Soilless Culture Technology

无土栽培技术

颜志明　张　更◎主编

中国财富出版社有限公司

图书在版编目（CIP）数据

无土栽培技术 = Soilless Culture Technology : 英文 / 颜志明，张更主编 . — 北京：中国财富出版社有限公司，2023.12

ISBN 978-7-5047-8036-2

Ⅰ. ①无…　Ⅱ. ①颜… ②张…　Ⅲ. ①无土栽培—教材—英文　Ⅳ. ① S317

中国国家版本馆 CIP 数据核字（2023）第 238066 号

策划编辑	刘静雯	**责任编辑**	刘静雯	**版权编辑**	武　玥
责任印制	荀　宁	**责任校对**	张营营	**责任发行**	敬　东

出版发行	中国财富出版社有限公司		
社　　址	北京市丰台区南四环西路188号5区20楼	**邮政编码**	100070
电　　话	010-52227588 转 2098（发行部）		010-52227588 转 321（总编室）
	010-52227566（24小时读者服务）		010-52227588 转 305（质检部）
网　　址	http: //www. cfpress. com. cn	**排　　版**	宝蕾元
经　　销	新华书店	**印　　刷**	北京九州迅驰传媒文化有限公司
书　　号	ISBN 978-7-5047-8036-2/S · 0059		
开　　本	710mm × 1000mm　1/16	**版　　次**	2025年4月第1版
印　　张	29.25	**印　　次**	2025年4月第1次印刷
字　　数	510千字	**定　　价**	69.80 元

编写人员名单

主　编：颜志明　张　更

副主编：刘建平　贾思振　王喜艳　张　瑜

编　者：王全智　王其传　王喜艳　王媛花　冯英娜

刘建平　李永金　张　瑜　张成尧　孟晓慧

钟　华　贾思振　曹维荣　解振强　袁玉涛

巩子毓　史　翔　李艳艳　闫征南　徐丽娟

Foreword

This textbook is designed based on the cognitive and developmental patterns of students and incorporates the modern vocational education concept, which emphasizes a student-centered, competency-based, and industry-oriented approach. It fully reflects the direction of vocational education reform in the new era. The textbook is structured into 10 projects, covering the basic principles and methods of soilless culture and their practical applications in home gardening. It encompasses the learning process, starting with foundational theoretical knowledge and progressing to familiarity with the theory, skill training, mastering operational procedures, and acquiring practical skills. It strongly emphasizes integrating theory with practical application, and the content is organized in a practical, targeted, and operationally robust manner. The main features are as follows:

On the one hand, it firmly adheres to the correct political orientation and actively cultivates and promotes core socialist values. It considers moral education as a fundamental task and emphasizes the seamless integration of ideological and political education with professional knowledge. By implementing comprehensive education for college students in the new era and under the new circumstances, it aims to cultivate a new generation of agricultural professionals who are capable of shouldering the responsibility of national rejuvenation.

On the other hand, the content is determined based on the professional competency requirements of soilless culture. It adheres to the principle of providing a balanced and adequate amount of theory while highlighting vocational skills and strengthening practical components. The textbook incorporates new knowledge, technologies, processes, and skills in the field of soilless cultivation. It follows a task-based curriculum structure, including project introduction, learning objectives, pre-class preparation, task implementation, project summary, skill training, and extension tasks. Through comprehensive and integrated actions, students learn how to learn, think critically, and master skills, embodying the concept of developing high-quality technical and skilled talents in higher vocational education.

The publication of this textbook has been supported by the "Second Batch of National Vocational Education Teaching Innovation Team Research Projects"(Project No.: ZH2021100201), the "Jiangsu Higher Education Teaching Reform Research Project"(Project No.: 2023JSJG539), and the "Qinglan Project" of Jiangsu Colleges and Universities.

Due to the limitations of the editors' expertise and time constraints, there may be inadequacies or omissions in the textbook. We wholeheartedly welcome readers' feedback, criticisms, and corrections.

Editors

March, 2025

Contents

Module 1　Introduction to Soilless Culture

[Learning Objectives]

I. Knowledge Objectives

(1) Master the concept, types, and characteristics of the soilless culture technique and its application in production;

(2) Understand the development process, present situation, and trend of the soilless culture technique.

II. Skill Objectives

Learn to consult relevant information, and get familiar with the concept, types, and characteristics of soilless culture techniques and their application in production.

[Preparation for Learning]

I. Required Resources

(1) A paper library and a periodical reading room;

(2) A digital reading room and an online resource library;

(3) A multimedia classroom;

(4) Standardized soilless culture production base.

II. Background Knowledge

(1) Master the basic knowledge related to soilless culture technique;

(2) Master basic methods for literature review;

(3) Learn about some basic principles of soilless culture technique;

(4) Visit the soilless culture production demonstration base.

[Learning Tasks]

Task 1 Soilless Culture and Its Classification

I. Concept of Soilless Culture

Soilless culture is a method of growing plants in any nutrient solution or substrate other than natural soil. It provides growth conditions such as moisture and nutrients for crops to make them grow normally and complete their entire life cycle. In short, soilless culture is the way to grow plants without natural soil. It used to be called solution culture or hydroponics in the early stage as nutrient solutions were used early and long for culture.

As defined by the International Society of Soilless Culture (ISOSC), soilless culture refers to all methods that use balanced nutrient solutions (with or without substrates) rather than natural soil to supply nutrients for plant growth and development, enabling the plants to successfully complete the entire life cycle.

II. Types of Soilless Culture

Soilless culture has a history of more than 140 years since the beginning of early laboratory studies. In the process of moving from laboratory to large-scale commercial production and application, soilless culture has evolved from the original basic technique developed by German scientists Sachs and Knop in the mid-19th century to a wide range of techniques at present (Fig. 1-1).

(I) Liquid Substrate Culture

Liquid substrate culture is also known as solution culture. It refers to the method in which the root system is not fixed in solid substrates but grows in nutrient solution or humid air containing nutrients. Solid substrates are generally

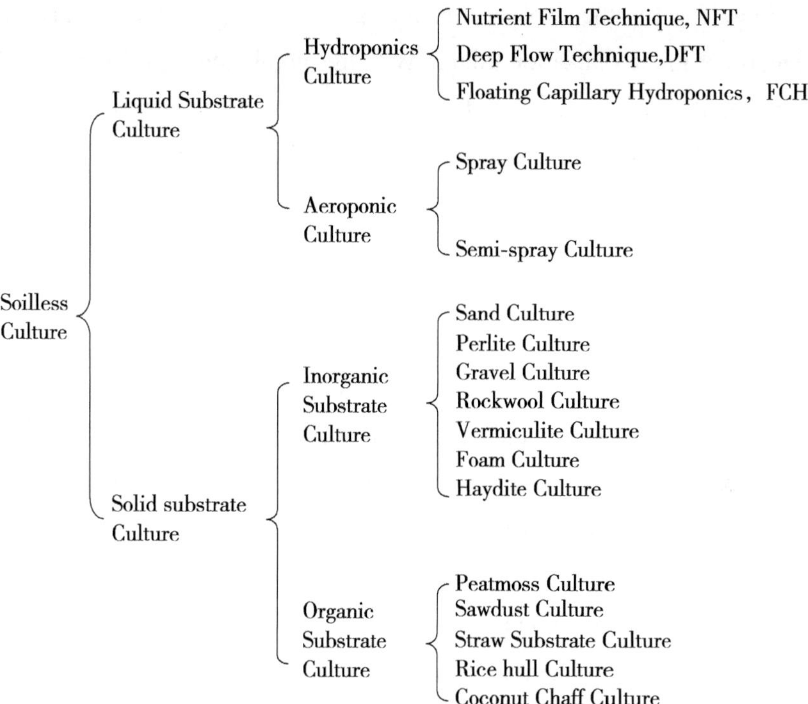

Fig. 1-1 Classification of Soilless Culture Technique

not used in the rhizospheric environment except in seedling cultivation. The method is called hydroponics if the root system grows in a nutrient solution and aeroponics if the root system grows in humid air containing nutrients. Hydroponics is further classified according to the depth of the nutrient solution.

1. Nutrient Film Technique (NFT)

NFT is a technique where the nutrient solution flows at a depth of 1–2 cm. NFT facilities include plantation troughs, reservoirs, nutrient solution circulation systems, and some auxiliary facilities (Fig. 1-2).

Advantages:

(1) Low investment in facilities, easy and convenient construction: The NFT plantation trough is made of lightweight plastic film or spliced with corrugated tiles. Light and simple in structure, it is easy to assemble and disassemble and requires low investment.

(2) Shallow and flowing stream of solution: The root systems of the crop are

partially immersed in the shallow stream of nutrient solution and partially exposed to the moisture in the plantation trough. With the circulation of the nutrient solution in the trough, the oxygen demand of the root systems can be better met.

(3) Easy for automatic management of the production process.

Disadvantages:

(1) Although the NFT facilities require less investment and are easy to construct, subsequent investment and frequent maintenance are needed owing to their poor durability.

(2) The shallow stream of solution and intermittent supply in NFT facilities can better satisfy the oxygen demand of the root system. However, the poor stability of the rhizospheric environment requires highly qualified management persons and higher-performance equipment.

(3) To streamline management, automatic control devices are essential, resulting in increased equipment and investment, which restrains promotion.

(4) Since the NFT system is a closed circulatory system, a disease that breaks out in the root system is more likely to spread throughout the whole system. Therefore, strict requirements must be followed in the cleaning and disinfection of facilities before use.

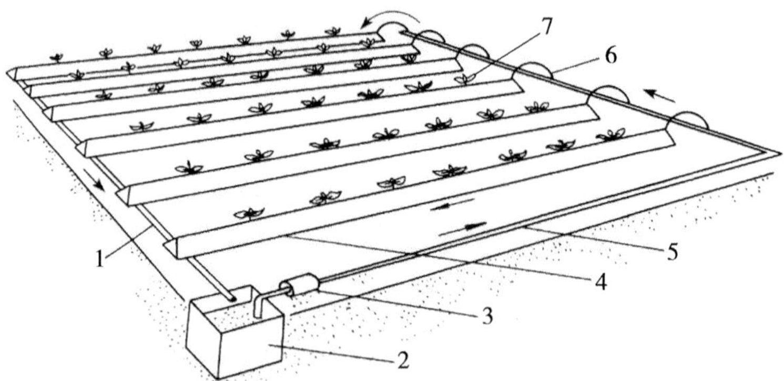

Fig. 1-2 Composition Diagram of Nutrient Film Hydroponic Facilities

1. Return pipe 2. Reservoir 3. Pump 4. Plantation trough 5. Liquid supply main pipe 6. Liquid supply branch pipe 7. Plant

2. Deep Flow Technique (DFT)

DFT is a culture technique that requires the depth of nutrient solution to be above 5 cm. Given the stable liquid temperature and the fact it is not affected by power outages and water cutoffs, this technique may be promoted and applied in subtropical and tropical areas(Fig. 1-3).

Culture characteristics of DFT: a. Deep plantation troughs and nutrient solution: With an adequate amount of nutrient solution in the plantation trough, the plants will witness little change in the composition, concentration, pH level, moisture, and temperature of the nutrient solution during growth, which is good for maintaining a stable environment for the growth of root systems; b. Circulating nutrient solution: The circulation of the nutrient solution can increase the concentration of dissolved oxygen in the nutrient solution, eliminate the locally deposited harmful metabolites and supplement for nutrient deficiency in the root zone, and promote the re-dissolution of nutrients that have become ineffective due to precipitation; c. Partially hanging plants in the nutrient solution: The root collar is hung above the liquid level, which can avoid decay or even the death of the plants caused by hypoxia during growth.

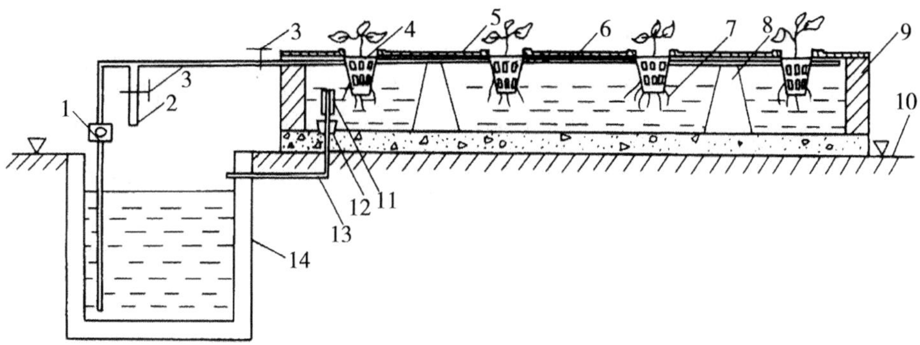

Fig. 1-3 Composition of Deep Flow Hydroponic Facilities

1. Water pump 2. Oxygen-increasing branch pipe 3. Flow regulating valve 4. Planting cup 5. Planting plate 6. Liquid supply pipe 7. Nutrient solution 8. Supporting pier 9. Plantation trough 10. Ground 11. Liquid layer control pipe 12. Rubber pipe 13. Return pipe 14. Reservoir

Advantages:

(1) Culture by hanging: The plants are fixed on a planting plate, with part of the root system inserted into the nutrient solution, and part of the root system aerated in the gap between the solution surface and the plate, making them semi-hydroponic and semi-aerial. This makes it easier to balance the water and air supply of the root system.

(2) Deep stream of solution: The root system stretches into the deep flow, ensuring an adequate supply of solution for a single plant. The large volume and deep stream of solution contribute to little sudden change in the concentration of the nutrient solution (including total salt and nutrients), dissolved oxygen, pH level, temperature, and moisture, providing a stable environment for the growth of the root system. This is the outstanding advantage of deep-flow hydroponics.

(3) Circulating nutrient solution: The circulation of the nutrient solution can increase the content of dissolved oxygen in the nutrient solution; eliminate the locally deposited harmful metabolites on the root surface (the most obvious impact is on the pH level); eliminate the difference in the concentration of solutions at the root and in other places to allow nutrients to be timely sent to the root surface to fully meet the growing needs of the plant; promote the re-dissolution of nutrients that have become ineffective due to precipitation to prevent the occurrence of nutrient deficiency. Therefore, even if it is to cultivate marsh plants or plants that can form oxygen-conducting tissue, it is necessary to circulate the nutrient solution.

(4) There are many types of crops suitable for cultivation with this technique. Except for root and tuber crops, almost all fruit and leafy vegetables can be cultivated.

Disadvantages: large initial investment.

3. Floating Capillary Hydroponics (FCH)

FCH is a hydroponic technique in which plants grow on moist non-woven fabric placed on a piece of foam plastic floating in a nutrient solution that is about 5–6 cm deep. This technique is widely used for the cultivation of vegetables such as tomatoes, cucumbers, cantaloupe, and head lettuce. The FCH facility consists of a plantation trough, an underground reservoir, circulation pipes, and a control system.

Except for the plantation trough, the other three parts are basically the same as those of the NFT facility(Fig. 1-4).

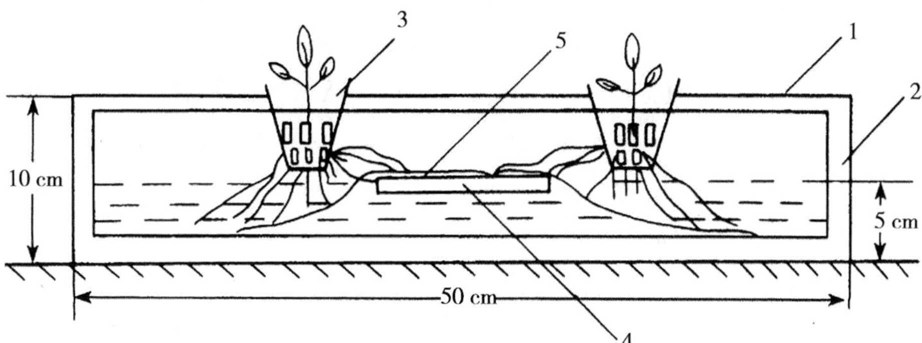

Fig. 1-4 Schematic Diagram of Cross Section of Plantation Trough for Floating Capillary Hydroponics

1. Planting plate 2. Plantation trough 3. Planting cup 4. Floating plate 5. Non-woven fabric

Advantages:

(1) This technique works with a floating board and capillary mat in the culture bed to create an oxygen-rich environment for cultivating roots in moisture, balancing the water and air supply.

(2) A large amount of nutrient solution is stored in a horizontal culture bed to ensure an adequate and stable supply of fertilizer and water during power outages; a heating coil is used to warm the culture bed in winter and deep-well water is used for cooling in summer to ensure stable rhizospheric temperature and humidity.

(3) This technique features low investment in equipment, low power consumption, and convenient installation, operation, and maintenance. The facility is composed of a culture bed, a reservoir, a circulatory system, and a control system. The nutrient solution is transferred into the culture bed via a pipe (with an air mixer) by a timer-controlled water pump and then returns to the reservoir. The closed-loop circulation of nutrient solution is rarely affected by external environmental conditions, ensuring little change in rhizospheric temperature and humidity, which is suitable for the growth of various plants.

4. Aeroponics

Aeroponics is a soilless culture method in which the plant root system is suspended in a container, and the nutrient solution is transferred by a water pump and sprayed through the inner nozzle of the container to the root system surface at intervals so as to meet the needs of the plant growth. Aeroponics is further classified into spray culture and semi-spray culture. The former is aeroponics, and the latter is a method in which part of the root system is in the nutrient solution, while the other part grows in the fine mist of the solution.

Advantages:

(1) This technique can well solve the problem of oxygen supply to the root system, and almost no poor growth will occur due to hypoxia.

(2) With a high utilization rate of nutrients and water, a rapid and effective supply of nutrients occurs.

(3) The space in the greenhouse can be fully utilized to improve the planting quantity and yield per unit area. The utilization rate of greenhouse space is twice to three times higher than that of traditional plane culture.

(4) This technique makes it easier to realize the automation of culture management.

Disadvantages:

(1) Large investment in production equipment and high reliability of equipment is required. Otherwise, it is easy to cause such problems as nozzle blockage, uneven spray, and excessively large droplets.

(2) There are high requirements on the management technology as the concentration and composition of the nutrient solution are prone to large variations during planting.

(3) The spray device would fail in case of a short-time power outage, which can easily cause damage to the plant.

(4) Owing to a closed system, diseases in the root system can spread easily if not controlled properly.

(II) Solid Substrate Culture

Soilless culture with solid substrates is referred to as substrate culture. It is about growing plants in an environment with a wide variety of natural or synthetic materials as substrates to fix the root system of such plants and maintain their supply of nutrients and oxygen. In this way, the plants can grow in a rhizospheric environment with a stable and coordinated supply of water, air, and fertilizer to normally complete their life cycle. Substrate culture can well balance the supply of water and air in the rhizospheric environment and requires less investment, which is convenient for production with local resources. Substrate culture is further classified into two types: organic and inorganic substrate culture.

(1) Organic substrate culture mainly refers to peat moss culture, sawdust culture, straw substrate culture, rice hull culture, and coconut chaff culture.

(2) Inorganic substrate culture mainly refers to sand culture, perlite culture, gravel culture, rockwool culture, vermiculite culture, foam culture, and haydite culture. In general, placing the substrate into plastic bags or culture troughs for cultivating crops has a certain buffer function and is safe for application.

Task 2 Characteristics of Soilless Culture

I. Advantages of Soilless Culture

(I) High Yields, High Quality, and High Commodity Rate

As in soilless culture, the growth needs of crops can be adjusted artificially to achieve higher yields per unit than that through soil cultivation. In addition, soilless culture can be applied for production all year round with a high annual yield. Vegetables cultivated in this way are large in size and excellent in quality. Soilless culture can reportedly increase the content of vitamin C in tomatoes by 30%.

(II) Improved Land and Space Utilization Rate

Through soilless culture, the land that is not suitable for cultivating crops, such as saline-alkali land, barren mountainous land, wasteland, and island, can be made full use of. Particularly, it can help solve the problem of increasing diseases and pests in the greenhouse and polytunnel due to continuous cropping for years, as well as the aggravated problem of secondary salinization. Moreover, the space of the greenhouse can be utilized to raise the yield per unit and increase the income of farmers.

(III) Time- and Labor-saving and High Utilization Rate of Resources

Soilless culture techniques, once put into use, can save the tedious labor such as intertillage, fertilization, and weeding, bringing high yields, high output, and high productivity.

II. Precautions in Soilless Culture

(I) Source, Disposal, and Disinfection of Substrates

The substrates used currently come in a wide variety with different sources and different physicochemical properties. It is required to mix them at a ratio and disinfect them during use, and their disposal and disinfection after use are also very troublesome. All these restrict the application of substrate culture to some extent.

(II) Prevention & Control of Diseases

Hydroponics features a rapid spread of pathogens due to the circulation of the nutrient solution. Once infected with pathogens, all plants may be at risk of being infected. Therefore, more research should be conducted on nutrient solution disinfection equipment and effective prevention and control agents.

(III) Regulation of Greenhouse Environment

At present, common plastic greenhouses and solar greenhouses are mainly used in China for the soilless culture of vegetables without corresponding regulating equipment, resulting in a low level of greenhouse environment regulation. The cost is also high for introducing equipment from advanced countries.

(IV) Breeding of Special Varieties

Currently, there is almost no vegetable variety specifically suitable for soilless culture. Due to the particularity of soilless culture, there is an urgent need for special high-quality and high-yield varieties that are resistant to low temperatures and diseases in the root system and are adaptable to weak light.

(V) High Requirements for Growers in Soilless Culture

Growers should master agricultural production technologies, physiological and biochemical knowledge about vegetables, and mechano-electronic technologies. At present, there are few technicians who have mastered these techniques and are engaged in the soilless culture in China. Therefore, agricultural colleges and universities and some agricultural research institutions should strengthen the training of professional technicians and promote relevant technologies to improve production efficiency.

[Skill Training]

Skill Training 1-1 Investigation on Types of Soilless Culture

I. Purposes and Requirements

Through field investigation of soilless culture types in the area, combined with viewing image materials and problem inquiry, grasp the structural characteristics, performance, and application of main soilless culture types in the area, identify the components of soilless culture facilities, and make a reasonable evaluation.

II. Plan

1. Tools and equipment

(1) Field investigation: measuring tools such as tape measure, steel tape measure, and angular instrument (slope meter); recording tools such as pencil and ruler.

(2) Image materials and equipment: slides, CDs, and other image materials of different soilless culture types and structures; slide projectors, VCDs, and other imaging equipment.

2. Implementation plans

(1) Investigate the types and characteristics of local soilless culture facilities, and observe the site selection, facility location, and overall planning of various

types of soilless culture.

(2) Determine and record the structural specifications, models, performance characteristics and applications of different soilless culture types.

① Record the types and materials of hydroponic facilities, the size of the plantation trough, the specifications of the planting plate, the model of the planting cup, the volumes of the liquid supply system and reservoir, the type of cultivated crop, etc.

② Record the substrate type, facility structure, and liquid supply system of substrate culture.

(3) Investigate the main cultivation seasons, varieties of cultivated crops, yield, quality, cropping pattern, and annual utilization of different soilless culture types in the area.

III. Implementation

(1) Work in groups to develop investigation plans for soilless culture types.

(2) Design investigation forms, conduct field investigations, and keep records.

(3) Analyze the similarities and differences of different types and structures of soilless culture, as well as the performance, cost composition, and economic benefits.

(4) Write an investigation report. State the investigation time, method, and invested enterprises; summarize the types, structures, performance, and application of soilless culture in the area; draw (or reflect with photos) the structural diagram of main facilities and types; indicate the name and size of each part,and point out the advantages and disadvantages; analyze the characteristics and formation reasons of the main soilless culture types in the area, and put forward reasonable suggestions for the development of soilless culture.

(5) Prepare slides based on the investigation report and make a presentation in class.

Module 2 Basis of Plant Physiology of Soilless Culture

[Learning Objectives]

I. Knowledge Objectives

(1) Grasp the plant physiological principles of the soilless culture technique;

(2) Understand the physiological changes and nutritional requirements of plant growth in the process of soilless culture.

II. Skill Objectives

(1) Learn to consult relevant information, and get familiar with plant physiological characteristics, growth characteristics, and their application in soilless culture production;

(2) Learn about the diagnosis and prevention methods for plant malnutrition.

[Preparation for Learning]

I. Required Resources

(1) Consult relevant plant physiological knowledge;

(2) A digital reading room and an online resource library;

(3) A multimedia classroom;

(4) Consult relevant journals of plant physiology.

II. Background Knowledge

(1) Master the basic knowledge of plant morphology and structure;

(2) Master basic knowledge of literature review;

(3) Learn about the basic knowledge related to plant growth and development;

(4) Learn about the basic knowledge related to plant growth environment.

[Learning Tasks]

Task 1 Structure and Function of the Plant Root System

I. Type of Roots and Root Systems

The root is the product of plants adapting to terrestrial life in the long-term evolution process, one of the important vegetative organs of terrestrial plants, and also the main organ for plants to absorb water and mineral nutrients from the medium.

The growth and distribution of the root system have a certain correlation with the growth and development of the aerial parts of the plant. The depth of root distribution varies at different growth and development stages of the plant. The root system is shallow at the seedling stage, deep in adult plants, shallow in herbaceous plants, and deep in woody plants. Most of the roots of crops are distributed in the loose tillage layer of the soil. Moderate deep tillage can promote the development of roots, increase the depth and breadth of the root system distribution in the soil, increase the absorption area, and achieve a high yield of the crop

(I) Normal Root and Adventitious Root

With the further growth of the main root, many new roots usually develop at certain positions on the main root or other parts of the plant to coordinate the growth of the plant. According to the different positions at which the roots develop, roots can be divided into two categories: normal root and adventitious root.

Normal root refers to the root that develops at certain positions on the plant, including the main root and lateral root. The main root comes from the radicle and the lateral root develops from the cells at certain positions on the pericycle of the main root. Aside from developing normal roots, many plants can develop roots from stems, leaves, old roots, or hypocotyls. These roots are called adventitious roots since they develop at unfixed positions. Adventitious roots can also continuously give rise to branched roots, that is, lateral roots. The main root of gramineous plants developed during seed germination has a short survival period and is mainly replaced by adventitious roots developed on hypocotyls or basal nodes of stems. The vegetative propagation techniques, such as cutting and layering in production, are carried out by making use of the ability of shoots, leaves, or subterraneous stems to produce adventitious roots.

(II) Tap Root System and Fibrous Root System

The subterraneous roots of a plant are collectively called the root system. The root system is gradually formed during the growth and development of the plant. According to the composition characteristics of the root system, it can be divided into tap root system and fibrous root system(Fig. 2-1). The tap root system consists of an obviously developed main root and its lateral roots at all levels. Since the main root is developed and deeply buried and the lateral roots at all levels are short, the tap root system is generally distributed in the shape of a gyro. The root systems of most dicotyledons belong to this type.

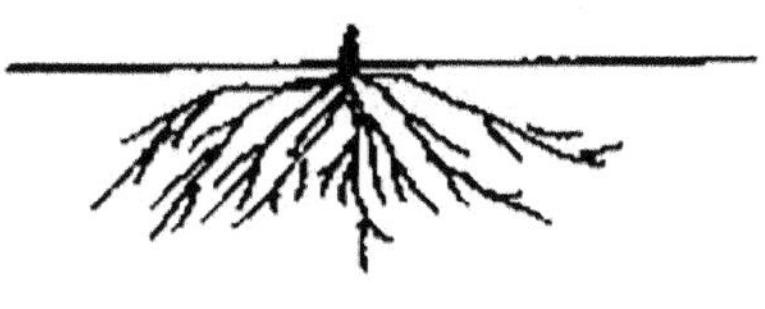

a. Fibrous root system

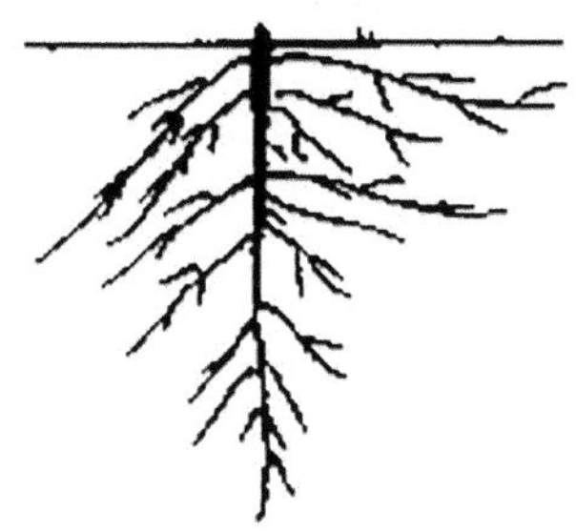

b. Tap root system

Fig. 2-1 Schematic Diagram of Fibrous Root System and Tap Root System

The fibrous root system is mainly composed of adventitious roots and their lateral roots. Some fibrous root systems are entirely composed of adventitious roots and their lateral roots. It is called the fibrous root system because its main root is underdeveloped and roots have a similar thickness and length, its burial is shallow, and it is tufted or beard-like. The root systems of most monocotyledons belong to this type.

II. Main Functions of Plant Root System

(I) Support and Fixation

Angiosperms have a huge root system, and their distribution range and root penetration depth are comparable to the aerial part, so as to support the tall, branched stem-leaf system and firmly fix it in the terrestrial environment, thereby facilitating their respective physiological functions.

Some plants can grow some adventitious roots from stems or near-surface stem nodes, which can support the vertical growth of plants and are hence called prop roots. This phenomenon can be seen in corn, sugarcane, banyan, etc. Some woody vines, such as *Hedera helix, Campsis grandiflora,* and *Euphorbia humifusa*, can develop an adventitious root from the stem that can grow on the surface of other objects, which is called the climbing root. With the help of these climbing roots, plants can adjust their spatial position, so that the slender and weak stems can grow upward by clinging to or climbing on other objects, and thus grow and develop better.

For different soilless culture methods, the supporting functions of roots are also different. For example, for hydroponics, aeroponics, and nutrient film techniques, the roots float in the nutrient solution or are exposed to humid air, so the fixation and support of plants are realized by artificial measures rather than by roots. In substrate culture, such as sand culture, gravel culture, vermiculite culture and rockwool culture, the fixation and support functions of roots are as important as in soil culture.

(II) Absorption and Conduction

The water and nutrients needed by plants are mostly obtained from the medium, besides those absorbed from the air by leaves or young stems. The main function of roots is to absorb water from the medium, and carbon dioxide and inorganic salts dissolved in water. This is mainly done by the root hairs and tender epidermis at the root tip. The part above the root tip often cannot absorb water from the medium due to the suberized cells of epidermis or exodermis or the formation of cork layer.

Inorganic salts are also absorbed from soil solutions, such as sulfate, phosphate, and nitrate, which are all absorbed in an ionic state by roots. They are indispensable to plant life, such as nitrogen, phosphorus, potassium, and other inorganic salt ions.

Root absorption is accompanied by conduction. The water and inorganic salts absorbed by root hairs and epidermal cells are transported to stems and leaves through the vascular tissues of roots. The organic nutrients produced by leaves are transported to roots through stems and then to all parts of roots through the vascular tissues of roots, so as to maintain their growth and life.

(III) Synthesis and Secretion

The plant root system can carry out many complex biochemical reactions and synthesize a variety of bioactive substances to regulate the growth and development of plants. Experiments show that a variety of essential amino acids, plant hormones (cytokinins), and alkaloids can be synthesized in the roots, which play an important role in regulating the growth and development of the aerial parts of plants.

Roots can secrete nearly 100 kinds of substances, including sugar, amino acid, organic acid, sterol, biotin, and other growth substances, as well as nucleotides and enzymes(Fig. 2-2). Some of these secretions can reduce the friction between roots and soil during growth; some can make the roots form a surface that promotes absorption; some are growth irritants or toxins to other organisms; some are resistant to diseases and pests, such as allicin, the secretion of garlic roots, which has a strong inhibitory effect on many kinds of bacteria and fungi; some plants can secrete autotoxic substances, which is one of the reasons for the avoidance of continuous cropping, such as Solanaceae crops.

(IV) Storage and Propagation

The roots of some plants (such as radish, carrot, beet, and sweet potato) are often succulent and store a lot of nutrients. The roots of some plants (such as sweet potato and jujube) also have special reproductive functions, which can produce adventitious buds.

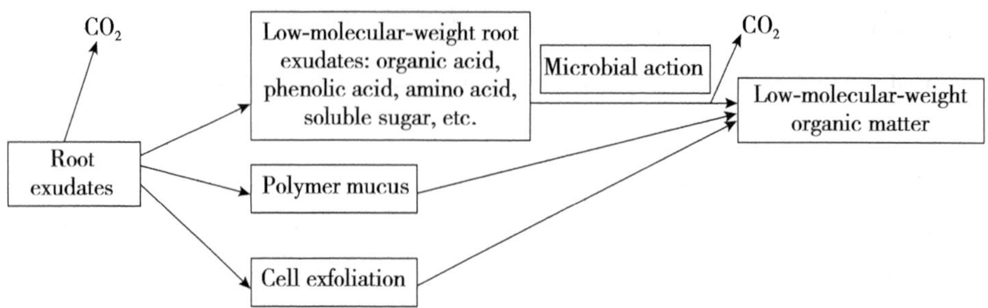

Fig. 2-2 Composition of Plant Root Secretion

Task 2 Nutrient Elements Needed by Plants

The existence of a chemical element in plants does not necessarily mean that such an element is essential for the normal growth and development of plants. Essential elements for plants refer to indispensable nutrient elements for the normal growth and development of plants. There are generally three criteria for determining whether an element is essential for crops: indispensability, irreplaceability, and direct functionality.

I. Indispensability

This chemical element is indispensable for the growth and development of all plants. Without this element, it would be impossible for plants to go through their life cycle. For higher plants, the life cycle refers to the process from seed germination to seed reproduction.

II. Irreplaceability

Without this element, plants would show unique symptoms, and no other chemical element can take its place. Only by supplementing this element can the symptoms be alleviated or eliminated.

III. Direct Functionality

This element is directly involved in the metabolism of plants and directly provides nutrition for plants, rather than indirectly improving the environment.

Only the nutrient elements that meet these three criteria can be determined as essential nutrient elements for plants. So far, 16 kinds of nutrient elements have been identified as essential elements for the growth and development of plants, namely, carbon (C), hydrogen (H), oxygen (O), nitrogen (N), phosphorus (P), sulfur (S), potassium (K), magnesium (Mg), calcium (Ca), iron (Fe), manganese (Mn), zinc (Zn), copper (Cu), boron (B), molybdenum (Mo), and chlorine (Cl).

According to the content (dry matter) in crops, the 16 essential nutrient elements in crops can be divided into macroelements and microelements.

Macroelements (dry matter content $\geqslant 0.01\%$) include carbon (C), hydrogen (H), oxygen (O), nitrogen (N), phosphorus (P), sulfur (S), potassium (K), magnesium (Mg), and calcium (Ca).

Microelements (dry matter content $<0.01\%$) include iron (Fe), manganese (Mn), zinc (Zn), copper (Cu), boron (B), molybdenum (Mo), and chlorine (Cl).

With the development of science and technology, especially analytical chemistry technology, there is another category of nutrient elements besides the 16 essential nutrient elements. They have positive effects on the growth and development of certain plants or are necessary for certain plants under certain conditions, but they are not generally essential for higher plants. These nutrient elements are called beneficial elements, which mainly include silicon (Si), titanium (Ti), cobalt (Co), selenium (Se), sodium (Na) and nickel (Ni).

Usually, among the nutrient elements needed by plants, carbon, hydrogen, and oxygen come from air and water, while nitrogen is mainly absorbed by plants from the soil through the root system, and part of it is absorbed from the soil air through the combined nitrogen fixation of rhizosphere microorganisms and symbiotic nitrogen fixation of rhizobia. Other elements, almost all come from the soil. In soilless culture, except for carbon, hydrogen, and oxygen, the remaining nutrient elements are all provided by the nutrient solution.

Task 3　Nutritional Diagnosis of Plants

All essential nutrient elements for plants have their special nutritional effects. The lack of them will affect various physiological and biochemical processes of plants. When the deficiency of certain nutrient elements reaches a certain level, plants will show certain symptoms in appearance. On the contrary, having an excessive amount of them may also lead to specific symptoms. These pathological features, known as physiological diseases, can be used as a morphological diagnostic basis for crop malnutrition, which is very important in formulating the nutrient solution.

I. Methods of Diagnosis

1. Morphologic diagnosis

Because of the different physiological functions of different nutrient elements, when a certain nutrient element in a crop is deficient or excessive, its exterior will show some characteristic symptoms. Therefore, we can determine the abundance or deficiency of a certain nutrient element by observing the seedling stage, that is, morphological diagnosis.

2. Chemical analysis of plants

On the basis of morphological diagnosis, take the tissues of abnormal parts (such as leaves) of abnormally growing plants and the tissues of normally growing plants (the same parts as abnormal plants) to carry out a chemical analysis of possible abnormal nutrients, and determine which kind (or kinds) of nutrient elements are insufficient or excessive by comparison.

3. Chemical analysis of substrate

Carry out a chemical analysis of the substrate to see if one or several kinds of nutrients are accumulated to cause nosotoxicosis or affect the absorption of other elements and cause nutrient deficiency symptoms. At the same time, measure the pH value of the substrate and analyze its effect on nutrient absorption.

4. Fertilization diagnosis

After diagnosis and analysis, if there is doubt about the imbalance of a certain nutrient element, use a few plants for fertilization verification. In the case of nutrient deficiency, double the nutrient element in the nutrient solution, spray the nutrient element on the foliage, or adjust the nutrient solution and spray on the foliage at the same time. When a plant is poisoned, halve that nutrient element in the nutrient solution and observe the change of the plant. After the correct result is obtained, the same measures can be taken immediately for a large area of crops.

II. Techniques of Morphological Diagnosis

The 16 essential nutrient elements for plants can be divided into mobile and immobile nutrient elements. The mobile nutrient elements include nitrogen, phosphorus, potassium, magnesium, and zinc. When these elements are in deficiency, they can move from the old leaves to the new ones, thus making the old leaves show a nutrient deficiency symptom. Immobile nutrient elements include calcium, iron, sulfur, boron, copper, and manganese. These elements cannot move in plants, so the symptoms of deficiency of these elements mostly appear on young leaves.

The identification of crop nutrient deficiency symptoms should be carried out in three steps. Step 1: Check to see where symptoms appear. If the symptoms appear on the old leaves first, it indicates a deficiency of nitrogen, phosphorus, potassium, magnesium, and zinc; if the symptoms appear first on the new tissue, it indicates a deficiency of calcium, iron, boron, and sulfur. Step 2: Check whether there are disease spots on old leaves and whether the new leaves are withered. In the case of old leaves showing symptoms, if there are no disease spots on old leaves, it may be in deficiency of phosphorus or nitrogen; if there are any disease spots on old leaves, it may be in deficiency of potassium or zinc. In the case of the symptoms starting from new leaves, if the terminal bud dies easily, it may be in deficiency of boron or calcium; if the terminal bud is not easy to die, it may be in deficiency of iron, sulfur, manganese, molybdenum and copper. Step 3: Determine the deficient elements according to the specific symptoms.

III. General Symptoms of Malnutrition

1. Nitrogen

When nitrogen is deficient, less protein is produced, the cells are small and thick-walled, and especially the cell division is blocked, which makes the growth slow, so the plants are short, thin, and upright. Besides, nitrogen deficiency leads to a decrease in chlorophyll content, which turns the green leaves pale and even pale yellow in severe cases. The leaves are uniform in color after turning pale, and there are generally no spots on the leaves. The leaves are thin and straight, and form a small angle with the stem. The green color of stems will fade due to nitrogen deficiency. For some crops, such as tomato, oilseed rape, and corn, nitrogen deficiency will cause the accumulation of anthocyanins, and the stems, petioles, and old leaves will also turn red or dark purple. Because of the high mobility of nitrogen in plants, it can be transferred from the old leaves to the young leaves; the symptoms of nitrogen deficiency start from the old leaves and gradually spread to the upper leaves. The root system of such crops is white and slender compared with that of normal crops, and the root volume is small. The lateral buds of plants are dormant or dead, so the lateral roots of tillers are reduced. The crops are prone to premature senescence. Flowers and fruits are few in number, the grains mature in advance, and the seeds are small and not full, which could seriously affect the yield and quality of the crops.

Excessive nitrogen promotes the formation of amino acids, protein, and chlorophyll in plants, thereby affecting the formation of raw materials (cellulose and pectin) that make up cell walls and making crop tissues soft and unresisting to diseases, pests, and lodging. Since the leaf area increases and the dark green loose leaves cover each other, it could affect ventilation and light transmission, resulting in late ripening, more immature grains, lower yields, and lower quality.

2. Phosphorus

In the case of phosphorus deficiency, since various metabolic processes are inhibited, the plants will grow slowly and become short, thin, and upright; there will be few tillers and branches, the flower bud differentiation will be delayed, and the blossom and fruit drop will increase. Because sugar transportation in plants is blocked, sugar accumulates in stems and leaves, forming anthocyanin, which could make many crops turn purplish-red. The leaves of a few crops, such as rice and tobacco, will turn dark green, and this is the result of the increase of chlorophyll density caused by the blocked growth of leaf cells and shrinking cells. The symptoms of phosphorus deficiency appear first from the old leaves. When phosphorus is seriously deficient, the leaves will die and fall off. The symptoms generally start from the old leaves at the base of the stem and gradually develop upwards. The maturity of crops will be delayed, and the number of empty immature grains will increase.

Excessive phosphorus could make crop respiration too vigorous, consume a lot of carbohydrates, increase the ineffective tillering of cereal crops, make reproductive organs develop prematurely, inhibit the growth of stems and leaves, and cause premature senescence of the plants. The number of empty immature grains will increase. Excessive application of phosphate fertilizer will also lead to zinc deficiency, iron deficiency, and magnesium deficiency.

3. Potassium

When potassium is deficient, the plants will become short, the stems will become thin, and the leaves will become narrow. The leaves will first turn yellow at the tip and edge of the old leaves, and then turn brown and become shriveled as if they are "burnt". The spots or patches on the leaves will gradually increase in number, but the middle of the leaves, the veins, or the places near the veins will remain green. With the aggravation of potassium deficiency, the whole leaf will become reddish-brown or withered, and die off. Generally, when potassium is deficient, the leaves are prone to wrinkling. All the above symptoms start from the old leaves and then spread to the young leaves. The root system will also be significantly damaged: the root is short and small, which is prone to premature senescence, the vitality of the roots will decrease, and the roots will even rot in severe cases.

4. Calcium

Calcium can hardly move in plants, so when calcium is deficient, the growth of young parts will stagnate first, new leaves can hardly grow out, and the tips of young leaves will stick together and bend, resulting in deformity. In severe cases, the leaves will turn yellow, become shriveled, and die off. The root system will be stunted, the root tip will swell and turn brown, and in severe cases, it will secrete mucus, decay, and die. The flowers and flower buds will fall off in large numbers. The empty shell rate of peanuts will increase significantly. Tomato, watermelon and other fruits will suffer from top rot (navel rot).

5. Magnesium

Magnesium is a component of chlorophyll. It can move easily in crops. When magnesium is deficient, the mesophyll between veins of the lower leaves will first lose its green color, but the veins will remain green. Gramineous plants often show stripe-like chlorosis between veins, and sometimes the chlorosis part will appear discontinuous beading. The chlorosis part of dicotyledons will change its color from light green to yellowish-green until purplish-red spots appear. In severe cases, the whole leaf will turn yellow, shrivel, and fall off, and this phenomenon will spread to the upper leaves. The root system will also be significantly inhibited, and the fruit yield will decrease.

6. Sulfur

Sulfur, like nitrogen, is a component of protein. When sulfur is deficient, the symptoms of crops are similar to those with nitrogen deficiency, and the chlorosis and etiolation are quite obvious. However, because sulfur can hardly move in the plant, the chlorosis part is different from that with nitrogen deficiency, and it will appear first in the young part. In the case of sulfur deficiency, plants are short, and leaves become small, curled up, hardened, fragile, and falling off early. The stem is blocked in growth and becomes stiff. It will lead to late flowering and less fruiting and podding. If rice is sulfur deficient, it can hardly strike roots after transplanting, with few new roots and poor root growth.

7. Iron

Although iron is not a component of chlorophyll, it participates in the formation of chlorophyll, so iron deficiency could lead to chlorosis. Because iron can hardly move in plants, chlorosis will first appear between the veins of young leaves. If the symptoms aggravate, the veins of the leaves will also lose their green color, and the leaves will show a uniformly pale yellow color. In severe cases, the leaves will turn yellow, and white or brown spots will appear near the leaf edge.

8. Boron

It is also hard for boron to move in plants. When boron is deficient, the symptoms will first appear at the growing point, and the root tip and shoot tip will stop growing. In severe cases, the growing point will shrink and die. After the root tip dies, the lateral root will grow again and then die again, thus making the root system become shorter. Boron deficiency will cause the accumulation of certain phenolic substances and damage the crops, e.g. cauliflower terminal buds will suffer from brown rot, radish roots will suffer from heart rot, and celery will suffer from cracked stem. Boron deficiency has a more significant effect on reproductive organs. It could cause abnormal flowering and fruiting, easy falling off of buds and flowers, prolonged anthesis, and unfilled fruit and seeds. In severe cases, there could be buds without flowers, or flowers without fruits, or even if there are fruits, there are no kernels or many immature grains. The leaves will become thick, rough, wrinkled, and curled as if they are withered, sometimes with purple spots. The root system of leguminous crops has few root nodules or cannot fix nitrogen.

Excessive application of boron fertilizer could easily poison the crops: the edges of old leaves will turn yellow and become shriveled, and there will be necrotic spots on the leaves. In severe cases, it could cause the crops to die.

9. Zinc

Zinc is related to the formation of auxin. Zinc deficiency will stop plants from growing and significantly shorten the internodes, e.g. rice will suffer from the dwarf disease, and fruit trees will suffer from witches, broom disease, and little leaf disease. Zinc can affect the formation of chlorophyll, and zinc deficiency will lead to chlorosis. If the heart leaves of rice turn white, it will be more evident near the midvein, with narrow leaves and brown spots appearing at the lower leaf tips. The corn seedlings will turn white and suffer from white bud disease and white seedling disease. Zinc deficiency will hinder the development of reproductive organs.

10. Molybdenum

When different crops are deficient in molybdenum, the symptoms will be different: if vegetables are deficient in molybdenum, the leaves will be slender and deformed, twisted spirally, the old leaves will become thicker and shriveled; Leguminosae plants will face difficulty in forming root nodules, their leaf color will fade, many tiny grayish brown spots will appear on their leaves, and the leaves will become thicker, wrinkled, and curled.

11. Manganese

Manganese is involved in the formation of chlorophyll. When manganese is deficient, the mesophyll between veins of young leaves will turn yellow, but the veins will remain green and the vein pattern is clear. In severe cases, dark brown spots will appear on the leaves, and then they will increase and expand, spreading all over the leaves. The plants will grow thin, the flowers will be stunted, and the root system will become weak.

12. Copper

Copper is also related to the formation and stability of chlorophyll. When copper is deficient, the new leaves will turn yellow and become withered and dry, the tips of leaves turn white and become curled, the edges of leaves will turn yellow and white, necrotic spots will appear on the leaves, and the development of reproductive organs will be hindered. Generally, graminaceous crops are prone to copper deficiency.

13. Chlorine

Chlorine is mainly involved in the photolysis of water during photosynthesis. Palmae plants generally need relatively more chlorine. When chlorine is deficient, yellow spots will appear on the leaves. Coconut yield is closely related to the application of chlorine.

Some plants are very sensitive to chloride ions. When a certain amount of chloride ion is absorbed, it will significantly affect the yield and quality. These plants are usually called chlorine-sensitive crops. When there is a large amount of chloride ions, it is not conducive to the conversion of sugar into starch, and the starch content of root and tuber crops will decrease. Chloride ions can promote the hydrolysis of carbohydrates, thus reducing the sugar content of watermelons, beets, and grapes. Excessive chloride ions will affect the combustibility of tobacco and cause cigarettes to easily extinguish. Chloride will have an adverse effect on

Solanaceae crops. Soybeans and kidney beans have weak chlorine resistance.

14. Silicon

Gramineous plants need relatively more silicon. Silicon can silicify the cells of crop leaves, leaf sheaths, and stems; enhance the hardness of stems and leaves; resist lodging, diseases, and pests; and enhance photosynthesis, thus increasing the yield. When silicon is deficient, the leaves will hang like weeping willows, which will affect the ventilation and light transmission of the lower leaves.

IV. Prevention and Control of Plant Malnutrition

If the symptoms of malnutrition arising from the soilless culture can be diagnosed accurately in time, and the nutrient solution components can be adjusted immediately or foliar fertilization can be carried out, then the effect will be seen in a few days.

1. Nitrogen deficiency

Spray 0.2%–0.5% urea solution on the foliage and add calcium nitrate or potassium nitrate into the nutrient solution.

2. Phosphorus deficiency

Spray 0.2%–0.5% potassium dihydrogen phosphate solution on the foliage or add a proper amount of potassium dihydrogen phosphate into the nutrient solution.

3. Potassium deficiency

Spray 1% potassium sulfate solution on the foliage or add potassium sulfate into the nutrient solution.

4. Magnesium deficiency

Spray a large amount of 2% magnesium sulfate on the foliage or add magnesium sulfate into the nutrient solution.

5. Zinc deficiency

Spray 0.1%–0.5% zinc sulfate on the foliage or directly add zinc sulfate into the nutrient solution.

6. Calcium deficiency

Spray 0.75%–1.0% calcium nitrate or 0.4% calcium chloride solution on the foliage or add calcium nitrate into the nutrient solution.

7. Iron deficiency

Spray 0.02%–0.05% chelated iron (Fe-EDTA) solution on the foliage once every 3–4 days for 3–4 times in a row, or directly add the chelated iron (Fe-EDTA) solution into the nutrient solution.

8. Sulfur deficiency

Add a proper amount of sulfate into the nutrient solution, and potassium sulfate is preferred for the sake of safety.

9. Boron deficiency

Spray 0.1%–0.25% borax solution on the foliage, or directly add the borax solution into the nutrient solution.

10. Copper deficiency

Spray 0.1%–0.2% copper sulfate solution and 0.5% hydrated lime on the foliage.

11. Manganese deficiency

Spray a large amount of 0.1% manganese sulfate solution on the foliage.

12. Molybdenum deficiency

Spray 0.07%–0.1% ammonium molybdate or sodium molybdate solution on the foliage or directly add it into the nutrient solution.

[Skill Training]

Skill Training 2-1 Observation of Nutrient Deficiency Symptoms of Soilless Culture Plants

I. Purposes and Requirements

Study the physiological diseases of plants that are deficient in mineral elements such as N, P, K, Ca, Mg, and Fe;

Understand the important roles of various elements in organisms;

Master the basic agricultural research methods and research report writing skills.

II. Materials, Instruments, and Drugs

1. Experimental materials

Corn seedlings. Requirements: Make sure that seedlings are growing uniformly after the first true leaf is fully unfolded (be careful not to damage the root system and remove the endosperm during transplantation).

2. Experimental apparatus

Ten 5-mL pipettes, one 1-mL pipette, one 1,000-mL graduated cylinder, seven culture flasks (opaque), absorbent cotton, suction ball, pH indicator paper (pH meter), glass tube, ear washing ball, etc.

The above is the quantity allocated for each group.

3. Experimental drugs

Calcium nitrate, magnesium sulfate, potassium dihydrogen phosphate, potassium sulfate, sodium sulfate, sodium dihydrogen phosphate, sodium nitrate, calcium chloride, ferrous sulfate, boric acid, manganese chloride, copper sulfate, zinc sulfate, molybdic acid, hydrochloric acid, and ethylenediaminetetracetic acid disodium (EDTA-2Na). (All the above drugs are analytically pure.)

III. Experimental Steps

1. Seedling raising

Put a certain amount of quartz sand or clean river sand into an enamel tray, evenly arrange the corn seeds or other seeds that have been soaked overnight on the sand surface, cover them with a layer of quartz sand, keep them moist, and then place them in a warm place to germinate. After the first true leaf is fully unfolded, select the uniformly growing seedlings, carefully transplant them for later use, and be careful not to damage the root system during transplantation.

It needs to be prepared in advance.

2. Formulation of stock solution (mother solution) and culture solution

It is recommended that the teacher formulate the solutions in advance so that the students can use them directly. The solution formulation practice should be mastered in the follow-up courses.

Make the formulated stock solution into a complete culture solution and culture solution lacking a certain element (with distilled water) according to Table 2-1. After the nutrient solution is formulated, determine the pH value of the solution in each flask and adjust it to 5–6 with 0.1 mol/L NaOH or 0.1 mol/L HCl.

Table 2-1 Nutrient Solution Preparation Formula

Stock solution	Amount of stock solution per 1,000 mL of culture solution (mL)						
	Complete	N Deficiency	P Deficiency	K Deficiency	Ca Deficiency	Mg Deficiency	Fe Deficiency
$Ca(NO_3)_2$	5	–	5	5	–	5	5
KNO_3	5	–	5	–	5	5	5
$MgSO_4$	5	5	5	5	5	–	5
KH_2PO_4	5	5	–	–	5	5	5
K_2SO_4	–	5	1	–	–	–	–
$CaCl_2$	–	5	–	–	–	–	–
NaH_2PO_4	–	–	–	5	–	–	–
$NaNO_3$	–	–	–	5	5		
Na_2SO_4	–	–	–	–	–	5	–
Fe-EDTA	5	5	5	5	5	5	–
Microelements	1	1	1	1	1	1	1

3. Planting culture

Take seven 1,000-mL culture flasks (opaque) and fill them with 1,000 mL of formulated complete culture solution and various nutrient-deficient culture solutions, adjust the pH value to 6–7, and label and date them. Punch four round holes in the middle of the flask cap with a hole punch, remove the endosperm of the selected plant, rinse it with distilled water, wrap the stem base with cotton, and carefully fix it on the flask cap through the round hole, make sure that the whole root system is immersed in the culture solution, and adjust the rhizome base so that it is 1 cm away from the liquid level, and put three plants on each flask. Put a label on the flask, where the label should include treatment, class, group, culturist, and culture date.

Besides, insert a glass tube with a rubber ring into another small hole of the flask cap, make sure that the lower end of the glass tube is 1–1.5 cm away from the bottom of the flask for ventilation, and put the planted culture flasks in the greenhouse for cultivation.

Then, put the culture flasks in a place with sufficient sunshine and suitable temperature (20–25℃) and culture for 3–4 weeks.

4. Culture management and observation

After the start of the experiment, properly manage and observe the records, and carefully record the symptoms when the essential elements are deficient and the parts where the symptoms first appear. Take photos during each observation, which can be used for the before and after comparison, and accurately record and summarize the experiment.

Observe once every two days; add distilled water in time to maintain the level of the culture solution. Replenish the air in time; apply the suction ball to pump air through the glass tube, pump 15–20 times per flask, and be careful not to suck out the culture solution. Measure and adjust the pH value of the culture solution every 4–5 days to keep it between 5.5 and 6.5. Change the culture solution once a week. Record the results in the Table 2-2.

Table 2-2 **Experimental Observation Record Table**

Date	Observation item		Treatment (growth, nutrient deficiency symptom)						
			Complete	N Deficiency	P Deficiency	K Deficiency	Ca Deficiency	Mg Deficiency	Fe Deficiency
	Aerial part	Plant height							
		Number of leaves							
		Leaf color							
		Stem color							
	Subterranean part	Number of roots							
		Root length							
		Root color							
		Damage							

IV. Result Analysis

Observe and draw a conclusion on the difference between the corn plants cultivated with 6 kinds of nutrient-deficient culture solutions and the corn plants cultivated with complete culture solutions. Try to analyze the reasons.

V. Precautions

(1) The experimental containers must be cleaned to prevent contamination.

(2) When formulating the nutrient-deficient culture solution, first add a proper amount of distilled water into the container to prevent the stock solutions from reacting with each other to generate precipitation.

(3) When adding the stock solution, stir it at the right time, and use one pipette for a single solution to avoid cross-contamination.

(4) Remove the endosperm when placing the seedlings, and be sure not to damage the root system and that the root system should be shaded.

(5) Ensure ventilation during culture management.

Module 3 Nutrient Solution Formulation and Management Techniques

[Learning Objectives]

I. Knowledge Objectives

(1) Understand the composition, requirements, types, and confirmation methods of nutrient solution;

(2) Understand the meaning and types of nutrient solution formula;

(3) Master the composition principle, formulation method, operation procedure, and management measures of nutrient solution.

II. Skill Objectives

(1) Be able to use a conductivity meter to detect the concentration of nutrient solutions and to perform the water quality test correctly;

(2) Be able to proficiently formulate the mother solution and working solution;

(3) Be able to manage nutrient solutions scientifically and effectively.

[Preparation for Learning]

I. Required Resources

(1) Consult periodicals and agricultural journals;

(2) A digital reading room and an online resource library;

(3) Learn about chemicals in the laboratory;

(4) Attend a hydroponic vegetable or flower show before class.

II. Background Knowledge

(1) Master the basic knowledge related to soilless culture techniques as well as the basic principles of soilless culture techniques;

(2) Master the basic knowledge of literature review;

(3) Master the basic knowledge of plant nutrition;

(4) Master the basic knowledge of chemical analysis;

(5) Learn about agricultural meteorology and agricultural laws and regulations.

[Learning Tasks]

Task 1 Composition of Nutrient Solution

Nutrient solution is a solution formulated by having the compounds containing various nutritional elements essential for plant growth and development and a small amount of auxiliary materials that make certain nutritional elements more long-lasting dissolved in water according to scientific quantity and proportion. Nutrient solution is the main source of nutrients and water for the growth and development of crops in various soilless culture forms. The success of soilless culture largely depends on whether the formula and concentration of the nutrient solution are scientific and reasonable and whether the management of nutrient solutions can meet the needs of plants at different growth stages.

I. Raw Materials and Requirements of Nutrient Solution

The basic components of the nutrient solution include water, fertilizers (inorganic salt compounds), and auxiliary substances. Classic or generally considered appropriate nutrient solution formulas must take into account the local water quality, climatic conditions, and cultivated crop types. The types, application rates and proportions of fertilizers used to formulate the nutrient solution should be properly adjusted so as to maximize the use effect of the nutrient solution.

(I) Requirements for Water Source and Water Quality of Nutrient Solution

1. Requirements for Water Source of Nutrient Solution

The water used to formulate the nutrient solution is very important. Distilled water or deionized water should be used to study the new formula of nutrient solution and the nutrient deficiency symptoms during hydroponics; tap water and well water are generally used in soilless production. Using tap water as the water source, the water quality is guaranteed, but the production cost is high; using well water as the water source has a low production cost, but soft well water should be chosen. River water, spring water, lake water, reservoir water and rainwater may also be used to formulate the nutrient solution. No matter what kind of water source is used, it is necessary to pass the water quality test or obtain relevant information from the local water conservancy department before using it to determine whether the water quality is suitable or not, and if necessary, the water should be treated to meet the hygienic standards of drinking water. Water flowing through farmland, unpurified seawater, and industrial sewage shall not be used as the water source.

Soilless culture of crops requires sufficient water, especially in summer. If the water from a single water source is insufficient, tap water may be mixed with well water, rainwater, river water, etc.

2. Requirements for Water Quality of Nutrient Solution

Water quality has a great impact on soilless culture. Therefore, the water quality requirement of soilless culture is slightly higher than that of the *Standard for Irrigation Water Quality* (GB 5084–2021) issued by the Ministry of Ecology and Environment of the PRC, which is equivalent to that of drinking water meeting hygienic standards. Various ion contents, electrical conductivity, and pH values of soilless culture water must be tested and determined, which can be used as a reference for formulating the nutrient solution. Organic matter contained in natural water is often good for soilless culture, but the concentration of organic matter should not be too high,or it will reduce the pH value and the supply of trace elements. The main indicators for water quality of nutrient solution are as follows.

(1) Hardness: According to the amounts of calcium salt and magnesium salt in water, water can be categorized into soft water and hard water. Calcium salts in hard water mainly include calcium bicarbonate [$Ca(HCO_3)_2$], calcium sulfate ($CaSO_4$), calcium chloride ($CaCl_2$), and calcium carbonate ($CaCO_3$). Magnesium salts mainly include magnesium chloride ($MgCl_2$), magnesium sulfate ($MgSO_4$), magnesium bicarbonate [$Mg(HCO_3)_2$], and magnesium carbonate ($MgCO_3$). The contents of these salts in soft water are relatively low. The water hardness is uniformly expressed by the CaO content per unit volume, that is, each degree is equivalent to 10 mg CaO/L. Generally, the hardness of the formulated nutrient solution should not exceed 10°. If the water is too hard, or the pH value of the water increases, or the water is slightly alkaline, then the effectiveness of Fe, B, Mn, Cu, and Zn ions will be reduced, and the plants will have nutrient deficiency symptoms. If there are too many calcium ions in water, the absorption of potassium ions by plants will be inhibited.

(2) pH value: Generally, the pH value should be 5.5–8.5.

(3) Dissolved oxygen: The dissolved oxygen should be close to saturation before use, that is, 4–5 mg O_2/L.

(4) NaCl content: less than 2 mmol/L. If the NaCl content in water is too high, plants will grow poorly or die.

(5) Residual chlorine: mainly the residual chlorine from the disinfection of tap water and facilities. Chlorine is harmful to the plant root system. Therefore, it is better to put aside tap water for more than half a day before it enters the facility system and to leave the facility vacant for more than half a day after disinfection, so that residual chlorine can escape.

(6) Suspended matter: less than 10 mg/L. River water or reservoir water may be used as the water source only after being clarified.

(7) Contents of heavy metals and toxic substances: The contents of heavy metals and toxic substances in soilless culture water shall not exceed that under the national standard (Table 3-1).

Table 3-1 Content Standard for Heavy Metals and Toxic Substances in Soilless Culture Water

Name	Standard	Name	Standard
Mercury (Hg)	≤ 0.005mg/L	Copper (Cu)	≤ 0.10 mg/L
Cadmium (Cd)	≤ 0.01 mg/L	Chromium (Cr)	≤ 0.05 mg/L
Arsenic (As)	≤ 0.01 mg/L	Zinc (Zn)	≤ 0.20 mg/L
Selenium (Se)	≤ 0.01 mg/L	Iron (Fe)	≤ 0.50 mg/L
Lead (Pb)	≤ 0.05 mg/L	Fluoride (F^-)	≤ 3.00 mg/L
Hexachlorocyclohexane	≤ 0.02 mg/L	Phenol	≤ 1.00 mg/L
Benzene	≤ 2.50 mg/L	Escherichia coli	≤ 1,000 Nos/L
DDT	≤ 0.02 mg/L		

(II) Requirements for Fertilizers and Auxiliary Substances in Nutrient Solution

There are many kinds of nutrients used to formulate the nutrient solution in the soilless culture production, and different nutrients are used according to different nutrient solution formulas for different crops. In production, the type, application rate, and proportion of nutrients in the nutrient solution previously used or considered suitable may be adjusted appropriately according to the different local water quality, climate, and crop varieties. In order to flexibly and effectively manage the nutrient solution for soilless culture, it is necessary to have a good understanding of the nutrients and auxiliary materials used to formulate the nutrient solution.

Fertilizer selection requirements:

(1) Selecting fertilizers according to the cultivation purpose. If it is to study the new formula of nutrient solution and explore the nutrient deficiency symptom or conduct other tests, chemical reagents must be used. Unless clearly specified, the chemical pure grade is generally used. If it is used for soilless culture production of crops, except that microelements are supplied by chemically pure reagents or medical supplies, macroelements are mostly supplied by agricultural supplies so

as to reduce the costs. If there are no qualified agricultural raw materials, industrial supplies may be used instead, but the fertilizer costs will increase.

(2) Selecting fertilizers according to the special needs of crops. Ammonium nitrogen (NH_4^+) and nitrate nitrogen (NO_3^-) are good nitrogen sources for crop growth and development. Ammonium nitrogen is better used in summer when the photosynthesis of plants is fast or when plants lack nitrogen; nitrate nitrogen can be used under any conditions. The study shows that the effect of applying nitrate nitrogen in soilless culture is far greater than that of ammonium nitrogen. Nowadays, most nutrient solution formulas in the world adopt nitrate as the main nitrogen source. The reason is that the physiological alkalinity caused by nitrate is relatively weak and slow, and the plants have a certain resistance, so it is easy to control manually. However, the physiological acidity caused by ammonium salt is relatively strong and rapid, which is difficult for plants to resist, so it is very hard to control manually. Therefore, when formulating the nutrient solution, nitrate nitrogen or ammonium nitrogen should be selected according to the needs of the crops. Generally, nitrate nitrogen or two kinds of nitrogen source fertilizers mixed in an appropriate proportion will be selected as the source, which generally has a better effect than the use of nitrate nitrogen alone.

(3) Selecting fertilizers with high solubility. For example, calcium nitrate has a higher solubility than calcium sulfate, is easily soluble in water, and has a good effect. Therefore, when calcium is needed to formulate the nutrient solution, calcium nitrate is generally adopted. Although calcium sulfate is cheap, it is insoluble in water, so it is seldom used in production.

(4) The purity of fertilizer should be high, and industrial products should be used appropriately. This is because inferior fertilizers contain a lot of inert substances, which will produce precipitation when used to formulate the nutrient solution, thus blocking the liquid supply pipe and hindering the root system from absorbing nutrients. The application rates indicated in the nutrient solution formula are expressed by pure grade. When formulating the nutrient solution, the application rates of raw materials should be calculated according to the percentage

purity indicated for the raw materials of various compounds. Nutritional elements other than the raw materials are treated as impurities, but attention should be paid to whether the amount of such impurities can interfere with the balance of nutrient solution. On the premise of taking into account the cost, industrial products may be adopted appropriately.

(5) Suitable types of fertilizers. The selection of different fertilizers that provide the same nutrient element should be based on the principle of maximumly meeting the needs for the formulation of the nutrient solution. If calcium nitrate is used as the nitrogen source, there will be one more nitrate ion than potassium nitrate. The relative proportion of nutrient elements provided by a compound must be compared with the required quantity in the nutrient solution formula before being selected.

(6) Fertilizer safety. The fertilizer contains no toxic or harmful ingredients and is easy to buy at a low price.

II. Representation of Nutrient Solution Concentration

The representation of nutrient solution concentration is divided into direct representation and indirect representation.

(I) Direct Representation

In a certain weight or a certain volume of nutrient solution, the method of representing the concentration of nutrient solution by the amount of nutrient elements or compounds contained is collectively called direct representation. The direct representation of nutrient solution concentration includes compound weight/liter, element weight/liter, and mole/liter, where the first one is an operating concentration, which can be directly used in the nutrient solution formulation; the latter two are mostly used in the comparison of nutrient solution formulas, and they must be converted into compound concentration before being used in the nutrient solution formulation.

1. Compound weight/liter (g/L, mg/L)

It is the weight of a certain compound per liter (L) of nutrient solution. It is usually expressed in grams/liter (g/L) or milligrams/liter (mg/L).

2. Element weight/liter (g/L, mg/L)

It is the weight of a certain nutrient element per liter of nutrient solution. It is usually expressed in grams/liter (g/L) or milligrams/liter (mg/L).

3. Mole/liter (mol/L)

It is the number of moles of a certain substance per liter of nutrient solution. Such substance could be a compound (molecule), an ion, or an element. The amount of a substance per mole is equivalent to the molecular weight, ionic weight, or atomic weight of the substance, and its mass unit is gram (g).

(II) Indirect Representation

Indirect representation includes representation by osmotic pressure and electrical conductivity. The main focus here is on electrical conductivity(EC).

The inorganic salts used in the formulation of nutrient solutions are mostly strong electrolytes, which ionize into ions with positive and negative charges in water. Therefore, the nutrient solution can conduct electricity. Its ability to conduct electricity can be expressed by electrical conductivity. Electrical conductivity refers to the ability of a solution to conduct electricity per unit distance. It is usually expressed in milliSiemens/cm (mS/cm) or microSiemens/cm (μS/cm).

III. Composition Principle of Nutrient Solution

The composition of the nutrient solution not only directly affects the growth and development of crops but also involves the economical and effective utilization of nutrients. According to the plant species, water source, fertilizer source, and climatic conditions, pertinently determining and adjusting the composition of nutrient solution can play a role in the use of nutrient solution so as to meet the requirements of crop cultivation. The composition of a balanced nutrient solution formula should follow the following principles.

1. Complete range of nutrient elements

There are 16 kinds of nutrient elements essential for plant growth and development, among which C, H, and O are provided by air and water, and the remaining 13 nutrient elements (N, P, K, Ca, Mg, S, Fe, Mn, B, Zn, Cu, Mu, and Cl) are absorbed from the rhizospheric environment(Fig.3-1). Therefore, the formulated nutrient solution should contain these 13 nutrient elements. Because the water source, solid substrate, or fertilizer already contains the amount of certain trace elements needed by plants, it is generally unnecessary to add more trace elements when formulating the nutrient solution.

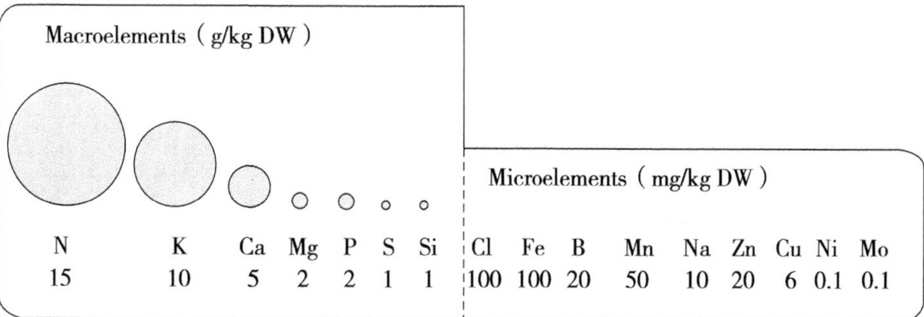

Fig. 3-1 Content of Mineral Elements in Plants

2. Nutrient elements can be absorbed by plants

The fertilizers used to formulate the nutrient solution should be mainly in a chemical state, have good solubility in water, and can be effectively used by crops. Usually, these fertilizers are inorganic salts, and some of them are organic chelates. Organic fertilizers that cannot be directly absorbed and utilized by plants should not be used as the fertilizer source of the nutrient solution.

3. Nutrition balance

The quantity ratio of each nutrient element in the nutrient solution should meet the requirements of normal growth and development of plants and should be physiologically balanced so that it can ensure the full play of the effectiveness of various nutrient elements and the balance of plant absorption. On the premise of ensuring the complete range of nutrient elements and meeting the formula requirements, the types of fertilizers used should be as few as possible (generally no more than four types) to prevent the compounds from bringing any unwanted excess ions or other harmful impurities into the plants (Table 3-2).

Table 3-2 Concentration Range of Elements in Nutrient Solution

Element	Concentration unit (mg/L)			Concentration unit (mol/L)		
	Minimum	Moderate	Maximum	Minimum	Moderate	Maximum
Nitrate nitrogen (NO_3^--N)	56	224	350	4	16	25
Ammonium nitrogen (NH_4^--N)	–	–	56	–	–	4
Phosphorus (P)	20	40	120	0.7	1.4	4
Potassium (K)	78	312	585	2	8	15
Calcium (Ca)	60	160	720	1.5	4	18
Magnesium (Mg)	12	48	96	0.5	2	4
Sulfur (S)	16	64	1440	0.5	2	45
Sodium (Na)	–	–	230	–	–	10
Chlorine (Cl)	–	–	350	–	–	10
Iron (Fe)	2	–	10	–	–	–
Manganese (Mn)	0.5	–	5	–	–	–
Boron (B)	0.5	–	5	–	–	–
Zinc (Zn)	0.5	–	1	–	–	–
Copper (Cu)	0.1	–	0.5	–	–	–
Molybdenum (Mo)	0.001	–	0.002	–	–	–

4. Appropriate total salinity and pH value

The total concentration in the nutrient solution shall meet the normal growth requirements of plants. Crops should not have nutrient deficiency symptoms because the concentration is too low, nor should they suffer saline injury because the concentration is too high. Although some fertilizers show physiological acidity or alkalinity or even strong physiological acidity and alkalinity because of the selective absorption of the root system after being dissolved, the pH value of the nutrient solution and its overall physiological acid-base reaction should be relatively stable and not exceed the range of pH changes required for normal growth of plants.

5. Long effective period of nutrient elements

All kinds of nutrient elements in the nutrient solution should keep their effective state for a long time during cultivation, and their effectiveness should not be reduced in a short time as a result of oxidation, root absorption and ion interaction.

6. Stable physiological acidity and alkalinity

The overall physiological acid-base reaction of the nutrient solution is relatively stable due to the selective absorption of all compounds in the nutrient solution formula by the root system during plant growth. In a nutrient solution formula, some compounds may show physiological acidity or alkalinity, and sometimes such physiological acidity and alkalinity are even relatively strong, but the overall physiological acidity and alkalinity of all compounds in the nutrient solution formula should be relatively stable.

Task 2 Formulation of Nutrient Solution

During the soilless culture of crops, it is necessary to formulate the nutrient solution correctly on the basis of the selected nutrient solution formula. In any balanced nutrient solution formula, there is a possibility that salts may precipitate from the solution. Only by adopting the correct method to formulate the nutrient solution can all kinds of nutrient elements in the nutrient solution be effectively supplied to meet the needs of crop growth, and achieve high yields and high quality of cultivation. However, incorrect formulation methods may make some nutrient elements invalid, or they may affect the balance of elements in the nutrient solution. In severe cases, they may damage the crop root system or even cause crop death. Therefore, it is necessary to master the correct formulation method of nutrient solution, which is the minimum requirement for the soilless culture of crops.

I. Formulation Principle of Nutrient Solution

The general principle for the formulation of nutrient solution is to ensure that no insoluble substances will precipitate after formulation and during the use of the nutrient solution. Every nutrient solution formula has the possibility of precipitating insoluble substances, which is closely related to the composition of the nutrient solution. Whether the nutrient solution will produce precipitation mainly depends on the concentration of the nutrient solution. Almost all balanced nutrient solutions contain cations such as Ca^{2+}, Fe^{3+}, Mn^{2+}, and Mg^{2+} and anions such as SO_4^{2-}, PO_4^{3-}, or HPO_4^{2-}, which may precipitate. When the concentration is high, these ions will interact with each other and precipitate. For example, Ca^{2+} interacts with SO_4^{2-} to produce $CaSO_4$ precipitation; Ca^{2+} interacts with phosphate (PO_4^{3-} or HPO_4^{2-}) to produce $Ca_3(PO_4)_2$ or $CaHPO_4$ precipitation; Fe^{3+} interacts with PO_4^{3-} to produce $FePO_4$ precipitation, and Ca^{2+} and Mg^{2+} interact with OH^- to produce $Ca(OH)_2$ and $Mg\ (OH)_2$ precipitations. In practice, guided by the law of solubility product of insoluble substances, the following two methods can be adopted to avoid precipitation in nutrient solution: ① Dissolve two kinds of salt compounds that are easy to precipitate, prepare and store them in separate tanks, and dilute and mix them before use; ② Add acid into the nutrient solution to reduce the pH value, and add alkali to adjust it to the normal level before use.

II. Formulation Technique of Nutrient Solution

(I) Preparation before the Formulation of Nutrient Solution

1. Correct selection and adjustment of nutrient solution formula

This is because different water quality and fertilizer purity in different areas can directly affect the composition of the nutrient solution. Different varieties and growth periods of cultivated crops require different proportions of nutrient elements, especially the proportions of N, P, and K. For cultivation methods, especially during substrate culture, both the adsorbability of the substrate and its own nutrients will change the composition of the nutrient solution. The use of different nutrient solution formulas also involves cultivation costs. Therefore, before preparing the nutrient solution, it is necessary to correctly select and flexibly adjust the nutrient solution formula according to plant species, growth period, local water quality, climatic conditions, fertilizer purity, cultivation methods, and costs, and apply it on a large scale after it is proved to be feasible.

2. Selecting appropriate fertilizers

The selection of fertilizers should take into account the concentration and proportion of available nutrient elements in the fertilizer, and the fertilizers with high solubility, high purity, few impurities, and low price should be selected.

3. Reading the relevant materials

Before preparing the nutrient solution, carefully read the instructions or packaging instructions on the fertilizers or chemicals, and pay attention to the molecular formula and purity of the fertilizer as well as the crystal water contained in the fertilizer.

4. Selecting the water source and conducting the water quality test

It can be used as a reference for the formulation of nutrient solution.

5. Prepare liquid storage tanks and other necessary items

Generally, the nutrient solution is prepared into a 100–1,000-fold concentrated mother solution, which requires 2–3 mother solution tanks. The volume of a small mother solution tank should be 25 L or 50 L. A dark opaque mother solution tank is preferred. In addition, it is necessary to prepare relevant testing equipment, dissolving and stirring apparatus, etc.

(II) Formulation Method of Nutrient Solution

The nutrient solution formulated in production is generally divided into mother solution (concentrated stock solution) and working solution. When formulating the mother solution, do not dissolve all fertilizers together because some of the anions and cations will interact with each other to precipitate after concentration. Therefore, three kinds of mother solutions (A, B, and C) are generally to be formulated. Mother solution A is mainly based on calcium salt. All salts that do not interact with calcium to produce precipitation can be dissolved together, and it is generally prepared into a 100–200-fold concentrated solution. Mother solution B is mainly based on phosphate, and all salts that do not interact with phosphate to produce precipitation can be dissolved together, and it is generally prepared into a 100–200-fold concentrated solution. Mother solution C is a mixture of iron salt and trace element compounds, and it is generally prepared into a 1,000–3,000-fold concentrated solution because of its small amount of use.

The formulation methods of the working solution include the mother solution dilution method and the direct formulation method. The mother solution dilution method is commonly used to formulate the working solution in production.

In practical production and application, the formulation method of nutrient solution can be as follows: Formulate the concentrated nutrient solution (or mother solution), and then use the concentrated nutrient solution to formulate the working nutrient solution. If the application rate is small, the working nutrient solution may also be directly formulated by directly weighing various nutrient element compounds. A formulation method may be selected according to actual needs. However, no matter which formulation method is selected, the general guiding principle is to avoid the precipitation of insoluble substances in the formulation process.

III. Dissolved Oxygen in Nutrient Solution

During the growth and development of the plant root system, oxygen is consumed in the process of respiration, so it is necessary to supply sufficient oxygen in order to make the root system grow normally. For soilless culture, especially hydroponic culture, whether sufficient oxygen is supplied in time often becomes a limiting factor for the normal growth of plants.

Supplemental sources of dissolved oxygen in nutrient solution include natural diffusion of oxygen from air to solution and artificial oxygen increase. Since the natural diffusion rate is slow and the increment is small, the natural diffusion method is suitable for use in the seedling stage only. The artificial oxygen increase method is adopted in hydroponic culture and most substrate cultures. Artificial oxygen increase methods mainly adopt mechanical and physical methods to increase the chance of contact between nutrient solution and air and enhance the diffusion ability of oxygen in the nutrient solution, thus increasing the oxygen content in the nutrient solution. Commonly used oxygen increase methods include spraying, stirring, compressing air, circulating flow, intermittent liquid supply, lowering the liquid temperature and nutrient solution concentration in summer, and using an oxygen-increasing device or a chemical oxygen-increasing agent (Fig. 3-2).

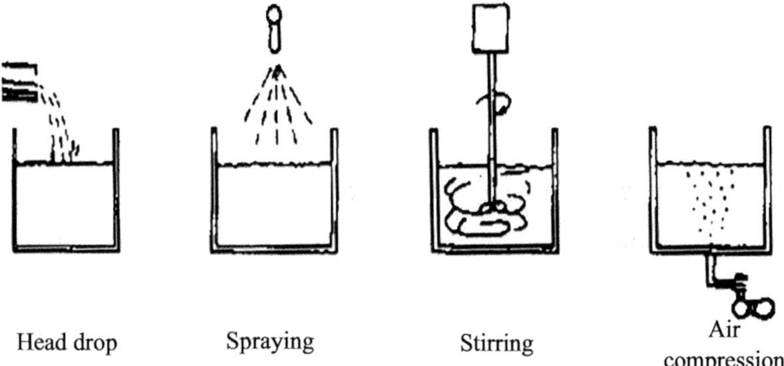

Fig. 3-2 Oxygen-increasing Methods of Nutrient Solution

1. Stirring of nutrient solution

Mechanically stirring the nutrient solution to break the air-liquid interface of the nutrient solution to dissolve the air in the nutrient solution has a good effect, but it is hard to carry it out concretely because there are a large number of roots in the nutrient solution where the plants are growing, and once stirred, the roots could be easily damaged, which will have adverse effects on the normal growth of plants.

2. Compressing air

It refers to the use of a compressed air pump to diffuse air directly into nutrient solution in the form of small bubbles to improve the dissolved oxygen content of the nutrient solution. This method has a good effect of increasing oxygen, but in large-scale production, ventilation pipes and bubblers need to be installed in many places of the plantation trough, which is difficult to construct and costly and hence is seldom used. This method is mainly used in small pot hydroponic culture for scientific research.

3. Reaction oxygen

It is a method of adding chemical oxygen-increasing agents into the nutrient solution to increase oxygen. In Japan, there is a device that can control hydrogen peroxide (H_2O_2) to slowly release oxygen. After this device is loaded with hydrogen peroxide, it can be put in the nutrient solution to increase the dissolved oxygen in the nutrient solution through the release of oxygen. Although this method has a good effect of increasing oxygen, it is costly and difficult to adopt in production. It is now mainly used in small household devices.

4. Circulating flow of nutrient solution

Use a water pump to pump the nutrient solution from the reservoir to the plantation trough, then let it flow in the plantation trough and finally flow back into the reservoir to form a continuous circulation. In the process of nutrient solution circulation, the dissolved oxygen content is increased by the impact and flow of water. This method has a good effect of increasing oxygen and is widely used in production. However, due to the different design of soilless culture facilities, different circulating times of water pumps, and different depths of nutrient solutions, the oxygen-increasing effect is also different. Therefore, producers should control the circulating time flexibly according to different facilities.

5. Head drop

When the nutrient solution circularly flows into the reservoir, a certain head drop is artificially caused to form a splash surface, which increases the contact area with the air. This method has a good effect and is widely used.

6. Spraying

Properly increase the pressure to make the nutrient solution disperse as much as possible to form a jet flow or mist spray when spraying from the plantation trough. This method has a good effect and is widely used.

7. Oxygen-increasing device

Install an oxygen-increasing device or air mixer at the water inlet of the circulation pipe to increase the dissolved oxygen content in the nutrient solution.

This method has been widely used in advanced hydroponic facilities.

8. Intermittent liquid supply

When the liquid supply stops, the nutrient solution flows back to the reservoir from the plantation trough, and the root system is exposed to the air to absorb oxygen. This method has a good effect and is widely used.

9. Drip irrigation

In soilless culture, such as substrate substitution culture, it can also ensure that the root system gets sufficient oxygen by controlling the flow rate and time of drip irrigation.

10. Intercropping

Intercrop xerophilous plants and aquatic plants whose root system secretes oxygen.

The combination of multiple oxygen-increasing methods can make the oxygen-increasing effect even more significant.

[Skill Training]

Skill Training 3-1 Formulation of Nutrient Solution

I. Purposes and Requirements

(1) Master the formulation methods of the mother solution and working solution;

(2) Understand the composition principle of nutrient solution;

(3) Be skilled in the operation and operate according to relevant procedures and standards;

(4) Formulate the nutrient solutions accurately without precipitation;

(5) Clearly identify the nutrient solutions and keep complete records.

II. Plan

1. Materials and chemical agents

Reagents or fertilizers are needed to formulate the mother solution according to the Japanese garden test formula (Table 3-3); 1M NaOH and 1M HNO_3 solution.

Table 3-3 Japanese Garden Test Nutrient Solution Formula

Name of salt compound	Application rate (mg/L)	Name of salt compound	Application rate (mg/L)
$Ca(NO_3)_2·4H_2O$	945	H_3BO_3	2.86
KNO_3	809	$MnSO_4·4H_2O$	2.13
$NH_4H_2PO_4$	153	$ZnSO_4·7H_2O$	0.22
$MgSO_4·7H_2O$	493	$CuSO_4·5H_2O$	0.08
Na_2Fe-EDTA	20	$(NH_4)_4MO_7O_{24}·4H_2O$	0.02

2. Instruments and tools

Tray balances or platform scales, electronic analytical balances (sensitivity 0.001 g), water pumps, acidimeters, conductivity meters, magnetic stirrers, black plastic buckets (50 L, 2), plastic beakers (500 mL, 1,000 mL) or plastic basins, black plastic liquid storage tanks (50L, 3), plastic water pipes, label paper, glass rods, short wooden sticks or plastic rods, pens, marker pens, registration form for mother solution formulation, and registration form for working solution formulation.

III. Implementation

1. Mother solution formulation(See Fig. 3-3 for the mother solution formulation process)

(1) Calculation: Determine the type, concentration ratio, and amount of nutrient solution formula and mother solution to be formulated, and then calculate the amount of reagent or fertilizer needed. The following mother

solutions are required to be formulated according to the requirements for the garden test formula: 10 L and 100-fold-concentrated mother solution A [$Ca(NO_3)_2 \cdot 4H_2O$ and KNO_3], mother solution B ($NH_4H_2PO_4$ and $MgSO_4 \cdot 7H_2O$), and 1 L and 1,000-fold-concentrated mother solution C (EDTA-Na_2Fe and various trace element compounds).

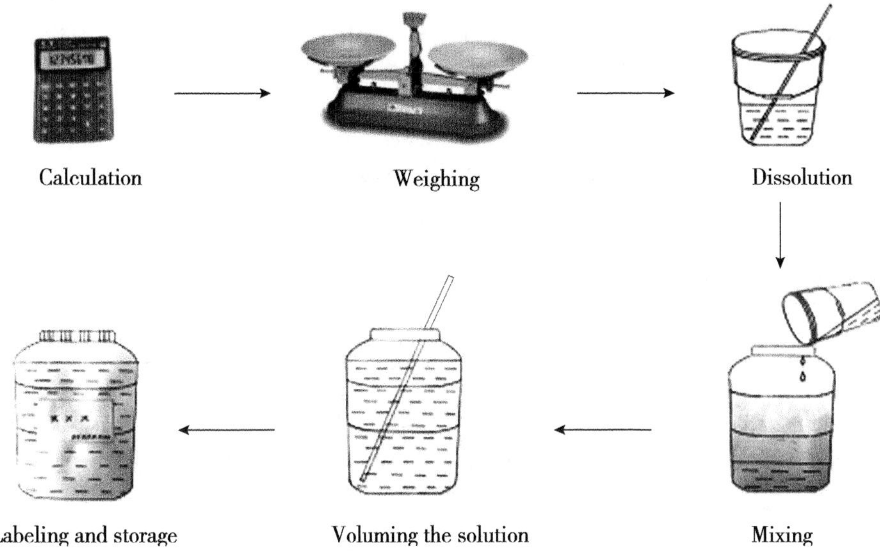

Fig. 3-3 Mother Solution Formulation Process

(2) Weighing: Weigh reagents or fertilizers with the platform scale, tray balance, or analytical balance, and place them in clean containers such as beakers and plastic basins. Ensure stable, accurate (to within ±0.1 g), and fast operation when weighing.

(3) Dissolution and mixing of fertilizers: Three kinds of mother solutions (A, B, and C) are to be formulated and held in three liquid storage tanks (A, B, and C), respectively. Formulate mother solution A in reservoir A by dissolving together the reagents or fertilizers that do not precipitate with calcium salts; Formulate mother solution B in reservoir B by dissolving together the reagents or fertilizers that do not precipitate with phosphate; formulate mother solution C in reservoir C by dissolving chelated iron salt and other trace element compounds respectively. If there is no readily available chelated iron reagent, $FeSO_4 \cdot 7H_2O$ and EDTA-Na_2

can be used. To formulate mother solution C, weigh 13.9 g of $FeSO_4{\cdot}7H_2O$ and 18.6 g of EDTA-Na_2, respectively, dissolve them with warm water, then slowly pour the $FeSO_4{\cdot}7H_2O$ solution into the EDTA-Na_2 solution, stir while adding to mix them well, then pour the mixture into reservoir C, then slowly pour the various trace element compound solutions dissolved respectively into reservoir C, stir while adding, and finally add water to the final volume to obtain the 1,000-fold mother solution C. To formulate mother solution A, first dissolve $Ca(NO_3)_2{\cdot}4H_2O$ and KNO_3 respectively and then mix them together. To formulate mother solution B, first dissolve $NH_4H_2PO_4$ and $MgSO_4{\cdot}7H_2O$ respectively and then mix them together. The mother solution C can be used in the soilless culture of any crop.

(4) Voluming the solution: Inject clear water into the liquid storage tanks A, B, and C to the volume required to be formulated, and mixed well.

(5) Labeling and storage: Affix labels or use a marker pen to indicate the name of the mother solution, mother solution number, concentration ratio or concentration, formulation date, and operator on the liquid storage tanks A, B, and C, and then store them in a cool and dark place. If the mother solution is stored for a long time, it should be acidified to prevent precipitation. Generally, it can be acidified to pH 3–4 with HNO_3.

(6) Making records: After the mother solution is formulated each time, the registration form for mother solution formulation should be carefully filled out. See Table 3-4 for the form format.

Table 3-4 Registration Form for Mother Solution Formulation

<table>
<tr><td colspan="2">Formula name</td><td></td><td>Used by</td><td></td></tr>
<tr><td rowspan="2">Mother solution A</td><td>Concentration ratio</td><td></td><td>Formulation date</td><td></td></tr>
<tr><td>Volume</td><td></td><td>Calculated by</td><td></td></tr>
<tr><td rowspan="2">Mother solution B</td><td>Concentration ratio</td><td></td><td>Reviewed by</td><td></td></tr>
<tr><td>Volume</td><td></td><td>Formulated by</td><td></td></tr>
<tr><td rowspan="2">Mother solution C</td><td>Concentration ratio</td><td></td><td rowspan="2">Remarks</td><td rowspan="2"></td></tr>
<tr><td>Volume</td><td></td></tr>
<tr><td colspan="2">Name and weighed amount of raw materials</td><td colspan="3"></td></tr>
</table>

2. Formulation of working solution

(1) Dilution of concentrated solution: This is a commonly used method to formulate the working solution in production. See Fig. 3-4 for the formulation method.

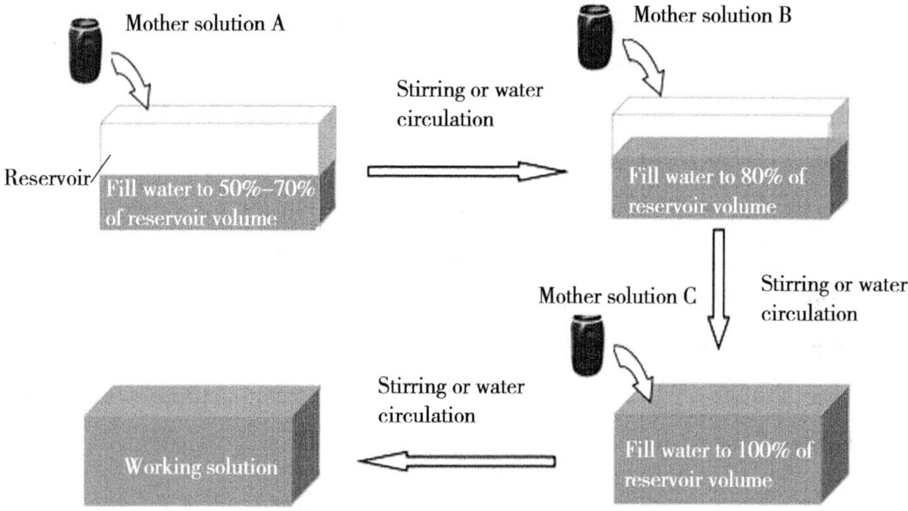

Fig. 3-4 Operating Procedure for Formulation of Working Solution

① Calculate the pipette volume of all three types of mother solutions. The formula for calculation of mother solution pipette volume:

$$V_2 \text{ (pipette volume of mother solution)} = \frac{V_1 \text{ (volume of working solution)}}{n \text{ (concentration ratio of mother solution)}}$$

For this training, 100 L of working solution is formulated with the above mother solution. According to the calculation results, 1 L of mother solutions A and B respectively, and 0.1 L of mother solution C should be pipetted.

② Fill water to 50%–70% of reservoir volume (volume of the nutrient solution to be formulated).

③ Pipette mother solution A and inject it into the reservoir; start the water pump to circulate the nutrient solution in there for 30 min or stir to allow it to spread evenly.

④ Pipette mother solution B, slowly inject it into the reservoir from the clear water inlet of the reservoir to dilute it by the running flow before it spreads in the reservoir; start the water pump to circulate the nutrient solution in the reservoir for 30 min or stir to allow it to spread evenly. The amount of water added in this process reaches 80% of the solution volume.

⑤ Pipette mother solution C, and add it into the reservoir in the same way as mother solution B. Circulate the solution through the water pump or stir it well. The water volume reaches 100% in this case.

⑥ Measure the pH and EC values of the nutrient solution with an acidimeter and a conductivity meter. If the measured pH value fails to meet the requirements for formula and crop cultivation, adjust it timely. After the pH value is adjusted for the nutrient solution, stand it for more than half an hour before use, then circulate it on the planting bed for about 5–10 min, and test the pH value again until it meets the requirements.

⑦ Fill in the registration form for working solution formulation for inspection. See Table 3-5 for the registration form for working solution formulation.

Table 3-5 Registration Form for Working Solution Formulation

Formula name		Used by	
Nutrient solution volume		Formulation date	
Calculated by		Reviewed by	
Formulated by		Water pH value	
EC value of nutrient solution		pH value of nutrient solution	
Name and measured (pipetted) amount of raw materials			

(2) Direct formulation: In production, if a large amount of working solution is required, it may be formulated directly after weighing raw materials containing macroelements or formulated by diluting mother solution C formulated in advance if the raw materials contain micronutrients. In this training, the 100 L of garden test formula nutrient solution can also be formulated directly. The specific procedures are as follows:

① Calculate the amount of reagent or fertilizer required according to the nutrient solution formula and the volume of nutrient solution to be formulated.

② Fill the reservoir with 50%–70% water.

③ Weigh corresponding amounts of reagents or fertilizers for the formulation of mother solution A, dissolve them in the plastic basin, and then pour them into the reservoir; start the water pump to circulate the nutrient solution for 30 min or stir it well.

④ Weigh corresponding amounts of compounds for the formulation of mother solution B, dissolve them in a plastic basin, and pour them into the reservoir from near the water inlet to allow them to be diluted by the running water before they spread in the reservoir; start the water pump to circulate the nutrient solution for 30 min or stir it well. The amount of water added in this process reaches 80% of the total solution volume.

⑤ Measure and dilute the pre-formulated mother solution C, then pour it slowly from the water inlet of the reservoir; start the water pump to circulate the nutrient solution for 30 min or until the solution is mixed well.

⑥ & ⑦ Same as the dilution of the concentrate solution.

IV. Precautions

(1) Check the calculated amount of reagents or fertilizers repeatedly to ensure accuracy. Ensure the accuracy of weighing and consistency between descriptions and real items.

(2) The following matters should be paid attention to in the calculation of the amount of reagents or fertilizers: ① The fertilizers used for soilless culture are mostly for agricultural and industrial uses, and often contain hygroscopic water and other impurities with low purity. The amount should be corrected according to the actual purity. ② Ca^{2+} and Mg^{2+} in the water should be deducted in hard water areas. For example, Ca^{2+} and Mg^{2+} in the formula are provided by $Ca(NO_3)_2{\cdot}4H_2O$ and $MgSO_4{\cdot}7H_2O$, respectively, and the actual amount of $Ca(NO_3)_2{\cdot}4H_2O$ and $MgSO_4{\cdot}7H_2O$ should be the formula amount minus the amount of Ca^{2+} and Mg^{2+} contained in the water. However, the amount of nitrogen in $Ca(NO_3)_2{\cdot}4H_2O$ is reduced after the deduction of Ca^{2+}, and the reduced amount can be supplemented with nitric acid (HNO_3), which not only plays a role in supplementing the nitrogen

but also neutralizes the alkalinity of the hard water. If the pH value of water is still not lowered to the ideal level after nitric acid is added, the amount of phosphate can be appropriately reduced, and phosphoric acid can be used to neutralize the alkalinity of the hard water. If the nutrient solution is too acidic, the amount of potassium nitrate can be increased to supplement nitrate nitrogen and reduce the amount of potassium sulfate accordingly. Deducting the amount of magnesium in nutrition, the actual amount of $MgSO_4 \cdot 7H_2O$ is reduced, so is the amount of SO_4^{2-}. However, since hard water contains a large amount of sulfate, additional supplementation is generally not required. If required, a small amount of H_2SO_4 can be added. In hard water areas, the amount of calcium nitrate is small, and the amount of phosphorus and nitrogen lacking is supplemented by nitric acid and phosphoric acid.

(3) Supplies, including weighed fertilizers for the formulation of the nutrient solution, should be placed on the formulation site in an orderly manner. Formulation may not begin until it is verified that nothing is missing. Do not rush to operate when materials are not ready.

(4) The container used for dissolving reagents or fertilizers should be swabbed with clean water, and the water used should be poured into the liquid storage tank or reservoir.

(5) To accelerate the dissolution of reagents or fertilizers, warm water can be used for dissolution or a magnetic stirrer can be used for stirring.

(6) Upon formulation of the working solution, avoid adding the mother solution too quickly, which may cause many precipitates due to the high local concentration. If the precipitates cannot be dissolved after long-term water pump circulation, the nutrient solution should be discarded and formulated again.

(7) Establish record files for inspection.

V. Answer Questions

(1) What is the purpose of mother solution formulation?

(2) What are the consequences of inaccurate nutrient solution formulation?

(3) What are the differences between the nutrient solution used for soilless culture and the medium used for tissue culture?

Module 4 Solid Substrate for Soilless Culture

[Learning Objectives]

I. Knowledge Objectives

(1) Master the properties, functions, disinfection methods, and selection principles of common substrates;

(2) Understand the physicochemical properties of solid substrates and the characteristics of common substrates;

(3) Learn about the requirements of substrate replacement and the development trend of substrate selection.

II. Skill Objectives

(1) Be able to identify different types of soilless culture substrates;

(2) Select the correct soilless culture substrate according to cultivated crops;

(3) Be able to disinfect the culture substrate.

[Preparation for Learning]

I. Required Resources

(1) Watch classic solid substrate videos before class;

(2) Visit an agricultural enterprise to observe the solid substrate culture;

(3) A multimedia classroom;

(4) Learn about the solid substrate through the Internet.

II. Background Knowledge

(1) Master the basic knowledge related to solid substrate for soilless culture;

(2) Master the basic knowledge of literature review;

(3) Learn about some applications of substrate culture;

(4) Learn about the relevant knowledge of soil and cultivation.

[Learning Tasks]

Task 1 Effect and Physicochemical Property of Solid Substrate

In soilless culture, solid substrates are widely used. A small amount of substrates is also needed to fix and support crops during seedling raising and planting. Commonly used substrates include sand, gravel, perlite, vermiculite, rock wool, turf, sawdust, carbonized rice husk, various foam plastics, and ceramsite.

I. Role of Solid Substrate

1. Supporting and fixing plants

Solid substrate can support and fix plants, make them rooted in the solid substrate without sinking and lodging, and provide the plant root system with a good growth environment that, for example, can facilitate the extension and adhesion of the plant root system.

2. Retaining moisture

All substrates have a certain moisture-retaining capacity, and the water retention capacity of substrates varies greatly. The substrate has a certain water-retaining property. Solid substrate absorbs water to prevent crops from being damaged by water loss during irrigation. For example, it can prevent plants from failing to absorb water and nutrients, drying up, and dying during the intermittent period of liquid supply and sudden power failure.

3. Air permeability

Air exists in the pores of the solid substrate, which can supply the oxygen needed by crop roots for respiration. The pores of the solid substrate are also the places where water is absorbed and held. Therefore, in the solid substrate, there is a unity of opposites between air permeability and water retention. It is required that the solid substrate should have a certain amount of both large pores and small pores in an appropriate proportion and can satisfy both water and oxygen requirements of the plant root system, so as to facilitate the growth and development of the root system.

4. Buffering

Buffering refers to the ability of a solid substrate to provide a stable environment for the growth of the plant root system. That is, when harmful substances or extraneous substances produced in the process of the root system growth are likely to harm the normal growth of plants, the solid substrate will reduce or even dissolve these hazards through some of its own physicochemical properties. Solid substrates with physicochemical absorption capacity, such as turf and vermiculite, have a buffering effect and are called active substrates. Substrates with weak or no buffering effect, such as river sand, gravel, and rock wool, are called inert substrates.

5. Providing nutrients

Organic solid substrates, such as peat, coconut fiber, smoked charcoal, and reed residue substrate, can provide certain mineral nutrients in the seedling stage or production stage.

II. Physical Properties of Solid Substrates

The quality of the substrate is first determined by its physical properties. The major indicators that reflect the physical properties of the substrate include particle size (particle diameter), bulk density, total porosity, and void ratio.

1. Bulk density

It refers to the weight per unit volume of dry substrate, generally expressed in g/L or g/cm^3. Bulk density is different from specific weight, which is the mass per unit volume of solid substrate (with substrate porosity not calculated) and is based on the volume of the substrate itself.

High bulk density indicates that the substrate is too compact and not loose enough with excellent water-holding capacity but poor air permeability. Low bulk density indicates that the substrate is too loose with high air permeability, which is conducive to the extension and growth of the root system, but with poor water-holding capacity, making it hard to fix the root system. The bulk density and specific gravity of several commonly used substrates are shown in Table 4-1.

Table 4-1 Bulk Density and Specific Gravity of Several Commonly Used Substrates

Substrate type	Bulk density (g/cm^3, approximate)	Specific weight (g/cm^3)
Soil	1.10–1.70	2.45
Sand	1.30–1.50	2.62
Vermiculite	0.08–0.13	2.61
Perlite	0.03–0.16	2.37
Rock wool	0.04–0.11	–
Turf	0.05–0.20	1.55
Bagasse	0.12–0.28	–
Bark	0.10–0.30	2.00
Pine needles	0.10–0.25	1.90

2. Total porosity

Total porosity refers to the sum of water-holding porosity and aeration porosity in the substrate, expressed as a percentage (%) of the substrate volume. The substrate with high total porosity has more space to hold air and water, and vice versa. Total porosity can be calculated as per the following equation:

$$\text{Total porosity} = \left(1 - \frac{\text{Bulk density}}{\text{Specific weight}}\right) \times 100\%$$

High total porosity means light and loose substrate which is conducive to the growth of the root system but is not beneficial for its fixation, making crops easier to fall over. For example, the total porosity of such substrates as bagasse, vermiculite, and rock wool is 90%–95% or higher(Table 4-2). The substrate with low total porosity is heavy, holding a small amount of water and air. For example, the total porosity of sand is about 30%. Therefore, in practice, to solve the problem that the total porosity of a single substrate is too high or too low, two or three substrates with different particle sizes are often mixed to make a complex substrate.

Table 4-2 Physical Properties of Several Commonly Used Substrates

Substrate	Bulk density (g/cm^3)	Total porosity (%)	Macropores (%) (aeration volume)	Micropores (%) (capillary volume)	Void ratio (Macropores / Micropores)
Vegetable garden soil	1.10	66.0	21.0	45.0	0.47
Sand	1.49	30.5	29.5	32.6	90.50
Coal cinder	0.70	54.7	21.7	33.0	0.66
Vermiculite	0.13	95.0	30.0	65.0	0.46
Perlite	0.16	93.2	53.0	40.0	1.33
Rock wool	0.11	96.0	2.0	94.0	0.02
Peat	0.21	84.4	7.1	77.3	0.09
Sawdust	0.19	78.3	34.5	43.8	0.79
Carbonized rice husk	0.15	82.5	57.5	25.0	2.30
Bagasse (retting for 6 months)	0.12	90.8	44.5	46.3	0.96

3. Void ratio (substrate gas-water ratio)

Gas-water ratio refers to the relative ratio of gas and water contained in the substrate in a certain period of time, usually expressed by the ratio of large pores to small pores of the substrate. It is expressed as the following equation:

$$\text{Void ratio} = \frac{\text{Air-filled pores (\%)}}{\text{Water-holding pores (\%)}}$$

The void ratio can reflect the distribution of air and water in the substrate and is an important indicator of substrate quality. It can comprehensively indicate the state of air and water in the substrate when used in combination with total porosity. A high void ratio means a high air-holding capacity and a low water-holding capacity. Generally speaking, a proper void ratio is in the range of 1 : 2–1 : 4, where the substrate has a high water-holding capacity and excellent air permeability, promising good crop growth and convenient management.

4. Particle diameter (particle size)

The particle size of the substrate directly affects the bulk density, total porosity, and void ratio. Particle size is expressed as particle diameter (mm): the larger the particle of the same substrate, the higher the bulk density, the lower the total porosity, and the higher the void ratio. In contrast, the finer the particle, the lower the bulk density, the higher the total porosity and the lower the void ratio. Therefore, the substrate particles should be moderate in size with rough but not angular surfaces and abundant and properly proportioned pores. The proper particle size varies among different types of substrates: It should be 0.5–2.0 mm for sand particles, within 1 cm for ceramsite, and whatever for substrates such as (blocky) rock wool as their particle size will not exert any of the above impacts.

III. Chemical Properties of Solid Substrates

The main chemical properties of the substrate that have a large impact on the growth of cultivated crops are the chemical composition of the substrate and the resulting chemical stability, pH value, physical and chemical absorption capacities (cation exchange capacity), buffering capacity, and electrical conductivity. Understanding the chemical properties and functions of substrates helps to well guide the selection of substrates and the formulation and management of nutrient solutions, thus improving cultivation management.

(I) Chemical Composition and Stability of Substrates

The chemical composition of the substrate refers to the type and content of chemical substances contained, including organic and mineral nutrients that can be absorbed and utilized by plants, as well as toxic and harmful substances. The chemical stability of a substrate is its chemical resistance. Different types of substrates have different chemical compositions, leading to different chemical stability (Table 4-3). Generally speaking, substrates mainly composed of inorganic substances, such as river sand and gravel, have high chemical stability, while those mainly composed of organic substances, such as wood chips and rice husks, have low chemical stability. However, turf shows stable chemical properties and is the safest to use.

The chemical stability of the substrates varies greatly with chemical composition. The substrates, such as sand and gravel that are composed of inorganic minerals, such as quartz, feldspar, and mica, have the highest chemical stability, followed by substrates made of hornblende, pyroxene, etc. The most unstable substrates consist of carbonate minerals such as limestone and dolomite. In soilless culture, the former two types will not affect the chemical equilibrium of nutrient solution by producing relevant harmful substances, but the latter will to a great extent by producing calcium and magnesium ions, which should always be noted.

Substrates composed of plant residues, such as peat, wood, rice husks, and bagasse, have complex chemical compositions and a great impact on the nutrient solution. Chemical composition can be roughly classified into three categories based on the impact on the chemical stability of the substrate: Category Ⅰ substances are those susceptible to microbiological decomposition, such as sugar, starch, hemicellulose, cellulose, and organic acid in carbohydrates. Category Ⅱ substances are toxic such as certain organic acids and phenols (tannin). Category Ⅲ substances are those that are not easily decomposed by microorganisms, such as lignin and humus. Substrates containing Category Ⅰ substances (such as fresh straw and bagasse) will cause strong biochemical reactions due to microbial activity in the early stage, which will seriously affect the equilibrium of the nutrient solution. The most

Table 4-3 Content of Nutrient Elements in Common Substrates

Substrate	Vegetable garden soil	Coal cinder	Vermiculite	Perlite	Rock wool	Cotton seed shells	Carbonized rice husks
Total nitrogen (%)	0.106	0.183	0.011	0.005	0.084	2.200	0.540
Total phosphorus (%)	0.077	0.033	0.063	0.082	0.228	0.210	0.049
Rapidly available phosphorus (mg/L)	50.0	23.0	3.0	2.5	–	–	66.0
Rapidly available potassium (mg/L)	120.50	203.90	501.60	162.20	1.34*	0.17*	6,625.50
Rapidly available calcium (mg/L)	324.7	9,247.5	2,560.5	694.5	–	–	884.5
Exchangeable magnesium (mg/L)	330.0	200.0	474.0	65.0	–	–	175.0
Rapidly available copper (mg/L)	5.78	4.00	1.95	3.50	–	–	1.36
Rapidly available zinc (mg/L)	11.23	66.42	4.00	18.19	–	–	31.30
Rapidly available iron (mg/L)	28.22	14.44	9.65	5.68	–	–	4.58
Rapidly available boron (mg/L)	0.425	20.300	1.063	–	–	–	1.290

Note: * is the percentage of total potassium (%).

noted consequence is a serious nitrogen deficiency. Substrates that contain many Category Ⅱ substances can directly harm the root system. Therefore, the substrate with more substances of Categories Ⅰ and Ⅱ cannot be used directly without treatment. Substrates containing mainly Category Ⅲ substances are the most stable and safest to use, such as peat, retted and decomposed wood chips, bark, and bagasse. Retting is to eliminate easily decomposed and toxic substances in the substrate and convert them into those that are difficult to decompose.

(II) Acidity and Alkalinity (pH) of the Substrate

The pH value indicates the acidity and alkalinity of the substrate. pH=7: neutral; pH<7: acidic; and pH>7: alkaline. The substrate itself has a certain acidity or alkalinity. Substrates with too-high acidity or alkalinity will affect the acidity or alkalinity of the nutrient solution, seriously disrupting the chemical equilibrium of the nutrient solution and hindering nutrient absorption of plants in severe cases. Therefore, a general understanding of the pH of a substrate is required before it is selected, so as to take corresponding measures to regulate the pH level. Although most ornamental plants are adaptable in the pH range of 5.5–6.5, a proper pH level of a substrate should be in the range of 6.5–7.0 to facilitate easy artificial regulation and avoid any influence on the effects of certain components in the nutrient solution and subsequent physiological disorders of the plant after the nutrient solution is supplied.

(III) Cation Exchange Capacity (CEC)

The cation exchange capacity (CEC) of the substrate is expressed as the milligram equivalents (me/100 g of the substrate) of absorbed cations exchanged per 100 g of substrate. The cation exchange capacity can not only reflect the capacity of the substrate to absorb and retain fertilizer nutrients and to protect the fertilizer ions from being leached by water and slowly release them for plant absorption, but also can buffer the acid-base reaction of the nutrient solution. In general, organic substrates with high cation exchange capacity have high buffering capacity which helps to resist nutrient leaching and excessive pH rise and fall. The cation exchange capacity of several commonly used substrates is listed in Table 4-4 to illustrate the differences.

Table 4-4 Cation Exchange Capacity of Several Commonly Used Substrates

Substrate type	Cation exchange capacity (me/100g)
High moor peat	140–160
Medium moor peat	70–80
Vermiculite	100–150
Bark	70–80
Inert substrates such as sand, gravel, rock wool,etc.	0.1–1

(IV) Electrical Conductivity of Substrate

The electrical conductivity of the substrate means the conductivity of the substrate itself before the nutrient solution is added. It can be determined with a conductivity meter, and reflects the amount of soluble salt originally contained in the substrate, which will directly affect the equilibrium of the nutrient solution. For example, sand affected by seawater often contains a high salt content and should be properly treated. The soluble salt content in the substrate should not exceed 1,000 mg/kg and preferably not exceed 500 mg/kg. The electrical conductivity of the substrate should be determined before use to facilitate leaching with freshwater or other appropriate treatment.

Based on the correlation between the electrical conductivity of the substrate and nitrate nitrogen, the conductivity value can be used to infer the nitrogen content of the substrate and determine whether nitrogen fertilizer is required. Generally, in flower cultivation, fertilizer must be applied when the electrical conductivity is lower than 0.37–0.5 mS/cm (equivalent to that of tap water); fertilizer is generally not applied when the conductivity reaches 1.3–2.75 mS/cm and salt leaching should be carried out where possible; the conductivity for cultivating vegetable crops should be greater than 1 mS/cm.

(V) Buffering Capacity of Substrates

The buffering capacity of a substrate refers to its ability to buffer pH changes after fertilizer is applied. The buffering capacity is mainly determined by cation exchange capacity and the content of weak acids and salts present in the substrate. Generally, high cation exchange capacity means high buffering capacity. Substrates containing rich calcium carbonate and magnesium salts show strong buffering capacity for acid, but no buffering capacity for alkali. Substrates containing rich humus show buffering capacity for both acid and alkali. Substrates ranking by buffering capacity are organic substrate > inorganic substrate > inert substrate > nutrient solution. Some of the commonly used mineral substrates show strong buffering capacity, such as vermiculite. Nevertheless, most of them have weak buffering capacity. Therefore, it is required to understand the buffering capacity of substrates so as to make the best use of their advantages and avoid their disadvantages.

(VI) Carbon-nitrogen Ratio

It is the relative ratio of carbon to nitrogen in the substrate. Substrates with a high carbon-nitrogen ratio (high carbon content to low nitrogen content) may cause nitrogen deficiency in plants as microorganisms compete for nitrogen during vital activities. Substrates with a high carbon-nitrogen ratio disallow plants to grow normally and develop properly, even if good culture techniques are adopted. Therefore, the dosage of organic substrates such as wood chips and bagasse shall not exceed 20% to prepare a mixed substrate, or 8 kg of nitrogen fertilizer should be added per m^3. The substrate should not be used before being composted for 2–3 months. In addition, the organic substrate with large particles has a slow decomposition rate as its surface area is smaller than its volume, and its effective carbon-nitrogen ratio is lower than that of the organic substrate with fine particles. Therefore, the substrates with coarse particles, especially those with a low carbon-nitrogen ratio should be used where possible.

According to general provisions, a carbon-nitrogen ratio of 200∶1–500∶1 is considered medium, lower than 200∶1 is considered low, and higher than 500∶1 is considered high. Generally, the carbon-nitrogen ratio should be medium to low rather than high. C∶N=30∶1 or so is more suitable for crop growth.

Task 2 Classification and Characteristics of Solid Substrates

I. Inorganic Substrates and Organic Substrates

Inorganic substrates mainly refer to some natural minerals or their products after being treated at a high temperature that are used as substrates for soilless culture, such as sand, gravel, ceramsite, vermiculite, rock wool, and perlite. Their chemical properties are stable, usually with a low cation exchange capacity, and their fertilizer storage capacity is poor.

Organic substrates are mainly culture substrates composed of organic residues containing C and H and their derivatives, such as turf, coconut chaff, bark, sawdust, and fungus residue. The chemical properties of organic substrates are often unstable. They usually have a high cation exchange capacity and relatively strong fertilizer storage capacity.

II. Complex Substrate

Complex substrate, also known as mixed substrate, refers to a culture substrate prepared by mixing two or more substrates in a certain proportion. This kind of substrate is produced by mixing two or more substrates in order to overcome the disadvantages such as being underweight, overweight, or poor or excessive ventilation caused by a single substrate in production. The mixed substrate combines substrates with different characteristics so that their components can complement each other, thereby enabling each performance indicator of the substrate to meet the required standard, so it is more and more widely used in production. Theoretically speaking, the more types of substrates to be mixed, the better the effect. However, due to the high labor cost when mixing substrates, the types of substrates to be mixed should be minimized for practical reasons. Generally, it's 2–3 types of substrates should be mixed in production.

III. Principles for the Selection of Solid Substrates

The substrate is an important constituent material in soilless culture. Therefore, the selection of substrates is very important. It requires that the substrate not only provide good nutrition and environmental conditions for the plant root system, just like soil, but also provide more convenient conditions for improving management measures. Therefore, the substrate should be carefully selected according to the specific situation. The principles for substrate selection can be based on the following four considerations: the applicability of the plant root systems, the economic efficiency of the substrate, marketability, and environment-friendliness.

Task 3 Examples of Solid Substrate Selection

I. Inorganic Culture Substrate

Inorganic substrate, as a major kind of substrate, is widely used in production. Rock wool, sand, gravel, and vermiculite are commonly used inorganic substrates. Although these substrates belong to the same category, their physicochemical properties are different. In order to give full play to the potential of inorganic substrates in soilless culture and obtain good benefits, it is particularly important to learn about the characteristics of the substrates.

1. Gravel

Gravel mainly comes from river stones or rock debris in quarries. Because of its different sources, its chemical composition and properties are very different. Generally, non-calcareous gravel, such as granite gravel, should be used in soilless culture. If calcareous gravel needs to be used as a last resort, the phosphate treatment method described above can be used to treat the gravel surface.

The particle size of gravel should be in the range of 1.6–20 mm, and the diameter of gravel accounting for half of the total volume should be about 13 mm. Gravel should be relatively hard and not easily broken. It is better to select gravel with fewer sharp edges and corners. Especially for tall plants or in the open air with strong wind, gravels with blunt edges and corners should be selected; otherwise, the stems of the plants will be scratched. Gravel has no cation exchange capacity. It has good ventilation and drainage performance but has poor water retention capacity.

Because of the large volume of gravel (1.5–1.8 g/cm^3), it brings great problems to daily management such as handling, cleaning, and disinfection. Moreover, when soilless culture is carried out with gravel, a solid water trough (usually made of cement bricks) should be built to circulate the nutrient solution. These shortcomings have made gravel culture less and less used in modern soilless culture. Especially in the recent 20–30 years, some lightweight synthetic substrates, such as rock wool and haydite (porous ceramsite), have been widely used, and they have gradually replaced sand and gravel as substrates. However, gravel played an important role in the early soilless culture production, and in today's deep flow technique, it is still very suitable to be used as an object to fix the plant in a planting cup.

2. Rock wool

Rock wool is a synthetic inorganic substrate, which is considered to be one of the best substrates for soilless culture and has been widely used in the world. It can provide a good rhizospheric environment with fertilizer and water conservation, sterility, and sufficient air supply for plants. In soilless culture, rock wool is mainly used in three areas: raising seedlings with rock wool, fixing the plants in the recirculating nutrient solution culture (such as NFT), and bag culture drip irrigation technique of rock wool substrate.

Since rock wool is easy and convenient to use, low in cost, and excellent in performance, rock wool culture is widely used all over the world. In soilless culture, the rock wool culture area ranks first. However, rock wool culture requires drip irrigation facilities and good cultivation techniques. When mixed with soil or other substrates, it is better to adopt granular rock wool since it is difficult to shred block or slab rock wool into small pieces and mix them evenly. In addition, the waste block or slab rock wool is extremely difficult to decompose in the soil and can damage the tillage characteristics of the soil, so it is regarded as a pollutant.

The physicochemical properties of rock wool are as follows.

(1) Excellent physical properties: Rock wool is a white or light green filament, and it is light in texture. It does not decay and decompose, and its bulk density is generally 70–100 kg/m^3. The total porosity is large, up to 96%–100%, where the large pores account for 64.3% and the small pores account for 35.7%, and the gas-water ratio is 1 : 0.55. It has good air permeability and strong water absorption ability, and can absorb an amount of water equivalent to 13–15 times its own weight. After rock wool absorbs water, the water content will decrease from bottom to top because of its varying thickness, and the air content will increase from bottom to top. See Table 4-5 for the vertical distribution of water and air in rock wool block.

Table 4-5 Vertical Distribution of Water and Air in Rock Wool Block

Height (cm) (from top to bottom)		Dry matter volume (%)	Pore volume (%)	Water-holding volume (%)	Air volume (%)
Top ↓ Bottom	1.0	3.8	96	92	4
	5.0	3.8	96	85	11
	7.5	3.8	96	78	18
	10.0	3.8	96	74	22
	15.0	3.8	96	74	42

(2) Stable chemical properties: Rock wool is composed of silica and some metal oxides and has the characteristics of low carbon-nitrogen ratio and low cation exchange capacity. It contains 0.084% total nitrogen and 0.228% total phosphorus. In the mineral composition, silicon dioxide accounts for 35.5%–47.0% and aluminum, calcium, magnesium, iron, manganese, sodium, potassium, and sulfur account for 53.0%–64.5% (Table 4-6). Most of these main components are inert substrates that plants cannot absorb and utilize. The pH value of new rock wool is relatively high, generally 7–8. It needs to be rinsed with clean water or have a small amount of acid added to it before use. The adjusted pH value of agricultural rock wool is relatively stable.

(3) Rock wool fiber does not adsorb element ions in the nutrient solution, and

the nutrient solution can be fully supplied to the crop root system for absorption.

(4) Rock wool can be completely sterilized at high temperatures with no pathogenic bacteria and thus can be used directly.

Table 4-6 Chemical Composition of Rock Wool

Component	Content (%)	Component	Content (%)
Silicon dioxide (SiO_2)	47	Sodium oxide (Na_2O)	2
Calcium oxide (CaO)	16	Potassium oxide (K_2O)	1
Aluminum oxide (Al_2O_3)	14	Manganese oxide (MnO)	1
Iron oxide (FeO)	8	Titanium oxide (TiO)	1

3. Vermiculite

Vermiculite is a secondary siliceous mineral of mica, which is a hydrous silicate of aluminum, magnesium, and iron. It is composed of laminated layers of thin sheets. It is a spongy substance formed after being heated in the furnace at 800–1,100℃. It has a light weight of 80–160 kg per cubic meter, a small bulk density (0.07–0.25 g/cm^3), a total porosity of up to 95%, a void ratio of about 1∶4, and a gas-water ratio of 1∶4.34. It has good air permeability and water-retaining properties, and its electrical conductivity is 0.36 mS/cm. Vermiculite has a strong water absorption capacity and can absorb 100–650 kg of water per cubic meter. It has a low carbon-nitrogen ratio, high cation exchange capacity, strong fertility preservation capacity, and buffering capacity. Vermiculite contains relatively more calcium, magnesium, potassium and iron, which can be absorbed and utilized by crops. The pH value of vermiculite varies slightly depending on different producing areas and different components. It is generally neutral to slightly alkaline (with a pH value of 6.5–9.0). It can be mixed with acidic substrates such as peat without any problem. If it is used alone, because the pH value is too high, a small amount of acid needs to be added for neutralization.

The particle size of vermiculite for soilless culture should be greater than 3 mm, and that of vermiculite for seedling culture can be slightly smaller

(0.75–1.0 mm). After a period of use, it could be easily broken due to collapse, decomposition, settlement, or other reasons, resulting in the structure being destroyed and thinned and the porosity being reduced, which will affect the air permeability and water drainage, so it should not be subjected to heavy pressure during transportation and planting. Generally, its structure will deteriorate after being used 1–2 times, so it is generally not suitable to be used as a substrate for long-term potted plants and needs to be replaced. After use, it can be used as fertilizer or applied to soil.

4. Perlite

Perlite is a kind of gray volcanic rock (aluminosilicate). When heated to 1,000℃, the rock particles will expand to form grayish-white, lightweight, porous closed-cell loose nuclear particles with uniform texture and a diameter of 1.5–1.4 mm, which is also known as expanded perlite or "spongolite". Perlite has a small bulk density of 0.13–0.16 g/cm^3, a total porosity of 60.3% and a gas-water ratio of 1 : 1.04. It can hold water of 3–4 times its own weight and can drain and ventilate easily. Perlite has a stable chemical property. It contains silicon, aluminum, iron, calcium, manganese, potassium and other oxides, and most of its nutrients cannot be absorbed and utilized by plants. Its electrical conductivity is 0.31 mS/cm. It is neutral, low in carbon-nitrogen ratio, small in cation exchange capacity, non-buffering, and difficult to decompose but easy to break when subjected to collision. Perlite can be used alone, but it is lightweight and could cause heavy dust pollution. Before use, it is better to wear a mask and spray it with water to avoid dust. Over-watering could make it float easily, which is not conducive to fixing the root system, so it is often used in combination with other substrates.

II. Organic Culture Substrate

The chemical properties of organic substrates are generally unstable. They usually have a high cation exchange capacity and relatively strong fertilizer storage capacity. In soilless culture, organic substrates generally have the advantages of good water-retaining property and strong fertilizer storage capacity and are widely used in practical soilless culture.

1. Peat

Peat, also known as turf, comes from the decomposition residues of peat moss, hypnum moss, bryophyte, and other aquatic plants, and is recognized as one of the best soilless culture substrates in the world so far. Especially in modern large-scale industrialized seedling culture, peat is mostly used as the main substrate, in which a certain amount of vermiculite, perlite, and other substrates are added to make peat pots (small pieces) containing nutrients or directly put in seedling trays to raise seedlings, and the effect is very good. Apart from being used for seedling culture, peat is often used as a substrate in drip irrigation of nutrient solution in bag culture or plantation trough culture, which can make plants grow well. The bulk density of peat is 0.2–0.6 g/cm^3 (high/low moor peat is lower/higher than this range), the total porosity is 77%–84%, the large pores account for 5%–30%, the water-holding capacity is 50%–55%, the electrical conductivity is 1.1 mS/cm, the cation exchange capacity is medium or high, the carbon-nitrogen ratio is low or medium, and the water content is 30%–40%. Turf is distributed in almost all countries around the world, but the distribution is very uneven, with more in the north and less in the south. The turf produced in the north of China is of good quality.

According to the different geographical conditions, plant species and decomposition degree of peat, peat can be divided into three categories: low moor peat, high moor peat, and medium moor peat. Low moor peat is distributed in low-lying swamps, so it should be applied directly as fertilizer rather than as a substrate for soilless culture. High moor peat is distributed in the high part of the terrain formed by low moor peat, and it mainly comes from moss plants. High moor peat should not be used as fertilizer directly but should be used as the adsorbent of fertilizer, and can be used as the raw material of complex substrate in soilless culture. Medium moor peat is a transitional type between high moor peat and low moor peat, which can be used in soilless culture. In the producing areas, peat is cheap because of its simple mining and processing. However, in non-producing areas, due to long-distance transportation and fine processing, the price of high-quality peat (organic matter: 95%, ash content: ≤ 5%, expansion rate: 200%) is relatively high. Peat is the second most frequently used substrate for soilless culture in the world, second only to rock wool. See Table 4-7 for the physical properties of peat from different sources.

Table 4-7 Physical Properties of Peat from Different Sources

Peat type	Bulk density (g/L)	Total porosity (%)	Air volume (%)	Easily available water volume (%)	Water absorption capacity (g/100g)
Moss peat (High moor peat)	42	97.1	72.9	7.5	992
	58	95.9	37.2	26.8	1,159
	62	95.6	25.5	34.6	1,383
	73	94.9	22.2	35.1	1,001
White peat (Medium moor peat)	71	95.1	57.3	18.3	869
	92	93.6	44.7	22.2	722
	93	93.6	31.5	27.3	754
	96	93.4	44.2	21.0	694
Black peat (Low moor peat)	165	88.2	9.9	37.7	519
	199	88.5	7.2	40.1	582
	214	84.7	7.1	35.9	487
	265	79.9	4.5	41.2	467

2. Reed residue

Make use of the paper mills' scraps—reed residues, add a certain proportion of chicken manure and other auxiliary materials, and compost, ferment, and synthesize high-quality environmentally-friendly organic reed residue substrate for soilless culture under the action of fermentative microorganisms. It was developed by Nanjing Agricultural University and other institutions and has been widely used in soilless culture and seedling culture, especially in the Yangtze River basin in southern China. Its physicochemical properties are as follows: bulk density: 0.20–0.40 g/cm^3, total porosity: 80%–90%, void ratio: 0.5–1.0, electrical conductivity: 1.20–1.70 mS/cm, pH value: 7.0–8.0, cation exchange capacity (CEC): 60–80 me/100g. Besides, it has a strong acid-base buffering capacity. It is basically comparable to natural peat, and especially it contains a variety of nutrient elements. The content of trace elements can basically meet the requirements of crop growth and development, so it is also known as artificial peat.

3. Bark

Bark is a scrap from wood processing. In places rich in wood, such as in Canada and the United States, bark is often used to replace peat as the soilless culture substrate. The chemical composition of bark varies greatly from tree to tree. The chemical composition of pine bark is as follows: The organic matter content is 98%, including wax resin (3.9%), tannin lignin (3.3%), starch pectin (4%), cellulose (2.3%), hemicellulose (19.1%), lignin (46.3%), ash (2%), etc. The C/N ratio is 135, and the pH value is 4.2–4.5.

Some bark contains toxic substances and cannot be used directly. Most bark contains relatively more phenolic substances, which are harmful to plant growth, and the C/N ratio of bark is relatively high, so direct use will cause competition between microorganisms for nitrogen. In order to overcome these problems, fresh bark must be retted, and the retting time should last for at least one month because the decomposition of toxic phenolic substances will take at least 30 days.

The retting not only decomposes the toxic phenolic substances and reduces the C/N ratio of bark but also increases the cation exchange capacity of the bark. The CEC can be increased from 8 me/100 g before retting to 60 me/100 g after retting. After retting, most of the pathogenic bacteria, nematodes and weed seeds contained in the bark will be killed, and no additional disinfection is needed during use.

The bulk density of bark is 0.4–0.53 g/cm^3. When bark is used as a substrate, its bulk density will increase, its volume will decrease, and its structure will be destroyed due to the decomposition of substances, resulting in poor ventilation and water accumulation. However, it will take about one year for the structure to deteriorate. Bark should not be used as a substrate if the oxide content in bark exceeds 2.5% and the manganese content exceeds 20 mg/kg.

The nature of bark is basically similar to sawdust, but it has stronger air permeability, weaker water-holding capacity, and higher resistance to decomposition. Before use, it should be broken into pieces of 1–6 mm in size, and it is better to be composted and thoroughly decomposed. Generally, it is used in combination with other substrates, and its application amount accounts for 25%–75% of the total volume. If it is used 100% alone, great care must be taken with watering and fertilizing due to excessive ventilation, so it is only used for planting orchids.

4. Bagasse

Bagasse, a by-product of the cane sugar industry, has a wide range of sources in southern China. The C/N ratio of fresh bagasse could be as high as 169, so it cannot be directly used as a substrate but may only be used after being retted. Two methods can be used for retting. The first method is to drench bagasse with water until the water content reaches 70%–80% (it is advisable to hold a handful of bagasse by hand until just a little water seeps out), and then stack it up. The second method is to add urea in the proportion of 0.5%–1.0% of the dry weight of bagasse. Adding urea can speed up the decomposition of bagasse and the reduction of C/N ratio. After a period of retting, the C/N ratio and physical properties of the bagasse change greatly (Table 4-8).

Table 4-8 Changes of Physicochemical Properties of Bagasse after Retting

Retting time	Total carbon	Total nitrogen	C/N ratio	Bulk density (g/L)	Aeration porosity (%)	Water-holding porosity (%)	Void ratio	pH value
Fresh bagasse	45.26	0.2680	169	127.0	53.5	39.3	1.36	4.68
Retting for 3 months	44.01	0.3105	142	118.5	45.2	46.2	0.98	4.86
Retting for 6 months	42.96	0.3613	119	115.5	44.5	46.3	0.96	5.30
Retting for 9 months	34.30	0.6058	56	205.0	26.9	60.3	0.45	5.67
Retting for 12 months	31.33	0.6375	49	278.5	19.0	63.5	0.30	5.42

Source: Laboratory of Crop Nutrition and Fertilization, South China Agricultural University, 1987.

If the bagasse retting time is too long (more than 6 months), bagasse will be poorly ventilated due to excessive decomposition, and its tolerance to additional available nitrogen is poor, so it is better to retted for 3–6 months in practical application. After retting and adding nitrogen fertilizer, bagasse can become a good substrate with the same planting effect as peat. This creates favorable material conditions for developing soilless culture in the southern region where sugarcane is produced. When bagasse is used as a culture substrate, bagasse should be fine, and the maximum particle size should not exceed 5 mm. When it is used for bag culture or trough culture, the particle size can be slightly coarse, but the maximum size should not exceed 15 mm.

5. Rice husk

Rice husk is a by-product of rice processing. Rice husk used in soilless culture is usually carbonized, which is called carbonized rice husk or carbonized rice chaff. The bulk density of carbonized rice husk is 0.15 g/m^3, the total porosity is 82.5%, of which the large pore volume accounts for 57.5%, the small pore volume accounts for 25%, the nitrogen content is 0.54%, the rapid available phosphorus content is 66 mg/kg, the rapidly available potassium content is 0.66%, and the pH value is 6.5. If carbonized rice husks are not washed with water before use, K_2CO_3 formed by carbonization will raise their pH value to above 9.0, so they should be washed with water before use.

Carbonized rice husks are carbonized at a high temperature, so they will not bear germs if they are free from external pollution. Carbonized rice husk contains abundant nutritional elements, and it is low-priced with good permeability. However, it has a small water retention porosity and poor water retention capacity, so it needs to be watered frequently during use. In addition, the rice husk should not be carbonized excessively; otherwise, it could easily break under pressure.

The carbonized rice husk is loose, with moderate air permeability and water absorbability, so it is not easy to be over-dried or over-wet (however, because of its strong hygroscopicity, if it is relatively thick and the liquid supply is sufficient, it is prone to wet injury). Although its ability to absorb nutrients is poor, it is rich in potassium, phosphorus, calcium, and other nutrient components, which can meet the needs of seedlings, so it is suitable for cutting and sowing. Because the pH value is slightly alkaline, the nutrients contained in it will interfere with the formulation of the nutrient solution, and the resources are not easy to obtain. It is generally not used as a substrate alone except for cutting and sowing, but is usually used in combination with other substrates to improve air permeability and soil fertility. In addition, the ash produced with rice chaff as fuel is called rice chaff ash or rice husk ash, which is actually not the same thing as carbonized rice husk. The former has a high burning degree, low carbon content, fine particles, and high ash content; the latter, on the contrary, has better performance than the former. Although some people confuse rice chaff ash with carbonized rice husk, the main substrate used in soilless culture is carbonized rice husk.

6. Coconut chaff

Coconut chaff, also known as golden coconut powder, is a compressed plant culture material, which is the waste from the processing of coconut husk. Coconut fruit is coated with a thick layer of fibrous material, which can be processed into coconut fiber. The coconut fiber can be made into ropes and other products. A large amount of powdery material, known as coconut chaff, can be produced during coconut fiber processing. Because of its coarse particles, strong absorption capacity, good air permeability and water drainage performance, its ability to retain water and maintain fertilizer is also strong. In addition, coconut fiber, after being cut into small pieces, or coconut husk, after being cut into blocks, may be used as the culture substrate. Coconut fiber that has not been finely cut and compressed contains filaments and has a fluffy texture. The finely-cut and compressed coconut fiber is brick-shaped, each piece weighing 450 g or 600 g. After being soaked in 3,000–4,000 mL of water, the volume can be expanded to 6,000–8,000 cm^3, and the wet bulk density is 0.55 g/cm^3. The pH value is 5.8–6.7. Its water absorption capacity is about 5–6 times its own weight. Because it is a plant-based organic substrate, and the ratio of nitrogen to phosphorus is relatively high, if there is only clear water poured on the plants, it is easy to show symptoms of nutrient deficiency, and especially the plant leaf will turn pale green or yellowish-green (possibly due to nitrogen deficiency). Because the pH value, bulk density, air permeability, water-holding capacity, and price are relatively moderate, after the coconut chaff, perlite, coal ash slag, and volcanic ash are mixed, the formulated pot substrate is relatively ideal, and especially for foliage plant substrate, the effect is very good. Coconut chaff resources in the South China Sea are abundant, and their development and utilization prospects are quite good.

III. Complex Substrate

Complex substrate, also called mixed substrate, is made by mixing two or more substrates in a certain proportion. The types and proportions of substrates vary with the species of plants to be cultivated. Each substrate has its own advantages and disadvantages when used in soilless culture, so there are many different problems existing in single substrate culture. Because of their complementary advantages, mixed substrates have ideal performance indicators.

1. General principles of substrate selection (mixing)

The general requirements for substrate mixing are appropriate bulk density, increased porosity, increased moisture content, and increased air content. Meanwhile, it is necessary to take into account the characteristics of the mixed substrate for culture and combine them with the crop nutrient solution formula, so as to give full play to its potential of high yields and high quality.

2. Substrate mixing method

When preparing the mixed substrate, a certain amount of fertilizer can be mixed in advance. It may be mixed with 0.25% NPK ternary compound fertilizer (15-15-15) in water or be added with potassium sulfate (0.5 g/L), ammonium nitrate (0.25 g/L), calcium superphosphate (1.5 g/L) and magnesium sulfate (0.25 g/L).

When the application rate of the mixed substrate is small, it can be stirred on the cement floor with a shovel; when the application rate is large, a concrete mixer should be used. Generally, dry turf is not easy to get wet, so it is necessary to spray water or add nonionic wetting agent one day in advance, and add 50 g of sodium hypochlorite into every 40 L of water to prepare a solution, which can wet 1 m^3 of the mixture. The turf blocks should be broken as much as possible during mixing; otherwise, it is not conducive to the growth of the plant root system.

In addition, a concrete mixer should be used during the substrate mixing if the application rate is large. Generally, the dry turf is not easy to get wet, so a nonionic wetting agent can be added. For example, if 50 g of sodium hypochlorite is added into every 40 L of water to prepare a solution, it can wet 1 m^3 of the mixture.

3. Mixed substrate for seedling culture and pot culture

Generally, turf is added into the seedling substrate. When the plant is taken out of the seedling-raising pot (tray), the substrate at the root of the plant is not easy to disperse. When the turf content in the mixed substrate is less than 50%, the substrate at the root of the plant could fall off easily. Therefore, care should be taken to prevent damage to the root system during transplanting. If turf is replaced by other substrates, limestone needn't be added to the mixed substrate. This is because limestone is mainly used to raise the pH value of the substrate. In order to make the seedlings grow strong, an appropriate amount of nitrogen, phosphorus, and potassium nutrients should be added during the mixing of the seedling substrate and pot substrate. The following are commonly used formulas for the seedling substrate and pot substrate.

(1) Mixed substrate of the University of California: 0.5 m^3 of fine sand (0.05–0.5 mm), 0.5 m^3 of crushed turf, 145 g of potassium nitrate, 4.5 kg of dolomite or limestone, 145 g of potassium sulfate, 1.5 kg of calcium limestone, and 1.5 kg of 20% calcium superphosphate.

(2) Mixed substrate of the Cornell University: 0.5 m^3 of crushed turf, 0.5 m^3 of vermiculite or perlite, 3.0 kg of limestone (preferably dolomite), 1.2 kg of calcium superphosphate (20% phosphorus pentoxide), and 3.0 kg of compound fertilizer (with a nitrogen, phosphorus, and potassium content ratio of 5 ∶ 10 ∶ 5).

(3) Pot substrate for soilless culture of the Institute of Vegetables and Flowers, Chinese Academy of Agricultural Sciences: 0.75 m^3 of turf, 0.13 m^3 of vermiculite, 0.12 m^3 of perlite, 3.0 kg of limestone, 1.0 kg of calcium superphosphate (20% phosphorus pentoxide), 1.5 kg of compound fertilizer (5-15-15), and 10.0 kg of sterilized dried chicken manure.

(4) Mixed substrate of turf minerals: 0.5 m^3 of turf, 700 g of calcium

superphosphate (20% phosphorus pentoxide), 0.5 m^3 of vermiculite, 3.5 kg of ground limestone or dolomite, and 700 g of ammonium nitrate.

Task 4 Disinfection and Replacement of Solid Substrates

I. Disinfection of Substrates

Many soilless culture substrates may contain some germs or worm eggs before use. Besides, germs and worm eggs may appear after long-term use, especially in the case of continuous cropping, which could easily cause diseases and insect pests. Therefore, it is necessary to disinfect most substrates before use or after harvest of each crop and before the next use, so as to eliminate any remaining germs or worm eggs. The most commonly used disinfection methods for substrates include steam disinfection, chemical disinfection, and solar disinfection.

(I) Steam Disinfection

Steam disinfection is simple, safe, and reliable, but it requires the use of expensive equipment that is not easy to operate. Put the substrate in a cabinet (box) (volume: 1–2 m^3), and carry out airtight disinfection with steam. Generally, germs can be killed by disinfection at 70–90℃ for 15–30 minutes. The substrate to be disinfected each time should not be too much; otherwise, the germs or worm eggs in the inner part of substrate cannot be completely killed. The water content of the substrate during disinfection should be controlled at 35%–45%. Too-wet or too-dry conditions may reduce the disinfection effect. When the amount of the substrate to be disinfected is large, the substrate may be piled up to a height of 20 cm, and the length depends on the conditions. Cover it with waterproof and high-temperature-resistant cloth, introduce steam, and disinfect it at 70–90℃ for 1 hour.

If a greenhouse heated by a steam boiler is used, the steam conversion device can be installed on the boiler, and the steam pipe can be directly connected to each planting bed to disinfect the substrate. If the steam disinfection is ineffective on the surface, a permanent tile pipe or any other hard pipe with holes can be installed at the bottom of the bed, so that the steam can enter the substrate through this pipe to achieve the purpose of disinfection.

(II) Chemical Disinfection

Commonly used chemicals include formaldehyde, chloropicrin, potassium permanganate, methyl bromide (bromomethane), metham, and bleaching agent.

1. 40% formaldehyde (formalin)

Formaldehyde is a good fungicide, but its insecticidal efficacy is poor. Generally, the 40% stock solution is diluted 50 times, and the substrate is sprayed wet evenly with a watering can. The required substrate content is generally 20–40 L/m^3. Cover and seal the substrate with plastic film for 24–48 hours, then uncover the film, spread out the substrate, air-dry it for 2 weeks or expose it to the sun for 2 days, and use the substrate until there is no formaldehyde smell in the substrate. Workers are required to wear masks and properly protect themselves.

2. Potassium permanganate

Potassium permanganate is a strong oxidant, which is generally used for disinfection of inert substrates such as gravel and coarse sand that have no absorption capacity and can easily be washed with clear water. However, it cannot be used to disinfect active substrates such as peat, sawdust, rock wool, and ceramsite that have a relatively large adsorption capacity or substrates that are difficult to wash with clear water because these substrates can adsorb potassium permanganate, directly poison the crops, or cause manganese poisoning of plants. When disinfecting the substrate, spray 0.1%–1.0% potassium permanganate solution on the solid substrate, mix it evenly with the substrate, embed the substrate in plastic for 20–30 minutes, and then rinse it with clear water.

3. Bleaching agent (sodium hypochlorite or calcium hypochlorite)

This disinfectant is especially suitable for the disinfection of gravel and sand. Generally, prepare a 0.3% –1% (available chlorine content) solution in a water tank, soak the substrate in the solution for more than half an hour, and then rinse it with clear water to eliminate the residual chlorine. This method is simple, rapid, and can be completed in a short time. Hypochloric acid can also be used to replace the bleaching agent for substrate disinfection.

(III) Solar Disinfection

Among the substrate disinfection methods, steam disinfection is relatively safe, but the cost is relatively high. The cost of agent disinfection is relatively low, but it is relatively unsafe and may even pollute the surrounding environment. Solar disinfection is a safe, cheap, simple, and practical disinfection method for substrates, and it is also suitable for solar greenhouse disinfection in China. The specific method is as follows: in the summer greenhouse or polytunnel leisure season, stack up the substrate to a height of 20–25 cm, with the length depending on the actual situation. While stacking the substrate, spray the substrate with water to make the water content exceed 80%, and then cover it with plastic cloth. In the case of trough culture, it may be directly watered in the pot and then covered with plastic film. In a closed greenhouse or polytunnel, exposure to the sun for 19–15 days can achieve the purpose of disinfection, and the effect is very good.

II. Substrate Replacement

After the substrate has been used for a period of time (1–3 years), various germs, crop root exudates, and rotten roots will accumulate in large quantities, and the physical properties will get worse, especially for the substrate with organic residues as the main material. Because of the decomposition of microorganisms, the fibers of these organic residues will break, resulting in reduced air permeability and high water retention of the substrate. These factors will affect the growth of crops, so it is necessary to replace the substrate.

Crop rotation is also advocated in substrate culture. For example, if tomato is planted among the preceding crops, eggplant and other solanaceous vegetables should not be planted among the succeeding crops, and melon vegetables may be planted instead. As most disinfection methods cannot completely kill germs and worm eggs, crop rotation or substrate replacement is the safer approach. The replaced old substrate can be regenerated by salt leaching, sterilization, ion reintroduction, oxidation, and other methods, and then be reused for soilless culture or be applied to farmland for soil improvement. Difficult-to-decompose substrates such as rock wool and ceramsite can be landfilled to prevent secondary pollution to the environment.

[Skill Training]

Skill Training 4-1 Determination of Physicochemical Property of Solid Substrate

I. Purposes and Requirements

(1) Understand the effect of the physicochemical property of the substrate on substrate culture;

(2) Master the determination method of physicochemical property of substrate;

(3) Ensure correct operation procedure and accurate calculation and weighing;

(4) Ensure standard and skilled operation.

II. Skill Training Preparation

(1) Materials and chemical agents

A certain amount of perlite, slag, vermiculite, sand, and other air-dried substrates, 1 mol/L HNO_3 solution, 1 mol/L NaOH solution, saturated $CaCl_2$ solution, and distilled water.

(2) Instruments and tools

Tray balance (or electronic analytical balance), steelyard, pH meter, conductivity meter, 500 mL canned bottle, 500 mL beaker, 50 mL beaker, 50 mL measuring cylinder, precision pH indicator paper, and gauze.

III. Methods and Steps

(I) Determination of Bulk Density

Weigh the 500 mL canned bottle with a steelyard, and mark it as m_1. After it is filled with the dry substrate to be measured, weigh it again, mark it as m_2, and let V be the volume of the canned bottle. Calculate the bulk density of the substrate according to the following formula (unit: g/L or g/cm^3):

$$\text{The bulk density of the substrate} = (m_2 - m_1)/V$$

(II) Determination of Total Porosity and Large/Small Porosity

Take a canned bottle with known volume (V) and mass (m_1), fill it with the substrate to be measured, weigh its total mass (m_2), then immerse it in water for 24 hours, and then weigh the mass (m_3) of the substrate and canned bottle after the substrate has absorbed enough water. When it is immersed in water, the total porosity of the substrate should be calculated according to the following formula:

$$\text{Total porosity (\%)} = [(m_3 - m_1) - (m_2 - m_1)]/V \times 100\%$$

Take a container with a known volume, measure the total porosity according to the above method, wrap the mouth of the canned bottle with a piece of wet gauze (m_4), then turn it upside down, let the water in the substrate seep out, and set it aside for 2 hours until no water seeps out of the container, and weigh it (m_5). Calculate the aeration porosity and water-holding porosity according to the following formulas:

$$\text{Aeration porosity (\%)} = [(m_3 + m_4 - m_5)/V] \times 100\%$$

$$\text{Water-holding porosity (\%)} = [(m_5 - m_2 - m_4)/V] \times 100\%$$

(III) Determination of pH Value and Buffer Capacity of Substrate

1. Determination of pH value of substrate

Weigh a certain volume of dry substrate and put it into a container, then add distilled water 5 times its volume, fully stir and filter, and then use a pH meter or precision pH indicator paper to determine the pH value of the substrate extract liquid; or weigh 10 g of dry substrate and put it into a 50 mL beaker, add 25 mL distilled water, shake it for 5 minutes, then set it aside for 30 minutes, and then filter it and use a pH meter or precision pH indicator paper to determine the pH value of the substrate extract liquid.

2. Determination of buffer capacity of substrate

Add 1 mL of 1 mol/L HNO_3 solution or 1 mol/L NaOH solution to the above different substrate extract liquids, set them aside for 30 minutes and then use a pH meter or precision pH indicator paper to determine the pH values of the different extract liquids, so as to compare the buffer capacity of different substrates.

(IV) Determination of Electrical Conductivity

Take 10 g of the air-dried substrate and put it into a 50 mL beaker, add 25 mL of saturated $CaCl_2$ solution, shake and extract it for 10 minutes, then filter it and use a conductivity meter to measure the electrical conductivity of the filtrate.

IV. Precautions

(1) The substrate must be in an air-dried state.

(2) The measurement of large/small porosity must be carried out after the water seeps out completely.

V. Answer questions

(1) How does the physicochemical property of the substrate affect the cultivation effect?

(2) What is the usual pH range for crops to grow well? Why?

Skill Training 4-2 Substrate Mixing and Disinfection

I. Purposes and Requirements

(1) Master the principles and methods of substrate mixing;

(2) Master the common methods of substrate disinfection;

(3) Evenly mix the substrate without impurities;

(4) Thoroughly disinfect the substrate.

II. Skill Training Preparation

(I) Materials and Chemical Agents

A certain amount of perlite, slag, vermiculite, turf, sand, and other common organic and inorganic substrates, 0.1%–1% potassium permanganate solution, and 50-fold dilution of 40% formaldehyde.

(II) Instruments and Tools

Tray balance, steelyard, shovel, spade, rubber gloves, watering can, plastic basin, bucket, and wide plastic.

III. Methods and Steps

(I) Substrate Preparation

(1) Pour all kinds of organic substrates and inorganic substrates into the plastic basin in advance, and pick out impurities and debris to make sure that the substrate granules are uniform in size and high in purity and cleanliness.

(2) Students work in groups to mix two complex substrates.

(II) Disinfection of Substrate Agent

Prepare 0.1%–1% potassium permanganate solution and 50-fold dilution solution of 40% formaldehyde in advance as disinfectants. Place the single substrate or complex substrate in a plastic basin or on a flat cement floor covered with plastic film. While mixing, spray disinfectant to the substrate with a watering can, and make sure that it is sprayed thoroughly. When potassium permanganate is used for disinfection, first spray the disinfectant on the substrate, and then cover the substrate with plastic film for 20–30 minutes before using it directly or temporarily bagging it for later use. When formaldehyde is used for disinfection, dilute the 40% stock solution into 50 times solution, spray the solution on the substrate evenly with a watering can accord to the application rate of 20–40 L/m^3, and then cover and seal it with plastic film for 12–24 hours. Before use, uncover the film and air dry the substrate for 2 weeks or expose it to the sun for 2 days to avoid the harm of residual agent.

(III) Solar Disinfection of Substrate

On the greenhouse or polytunnel indoor floor or outdoor flat cement floor covered with plastic film, stack up the substrate to a height of 25 cm, a width of about 2 m, and a length without limitation. While stacking, spray water on the substrate to make its water content exceed 80%, and then cover it with film. In the case of trough culture, directly water the substrate in the trough, and then cover it with film. After covering it with film, close and seal the greenhouse or polytunnel, and expose it to the sun for 10–15 days, during which turn over the stack to have it tedded once. After disinfection of the substrate, bag the substrate for later use.

IV. Precautions

(1) Select different disinfection methods according to different substrate types.

(2) After the substrates are mixed, visually observe the uniformity of the mixed substrate.

(3) It is better to carry out solar disinfection in the high-temperature season to achieve a quick disinfection effect and high disinfection quality.

V. Answer Questions

(1) Why is complex substrate often used in production?

(2) What are the possible consequences of incomplete disinfection of the substrate?

Module 5 Soilless Seedling Culture Technique

[Learning Objectives]

I. Knowledge Objectives

(1) Understand the meaning, advantages, and main methods of soilless seedling culture;

(2) Get familiar with soilless seedling culture equipment;

(3) Master the operation and management techniques of soilless seedling culture.

II. Skill Objectives

(1) Be able to skillfully operate and manage in a scientific way to cultivate healthy and evenly-growing seedlings;

(2) Be able to propose effective regulation measures for seedling culture.

[Preparation for Learning]

I. Required Resources

(1) A library, a periodical reading room, and agricultural newspapers;

(2) Master the seed treatment method before seedling culture in the laboratory;

(3) A multimedia classroom;

(4) Learn about industrialized seedling culture through field investigation.

II. Background Knowledge

(1) Master the basic knowledge related to soilless seedling culture technique;

(2) Master the basic knowledge of literature review;

(3) Learn about some basic principles of soilless seedling culture technique;

(4) Understand the agricultural meteorology and agricultural regulations.

[Learning Tasks]

Task 1 Sowing and Seedling Raising in Soilless Seedling Culture

Soilless seedling culture refers to the method of cultivating seedlings by using substrate and nutrient solution or using only nutrient solution instead of soil. Soilless seedling culture is an indispensable technical part of soilless culture of vegetables and flowers, and it is also one of the key techniques to ensure the quality and efficiency of soilless culture of horticultural crops. Accurately mastering the soilless seedling culture technique and skillfully cultivating the seedlings are the objective requirements for the follow-up soilless culture work.

I. Methods of Soilless Seedling Culture

Soilless seedling culture mainly includes three methods: seedling culture by sowing, seedling culture by cutting, and seedling culture by test tube (tissue culture), among which the seedling culture by sowing is the most commonly used. The following are the main methods of seedling culture by sowing.

1. Seedling culture by plastic pot

The seedling culture by plastic pot is widely applied, and the types of pots are diversified (Fig. 5-1). The pots include round pots and square pots in terms of shape, single pots and conjoined pots in terms of structure, and polyethylene pots and PVC pots in terms of material. At present, a single soft round pot made of polyethylene is mainly applied. The upper mouth diameter and the height of the pot are 8–17 cm, the lower mouth diameter is 6–12 cm, the volume is 200–800 mL, and there is one or more seepage holes at the bottom to facilitate drainage. During seedling culture, plastic pots of different specifications are selected according to crop types, seedling stage duration, and seedling size for sowing or seedling transplanting. Generally, pots with an upper mouth diameter of 8–10 cm are adopted for vegetable seedling culture, and pots with a larger diameter can be selected for flower and forest seedling culture. The crops that are subject to one-step seedling culture can be directly sown. The crops that require seedling separation should first be sown on the seedbed, and then be separated and put into the pots after the seedlings grow to a certain size (Fig. 5-2). The substrate filled in the pot can be a single or mixed substrate, and the liquid can be supplied by upper irrigation or bottom infiltration irrigation.

Fig. 5-1 Plastic Seedling-raising Pot

Fig. 5-2 Seedling Culture by Plastic Pot

2. Seedling culture by foam cube

This method is suitable for deep-flow hydroponics or nutrient film culture. The method is as follows: Lay special polyurethane foam cubes for seedling culture in a seedling tray, where the size of the seedling culture cube is about 4 cm square and the height is about 3 cm, cut an "X"-shaped gap in the center of each small cube, and embed the seeds that are easy to germinate into the gaps one by one (Fig. 5-3), and add nutrient solution to the seedling tray to allow the seeds to emerge and grow. After the seedlings are grown up, separate them one by one and plant them in the plantation trough.

3. Seedling culture by rock wool block

The seedling culture by rock wool blocks is widely applied to various soilless culture types. The specifications of commonly used rock wool blocks mainly include 3 cm × 3 cm × 3 cm, 4 cm × 4 cm × 4 cm, 5 cm × 5 cm × 5 cm, 7.5 cm × 7.5 cm× 7.5 cm, and 10 cm × 10 cm × 5 cm. Rock wool blocks of different specifications should be used according to the crop types and requirements of seedling age. Except for the upper and lower sides, the rock wool block should be wrapped with a milky white opaque plastic film around it to prevent water evaporation, salt accumulation, and algae breeding around it. During seedling culture, cut a small slit on the surface of the rock wool block, embed the seeds that have been germinated into the slit, and then densely place them in a box or trough containing nutrient solution. Wet it with low concentration nutrient solution, and keep the rock wool block wet. After seedling emergence, maintain a liquid layer below 0.5 cm at the bottom of the box or trough, and supply water and fertilizer by the capillary pipe at the bottom. Another method to supply liquid is to replace the nutrient solution layer at the bottom of the seedling culture block with a 2-cm-thick hydrophilic non-woven fabric, where the non-woven fabric is placed at about 1 cm on one side from the bottom of the seedling culture block, and supply liquid to the non-woven fabric through drip irrigation. The nutrient solution is transferred to the rock wool block by the capillary action of the non-woven fabric. In the case of the cultivation of big seedlings such as those of fruit vegetables, the "pot-in-pot" seedling culture method (Fig. 5-4) can be adopted. The method is as follows: Make a small square hole in the center of the big seedling culture block, and make sure that the size of the small square hole just coincides with the embedded small square. In the later stage of seedling culture, move the small rock wool blocks into the big seedling culture blocks, then line them up and gradually widen the distance between the seedling culture blocks as the seedlings grow up, so as to avoid mutual shading between the seedlings. After moving into the big seedling culture block, the nutrient solution layer can be maintained at a depth of 1 cm. This method is more effective than the watering method and immersion method.

4. Seedling culture by nutrient block

The nutrient block is a special nutrient substrate block for seedling culture prepared by adopting the compression-rebound technique, which is mainly made of

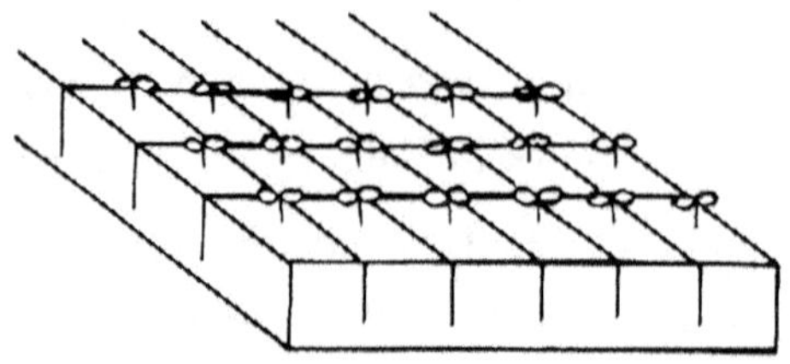

Fig. 5-3 Seedling Culture by Polyurethane Foam

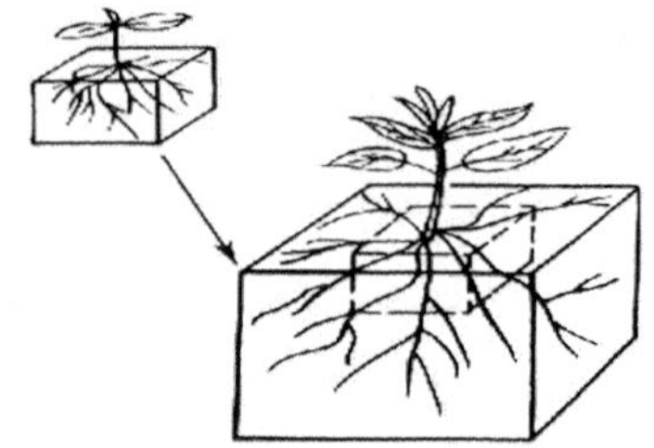

Fig. 5-4 "Pot-in-Pot" Seedling Culture by Rock Wool Block

high-quality peat, plus an appropriate amount of nutrient elements, water-retaining agent, curing agent, microorganisms, etc. The common Jiffy nutrient block is a compressed round-cake-shaped small seedling culture block made of 30% paper pulp, 70% peat, and some fertilizers and adhesives (Fig. 5-5). Its outside is covered with an elastic nylon net (some are not), with a diameter of 4.5 cm and a thickness of about 7 mm. It has the characteristics of aeration, strong water absorption, fertility, lightness, convenience in use and transportation, etc. It is mainly used for cultivating seedlings of fruits, vegetables, flowers, and trees. During seedling cultivation, first put the Jiffy nutrient block in a tray, pour water on it or make it absorb water at the bottom, so that it swells to a seedling culture block with a height of 4–5 cm, and then sow the seeds or transplant the seedlings. When the root system

Fig. 5-5 Seedling Culture by Seedling Culture Block

of the seedlings passes through the nylon net, plant them together with the seedling culture block. Generally, the fertilizer mixed in the seedling culture block can maintain the growth throughout the seedling stage, so no more fertilizer is required.

5. Seedling culture by plug tray

The plug tray is a seedling tray with many small pot-shaped holes made according to certain specifications (Fig. 5-6). Generally, it is molded from polyethylene sheet or polystyrene and polyurethane foam. The plug tray can be divided into square-cone-shaped plug tray, cone-shaped plug tray, and separable plug tray in terms of shapes; paper-based plug tray (such as that for rice seedling raising and throwing, which needs to be opened before use), polyethylene plug tray, and polystyrene plug tray in terms of fabrication materials; 50-hole, 72-hole, 128-hole, 200-hole, 288-hole, 392-hole, 512-hole, and 648-hole plug trays in terms of the number and size of holes, among which the 72-hole, 128-hole, and 288-hole plug trays are commonly used. Most of the plug trays used in the world are of the size of 27.8 cm × 54.9 cm, with the hole depth varying from 3 cm to 10 cm. According to the types, seedling age and purpose of the crops for seedling culture, plug trays of different specifications can be selected for one-step seedling culture or cultivation of seedlings for transplanting. Generally, the larger the seedling plant type or the longer the seedling age, the larger the hole diameter of the selected plug tray. The plug tray used for mechanized sowing is made according to the specification requirements of the automatic precision sowing production line. During seedling culture, first fill the substrate in the holes of the plug tray, then sow 1–2 seeds per hole, cover them with a small amount of substrate, compact it slightly, and finally water it. When the seedlings emerge, plant one seedling in each hole.

Fig. 5-6 Plug Tray for Seedling Culture

Other soilless seedling culture methods include seedling culture by seedling tray (box) and seedling culture by seedling barrel. In production, we can flexibly choose seedling culture methods according to specific conditions.

II. Soilless Seedling Culture Equipment

For the selection of seedling culture equipment, we can take a comprehensive consideration of the requirements, purposes, and conditions of seedling culture. For large-scale specialized seedling culture, the equipment for soilless seedling culture should be advanced with complete supporting facilities. For example, industrialized plug seedling culture requires sophisticated seedling culture facilities, equipment, and instruments, as well as modern measurement and control techniques, which are generally carried out in a multi-span greenhouse. For ordinary soilless seedling culture in small areas, the seedling culture equipment can be selected according to local conditions, which are mainly operated in the solar greenhouse, polytunnel, and other similar facilities. In addition, according to the conditions, other soilless seedling culture equipment can also be arranged. The main seedling culture equipment includes the following types.

1. Substrate disinfector

In order to prevent pathogenic microorganisms or nematodes from infecting the seedling culture substrate, the substrate can be disinfected by a substrate disinfector before use. In fact, the substrate disinfector is a small steam boiler that disinfects the substrate using the steam generated. According to the steam pressure and volume generated by the boiler, build a certain volume of substrate disinfection tank in the substrate disinfection workshop, connect the tank to a steam pipe with steam outlet holes, and properly design the inlet and outlet ports that are convenient for the substrate to enter and exit, close the ports, leave a small hole, insert a high-temperature thermometer, and observe the temperature in the substrate.

2. Substrate stirrer

Before the seedling culture substrate is sent to the feeding and loading machines, it is usually stirred with a stirrer. The purpose is to mix all the components in the substrate evenly and break the agglomerated substrate so as not to affect the quality of tray loading. The substrate stirrers can be individual or connected with feeders, and Korea-made individual stirrers are commonly used.

3. Automatic precision sowing production line

The device of plug tray automatic precision sowing production line (Fig. 5-7) is the core equipment for industrialized seedling culture, which is composed of five parts: plug tray placement machine, feeding and substrate loading machines, hole pressing and precise sowing machines, soil covering machine, and spraying machine. It mainly completes a series of operations such as substrate loading, hole pressing, sowing, covering, repression, and water spraying. These five parts are integrated into an automatic production line, and each part can work independently after being separated. In terms of working principle, the precision sowing machine can be divided into two types: mechanical rotary and vacuum pneumatic. The mechanical precision sowing machine has strict requirements on the seed shape, and the seeds need to be pelleted before they can be used. The pneumatic precision sowing machine does not have strict requirements on the seed shape, so the seeds need not be pelleted. One semi-automatic sowing machine can be purchased for a seedling culture farm with an annual output of less than 1 million commercial seedlings; two or three semi-automatic precision sowing machines can be purchased for a seedling culture farm with an annual output of 1 million to 3 million seedlings; highly automated precision sowing machines can be used for a seedling culture farm with an annual output of more than 3 million seedlings.

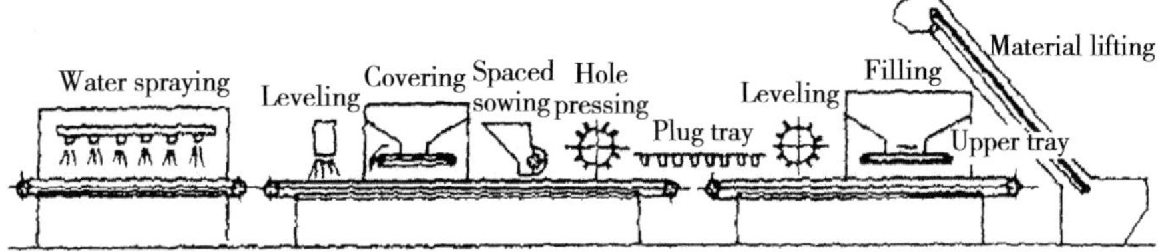

Fig. 5-7 Precision Sowing Production Line for Seedling Culture by Plug Tray

4. Plug tray for seedling culture

The plug tray for seedling culture is an essential container for seedling culture. According to different applications and vegetable types, the specifications and making methods of the plug tray are also different. The specifications of plug trays used for mechanized sowing are generally customized according to the specifications and models of the automatic precision sowing production line. The area and depth of each small hole in the plug tray for seedling culture depend on the seedling culture type.

5. Water spraying equipment and irrigation system

Water spraying equipment and irrigation systems are provided in the greening room or seedling cultivation facilities. Generally, the water spraying system for industrialized seedling culture adopts a walking spraying device, which can spray both water and pesticide, thereby saving labor, improving efficiency, and producing a good operation effect. When the seedlings are small, the amount of water to be sprayed into the substrate of each hole by the walking spray system is relatively uniform. When the seedlings grow to a certain size and the leaves are large, spraying water from above often causes an uneven amount of water to be sprayed in each hole. Therefore, it is better to adopt the groundwater supply method so that the water can be absorbed through the holes at the bottom of the plug tray.

6. CO_2 generator

There are many types of CO_2 generators, which take either coke and charcoal or kerosene and liquefied (petroleum) gas as raw materials, or make use of the chemical reaction between ammonium bicarbonate and dilute sulfuric acid to release carbon dioxide. Adding CO_2 to the seedling culture space can help the seedlings grow fast and strong.

7. Others

The facilities adopted for industrialized seedling culture are usually modern greenhouses or polytunnels with temperature regulation, humidity control, and ventilation devices, which are of high grade and have a high degree of automation. The facilities have large spaces which are suitable for mechanized operation. The indoor is provided with automatic drip irrigation, water spraying and pesticide spraying equipment, as well as a seedling germination accelerating room, greening room, seedling separation room, automatic intelligent grafting machine, and healing promoting device.

III. Main Soilless Seedling Culture Techniques

According to the scale and technical level of seedling culture, the soilless seedling culture can be divided into the ordinary soilless seedling culture and industrialized soilless seedling culture. Generally, ordinary soilless seedling culture is small in scale and low in cost, but the seedling culture conditions are poor, mainly relying on manual operation and management, which affects the quality and uniformity of seedlings. The industrialized soilless seedling culture is the large-scale mechanical production of seedlings according to certain technological processes and standardization techniques. It has the characteristics of large-scale seedling culture, good seedling culture conditions, labor saving, high efficiency, advanced management technique, energy saving, seed saving, area saving, convenience for long-distance transportation and mechanized planting, high quality and standardization of seedlings, etc. However, the cost of seedling culture is high, and it requires sophisticated seedling culture facilities and equipment, modern measurement and control techniques, as well as scientific automation, standardization, and intensification management.

(I) Ordinary Soilless Seedling Culture

Ordinary soilless seedling culture is similar to traditional soil seedling culture except for the relatively special seedling culture method, seedling culture facilities, nutrition supply and management, and rhizospheric environment of seedlings. The operation process of ordinary soilless seedling culture is shown in Fig. 5-8.

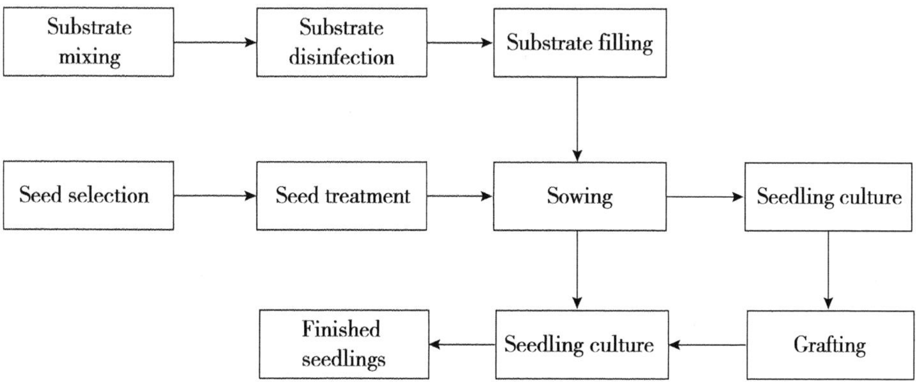

Fig. 5-8 Operation Process of Ordinary Soilless Seedling Culture

1. Seed selection

On the premise of ensuring a reliable source of seeds, the large plump seeds with complete structure and free from pests and diseases are selected as the seeds for seedling culture. The seeds are selected manually or mechanically.

2. Seed treatment

(1) Seed disinfection: As many crop germs lurk inside or attach to the surface of seeds, seed disinfection is an effective measure to reduce diseases at the seedling stage. The seed disinfection methods include soaking seeds in chemical agents, soaking seeds in warm water, scalding seeds in hot water, and dry heat treatment.

① Soaking seeds in chemical agents: Soaking seeds in chemical agents is to formulate the disinfectant into a solution with a certain concentration and then soak the seeds in it to kill the germs carried by the seeds. There are many kinds of chemical agents for seed disinfection, so different chemical agents should be selected according to the types of pathogenic bacteria. For example, to control bacterial diseases of solanaceous vegetables at the seedling stage, they can first be soaked in a 0.1%–0.3% mercuric chloride solution for 5 minutes, and then be soaked in a 1% potassium permanganate solution for 10 minutes. To control tomato and cucumber seedling blight, 70% dexon can be used for seed dressing (with a dosage of 0.3% of the seeding amount). To control the early blight of solanaceous vegetables, they can be soaked in a 1% formalin solution for 15–20 minutes, then covered with a wet cloth for 12 hours. Tomato mosaic virus can be inactivated by soaking the seeds in a 10%–20% trisodium phosphate or 20% sodium hydroxide solution for 15 minutes.

② Soaking seeds in warm water: The water temperature used for soaking seeds in warm water is 55℃ (the pathogenic temperature of germs), and the water amount is 5–6 times of the seed amount. During seed soaking, the seeds should be stirred constantly, and the water temperature should be kept at 55℃ for 10 minutes, and then the water temperature should be gradually reduced before carrying out general seed soaking. The water temperature of cold-resistant and semi-cold-resistant vegetables should be reduced to 20–25℃, that of thermophilic and heat-resistant vegetables should be reduced to 25–28℃, and the seed soaking time should be shortened by 1–2 hours compared with that at room temperature. For eggplant, loofah, wax gourd, and other vegetables' seeds with hard and thick seed coats or for seeds that are also fruits, it is difficult for them to absorb water, so they can be treated mechanically before soaking. For large melon seeds, the shell at the embryo end can be broken, or the peel can be rubbed with bricks or stones. Some seeds, such as eggplant seeds, are covered with relatively more sticky substances, which hinder air permeability and affect absorption and germination. They can be washed with a 0.2%–0.5% alkaline solution and then be scrubbed in water (the water needs to be changed) continuously during soaking until the seed coat is clean and not sticky.

③ Scalding seeds in hot water: It is generally used for soaking seeds that are difficult to absorb water (such as eggplant and wax gourd seeds) and seeds that should not be soaked for a long time (such as legume seeds). The water temperature can be up to 70–80℃ or even higher. For cryophilic vegetables with thin seed coats, such as cabbage and lettuce seeds, the water temperature should be lower. The application amount of water should not exceed 5 times that of seeds. Seeds should be dried thoroughly and scalded quickly. The time spent on scalding seeds in hot water should be less than half that of soaking seeds in warm water.

④ Dry heat treatment: It is a method to disinfect seeds in an incubator. Treating the dried seeds in a drying oven at above 70℃ for 2–3 days can inactivate the viruses attached to the seeds. Besides, it can also increase the vitality of the seeds, thereby promoting the seeds to germinate neatly and uniformly. For example, the seeds of melons, tomatoes, kidney beans, and other vegetables can be dry heat treated at 70–80℃ to kill the germs on the surface and inside of the seeds and reduce the diseases at the seedling stage. This method is suitable for heat-resistant vegetable seeds, such as melon and solanaceous vegetable seeds. However, during dry heat treatment, it should be noted that the seeds to be treated must be dry (generally, the water content is lower than 4%), and the treatment time should be strictly controlled; otherwise, the heat will pass through the seed coat and kill the embryo, making the seeds lose their germination ability.

(2) Seed soaking: Soak seeds after disinfection. The purpose of seed soaking is to let the seeds thoroughly absorb water in a short time, shorten the germination time of seeds, and make the seeds grow uniform and strong seedlings. The suitable soaking time is different for different crops (Table 5-1). Generally, seeds can be soaked in warm water at 25–30℃ for 4–12 hours. The soaking time should be longer for those with thick seed coats and shorter for those with thin seed coats. Seed soaking can be carried out in combination with chemical disinfection. The seed coats of canna, watermelon, and other seeds are hard. Before soaking, the seed coats are worn down by mechanical means or the seed shells at the embryo end are cracked, or they can also be soaked in sulfuric acid, so that the seed coats become soft and then the sulfuric acid is rinsed off with clear water immediately. If the seed coat is sticky, it can be scrubbed with 0.2%–0.5% alkali solution. It is advisable to soak the seeds to the extent that the water level is 2–3 cm higher than the seed level, and the seed level should not exceed 15 cm, so as to facilitate the respiration of the seeds and prevent the embryos from asphyxiating and dying. The seeds of Rosaceae flowers must be stratified in a low-temperature and humid environment before breaking dormancy.

Table 5-1 Suitable Soaking Time for Several Vegetable Seeds

Type	Tomato	Pepper	Eggplant	Kale	Celery	Wax gourd	Watermelon	Cucumber	Zucchini
Time (h)	6	12	24	4	12	12	12	4	8

(3) Accelerating seed germination: Accelerating seed germination can promote the rapid and uniform germination of seeds. The commonly used germination accelerating methods mainly include the following: ① Accelerating germination in an incubator or germination accelerating box: Put a gauze bag with seeds in a germination accelerating tray, and then put it in an incubator or germination accelerating box for accelerating germination. This is an ideal method of accelerating germination at present, and its temperature, illumination, and alternating temperature treatment can be automatically controlled. ② Conventional germination accelerating method: Pack the seeds in a well-washed coarse cloth or

gauze, put them in a wooden box or earthen pot with wet straw at the bottom, cover them with clean sacks, and place them near the flue or fire wall of the greenhouse for accelerating germination. The seeds should not be packed too full in the bag. It is better to be packed 60%–70% full so that the seeds are loose in the bag. For seeds that are not easy to germinate, sand can be added to accelerate germination, ensuring even humidity and temperature, thus achieving uniform germination. In the germination accelerating process, the seeds should be turned every 4–5 hours so as to heat the seeds and facilitate aeration and uniform germination. ③ Accelerating germination by sawdust: For seeds that are difficult to germinate, such as eggplant and pepper seeds, the method of accelerating germination by sawdust can have a better effect. Put steamed and disinfected fresh sawdust in a wooden box to a thickness of 10–12 cm, then spray water on it. After the water seeps down, pack half a bag of seeds in a coarse gauze bag, and spread them on the sawdust, where the seed thickness should be 1.5–2 cm, then cover them with steamed wet sawdust to a thickness of 3 cm, and put the wooden box near the flue or fire wall or on the heated kang, and keep it at a suitable temperature. In the process of accelerating germination, there is no need to turn the seeds frequently. It can germinate quickly and neatly at room temperature for 4–5 days.

Other methods include accelerating germination at low temperatures, accelerating germination by hormone treatment, and accelerating germination by alternating temperature treatment. The temperature and time for accelerating germination of several vegetable seeds are shown in Table 5-2.

Table 5-2 Temperature and Time for Accelerating Germination of Several Vegetable Seeds

Vegetable species	Optimal temperature (℃)	Early temperature (℃)	Late temperature (℃)	Number of days required	Germination control temperature (℃)
Tomato	24–25	25–30	22–24	2–3	5
Pepper	25–28	30–35	25–30	3–5	5
Eggplant	25–30	30–32	25–28	4–6	5

continued

Vegetable species	Optimal temperature (℃)	Early temperature (℃)	Late temperature (℃)	Number of days required	Germination control temperature (℃)
Zucchini	25–36	26–27	20–25	2–3	5
Cucumber	25–28	27–28	20–25	2–3	8
Kale	20–22	20–22	15–20	2–3	3
Celery	18–20	15–20	13–18	5–8	3
Lettuce	20	20–25	18–20	2–3	3
Cauliflower	20	20–25	18–20	2	3
Chinese chive	20	20–25	18–20	3–4	4
Onion	20	20–25	18–20	3	4

3. Selection of seedling culture substrate

The selection of a suitable seedling culture substrate is the basis for cultivating strong seedlings. Soilless seedling culture substrate should be loose, air permeable, water-retaining, and fertilizer-retaining, stable in chemical properties, free of germs, worm eggs, and weed seeds, and harmless to seedlings. In order to reduce the cost of seedling raising and ensure the seedling raising effect, it is better to make full use of local resources and local materials when selecting the seedling culture substrate, and it is advisable to mix 2–3 kinds of organic or inorganic substrates to achieve complementary advantages and improve the seedling raising effect. At present, the substrate ratio commonly used at home and abroad is roughly 50%–60% of turf, 30%–40% of vermiculite, and 10% of perlite. In order to make the seedlings grow strong, apart from the method of watering nutrient solution, an appropriate amount of inorganic fertilizer, biogas residue, biogas slurry, sterilized chicken manure is often mixed into the seedling culture substrate, and appropriate topdressing can be applied at the later growth stage. It only needs to pour clear water at ordinary times. Commonly used seedling culture substrate formulas at home and abroad are listed in Table 5-3.

Table 5-3 Commonly Used Seedling Culture Substrate Formulas

Formula code		A	B	C	D
Substrate type and added fertilizer	Fine sand① (m^2)	0.5	–	–	–
	Crushed Turf (m^2)	0.5	0.5	0.75	0.5
	Vermiculite(m^2)	–	0.5	0.13	0.5
	Dolomite②(kg)	4.5	3.0	–	–
	Perlite(m^2)	–	–	0.12	–
	Potassium nitrate(kg)	0.145	–	–	–
	Ammonium nitrate(kg)	–	–	–	0.7
	Potassium sulfate(kg)	0.145	–	–	–
	Calcium superphosphate③ (kg)	1.5	1.2	1.0	0.7
	Compound fertilizer(kg)	–	3.0④	1.5⑤	–
	Calcium limestone (kg)	1.5	–	3.0	3.5
	Sterilized dried chicken manure (kg)	–	–	10.0	–

Notes: A. Mixed substrate of the University of California; B. Mixed substrate of the Cornell University; C. Seedling culture and pot culture substrates of the Chinese Academy of Agricultural Sciences; D. Mixed substrate of turf minerals.

① The grain size of fine sand is 0.05–0.5 mm. ② Dolomite may also be replaced by limestone. ③ Calcium superphosphate contains 20% phosphorus pentoxide. ④ The N-P-K ratio in the compound fertilizer is 5 : 10 : 5. ⑤ The N-P-K ratio in the compound fertilizer is 5 : 15 : 15.

4. Sowing

(1) Determination of sowing period: Determination of the right sowing period is the fundamental guarantee for the planned seedling culture, which is generally estimated according to the number of days required for the seedling culture and planting period as well as factors such as cultivated varieties and cultivation seasons. The formula for calculating the number of days of seedling culture is as follows:

The number of days of seedling culture = The number of days of seedling age + The number of days of seedling hardening (7–10 days) + The number of flexible days (3–5 days)

(2) Determination of sowing amount and sowing area

① Sowing amount: The sowing amount can be calculated according to the following formula:

Sowing amount (g/mu) = (The number of seedlings planted or to be sold per mu + safety factor) × The thousand-seed weight ÷ Germination rate ÷ 1,000.

For example, 3,000 tomato plants are usually planted per mu, with a thousand-seed weight of 3.25 g and a germination rate of 85%. Then:

Sowing amount = (3,000 + 3,000 × 20%) × 3.25 ÷ 85% ÷ 1,000 ≈ 14 g/mu, i.e. 14 g of tomato seeds are required per mu.

② Sowing bed area (m^2) = [Sowing amount (g) × Number of seeds per gram × (3–4)] ÷ 10,000

Note: 3–4 refers to the average area of 3–4 cm^2 occupied by each seed, e.g. 3 for hot pepper, early-maturing cabbage, and cauliflower; 3.5 for tomato; and 4 for eggplant.

③ Separate seedbed area (m^2) = [Total number of separate seedlings × Nutrient area per plant (cm^2)] ÷ 10,000

Note: The nutrient area per plant is generally as follows: The nutrient area per plant for pepper (double planting), cucumber, watermelon, zucchini, eggplant, and tomato is 10 cm × 10 cm, and that for cauliflower is 8 cm × 6 cm.

Table 5-4 shows the sowing amount per unit area, sowing bed area and separate seedbed area for cultivation of several kinds of vegetable seedlings for reference.

Table 5-4 Sowing Amount and Seedbed Area Required for Cultivation of Several Kinds of Vegetable Seedlings (per mu)

Vegetable species	Sowing amount (g)	Area to be sown (m^2)	Seedbed area to be separated (m^2)	Remarks
Tomato	40–50	6–8	40–50	–
Pepper	100–150	6–8	40–50	–
Eggplant	50–80	3–4	20–25	–
Cucumber	150–200	–	40–50	The seedlings are generally not separated.
Zucchini	200–250	–	25–30	The seedlings are generally not separated.
Early-maturing cabbage	20–30	4–5	40–50	–
Cauliflower	20–30	3–4	20–25	–
Celery	50	25	–	The seedlings are not separated.
Asparagus lettuce	20	2.5–3	24–28	–

(3) Sowing methods: Different sowing methods are adopted for different soilless seedling culture methods. The commonly used sowing methods include spaced sowing, line sowing, and broadcast sowing. Generally, sowing is carried out on a windless, sunny morning. Before sowing, make a seedbed (electric hotbed in winter and spring, and low seedbed in summer), disinfect the seedlings with 0.1%–1.0% potassium permanganate, and spray the substrate with clear water. After sowing, cover the seed with the substrate to a thickness of 1–2 cm. After spraying a small amount of water, cover the substrate with film, increase the temperature, preserve moisture, and keep it moist during seedling emergence. Melon and legume crop seeds should be sown at 8–10 cm planting spacing (generally 2 seeds per hole). For solanaceous and leafy vegetables that need to separate seedlings, broadcast sowing can be adopted with a seedling spacing of 1–2 cm, and then the seedlings will be separated after 1–2 true leaves are developed. In industrialized plug seedling

culture, an automatic precision sowing production line is adopted to realize automatic sowing.

5. Seedling culture

When cultivating seedlings on a plug tray or seedling bed, it is necessary to irrigate the nutrient solution regularly or mix the fertilizer into the substrate in advance, and make sure that only water is poured in the seedling stage. For seedling culture by rock wool blocks or foam cubes, place the seedling block in a seedbed containing a shallow nutrient solution for circulating solution supply. No matter what kind of soilless seedling culture method is adopted, it should be adjusted according to the growth of seedlings and the change of nutrient solution. After reaching the seedling standard, it should be planted in time. One week before planting, the solution supply should be reduced, and the seedlings should be hardened in time. In order to improve the disease resistance, the seedlings of some crops, such as cucumber, should be grafted when they grow to a certain size. In addition, pest control and environmental regulation should be strengthened at the seedling stage.

(II) Industrialized Plug Seedling Culture

Industrialized soilless seedling culture mainly adopts the plug seedling culture method, which is a modern seedling culture system that takes turf, vermiculite, and other light substrate materials as seedling culture substrates and adopts industrialized precision sowing and one-time seedling cultivation methods. The operation process of industrialized plug seedling culture is shown in Fig. 5-9.

1. Seed selection and treatment

Seed selection is the same as that of common soilless seedling culture. However, the seed treatment is different from that of common soilless seedling culture as it requires seed coating and precision sowing before intensive germination. Coated seeds need not be soaked and disinfected.

2. Selection of substrate

The selection of seedling culture substrate is the same as that of common soilless seedling culture. The mixing, disinfection, and filling of the substrate are to be accomplished by the operation of machines such as a substrate mixer and

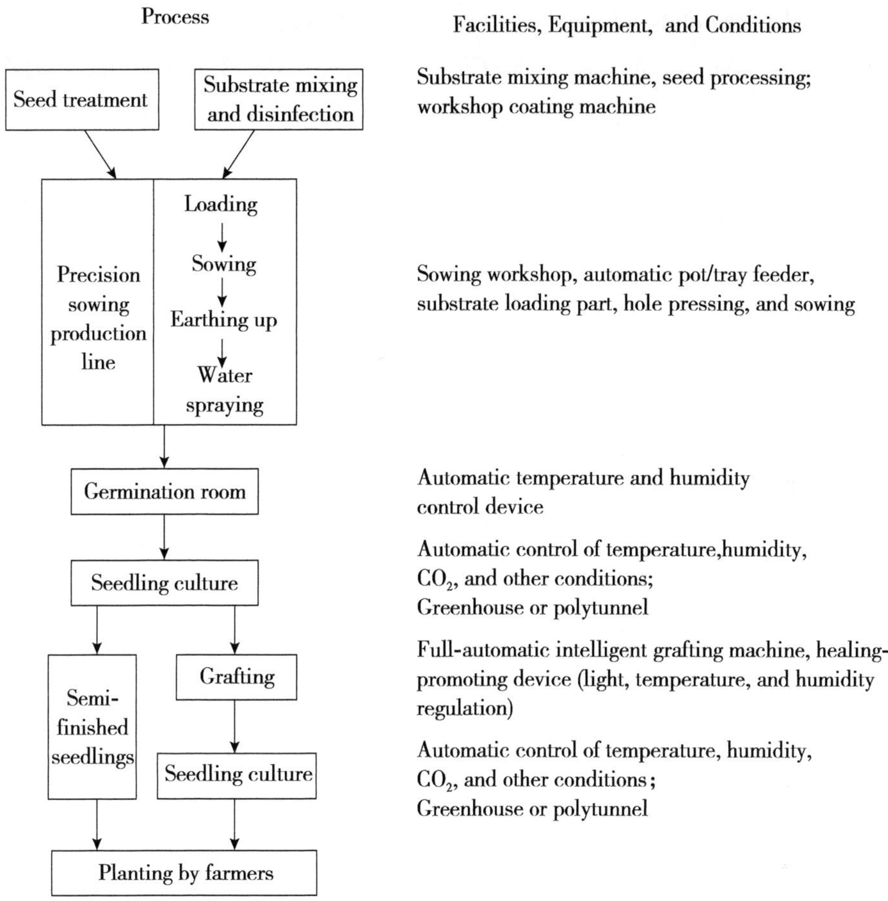

Fig. 5-9 Process, Facilities, and Operation Diagram of Plug Seedling Culture

substrate disinfector, which have high efficiency and good mixing and disinfection effects.

3. Precision sowing

The automatic precision sowing production line is used in the sowing workshop to realize a series of mechanized and programmed automatic assembly line operations such as tray loading, hole pressing, sowing, covering, repressing, and watering, which is convenient, quick, and efficient. The size of the plug tray used for industrialized plug seedling culture should meet the requirements of the automatic precision sowing production line. The number of plug trays used and the area of the plug seedbed are calculated as follows.

(1) The calculation formula for the number of plug trays:

The number of plug trays = (Sowing amount ÷ Thousand-seed weight × 1,000) ÷ [The number of holes per plug tray × Number of seeds sown in each hole(1–2)]

(2) The calculation formula for the area of the plug seedbed:

The effective area of plug seedbed (m^2) = The number of plug trays × The area of each tray (m^2)

4. Germination accelerating and greening

Put the plug trays neatly on the seedling cart after sowing. The seedling cart should be directly moved into the germination accelerating room to accelerate germination. After germination, the seeds should be placed in a greening room for greening in the light so as not to affect the growth and quality of the seedlings. The greening room generally refers to a multi-span greenhouse used for seedling culture, which has good light transmittance and heat preservation, so that the seedlings can be managed according to the predetermined requirements after they emerge from the soil. After the seedlings are greened, they will be subject to normal seedling management (some crops need to be grafted).

5. Seedling culture

The seedling management of industrialized plug seedling culture is roughly the same as that of ordinary soilless seedling culture, except that the industrialized plug seedling culture makes full use of advanced facilities and equipment, strengthens nutrient solution management and environmental regulation, effectively controls pests and diseases, and makes seedlings grow fast, neat and strong.

Task 2　Management of Soilless Seedling Culture

I. Soilless Seedling Culture Substrate and Nutrient Solution

(I) Seedling Culture Substrate

The seedling culture substrate is an important condition for fixing and supporting the seedlings, retaining moisture and nutrition, and providing a sound environment for the normal growth and development of the root system. The selection of suitable substrates is an important part of soilless seedling culture and the basis for cultivating strong seedlings. The soilless seedling culture substrate is required to have a large porosity, reasonable gas-water ratio, stable chemical properties, and be harmless to the seedlings. Besides, in order to reduce the cost of seedling culture, the substrate selection should be based on the principle of using local materials to make it economical and practical and make full use of local resources.

1. Types and properties of substrates

There are many kinds of substrates commonly used for soilless seedling culture, mainly including peat, vermiculite, rock wool, perlite, carbonized rice husk, slag, sawdust, cottonseed hull used for cultivating mushrooms, and bark. Different substrates have different physicochemical properties. These substrates may either be used alone or mixed according to a certain proportion. Generally, mixed substrates can produce a better seedling culture effect.

2. Selection of seedling culture substrate

The seedling culture substrate should have excellent physicochemical properties. It should be loose and air permeable, capable of retaining water and fertilizer, be slightly acidic, have stable chemical properties, and be free of germs, worm eggs, weed seeds, and substances harmful to seedlings. It is usually composed of two or more substrates that are mixed in a certain proportion. When preparing a complex substrate, usually 2–3 kinds of substrates need to be used, and locally abundant and inexpensive light substrates should be selected as far as possible, especially the organic or inorganic complex substrates that can produce a better effect. The culture substrate should also be conducive to root winding and root nodule formation. Physicochemical properties of several complex substrates are shown in Table 5-5.

Using complex substrates for seedling culture can realize complementary advantages and improve the effect of seedling culture. The substrate not only can fix the seedlings but also can provide the water and nutrients needed for their growth, so the nutritional conditions of the substrate have a great influence on their life activity. At present, most plug seedling culture methods in China adopt the "turf + vermiculite"

Table 5-5 Physicochemical Properties of Several Complex Substrates

Complex substrate	Bulk density (g/cm^3)	Specific gravity (g/cm^3)	Total porosity (%)	Aeration porosity (%)	Capillary porosity (%)	pH value	EC (mS/cm)	Cation exchange capacity (mmol/100g)
The ratio of turf, vermiculite, slag, and perlite is 2 : 2 : 5 : 1	0.67	2.29	70.7	17.1	53.6	6.71	2.62	13.77
The ratio of turf and vermiculite is 1 : 1	0.34	2.32	85.3	38.1	47.2	6.09	1.19	30.37
The ratio of turf, vermiculite, slag, and perlite is 4 : 3 : 1 : 2	0.41	2.22	81.5	25.3	56.2	6.44	2.82	29.03
The ratio of turf and slag is 1 : 1	0.62	1.93	67.9	17.7	50.2	6.85	2.43	21.50

complex substrate with a ratio of 2 : 1 or 3 : 1. Both turf and vermiculite contain a certain amount of major elements and trace elements, which can be absorbed and utilized by seedlings. However, for crops with a long seedling stage, the nutrients in the substrate cannot meet the needs of seedling growth. Therefore, apart from the method of irrigating the nutrient solution, different fertilizers are often added according to the nutrient content and crop requirements when formulating the substrate, and then appropriate topdressing is carried out at the later stage of growth. Clear water needs to be poured only at ordinary times, which is easy to operate. It can be seen from Table 5-6 that adding a certain amount of organic fertilizer and chemical fertilizer in the process of substrate formulation not only promotes the seedling emergence but also produces an effect that the physiological indexes of seedlings are better than those of seedlings with chemical fertilizer or organic fertilizer applied alone in the substrate.

Table 5-6 Seedling Culture Effect of Complex Substrate

Treatment	Plant height (cm)	Stem diameter (mm)	Number of blades	Leaf area (cm^2)	Dry weight of whole plant (g)	Sound seedling index
N-P-K compound fertilizer	14.2	3.1	4.5	27.96	0.124	0.121
Urea + potassium dihydrogen phosphate + deodorized chicken manure	17.6	3.6	4.9	39.58	0.180	0.181
Deodorized chicken manure	12.5	2.9	4.1	19.12	0.110	0.104

(II) Nutrient Solution

1. Selection of nutrient solution

The formula of nutrient solution for seedling culture is determined according to crop types. 1/3–1/2 dose of Japanese garden test formula and Yamazaki formula are commonly used in production. Besides, special formulas for seedling culture may also be used. The test shows that the formula containing 140–200 mg/kg of N, 70–120 mg/kg of P, and 140–180 mg/kg of K can be used for the leafy vegetable seedling culture. The formula for the solanaceous vegetable seedling culture contains 140–200 mg/kg of N, 90–100 mg/kg of P, and 200–270 mg/kg of K in the early stage, and 150–200 mg/kg of N, 50–70 mg/kg of P, and 160–200 mg/kg of K in the later stage. In addition, the formulated solution of N-P-K compound fertilizer (N-P-K content: 15-15-15) can also be used to spray irrigate the seedlings. The concentration at the cotyledon stage is 0.1%, and it can be increased to 0.2%–0.3% after one true leaf appears.

The general requirements of soilless seedling culture for nutrient solution are as follows: complete and balanced nutrients, safe use, and convenient formulation. Therefore, in the actual formulation process, we should reasonably select the fertilizer types, reduce the cost as much as possible, and control the pH value of the nutrient solution within the range of 5.5–6.8. If the concentration of ammonium nitrogen in the nutrient solution is too high, it could easily cause harm to the seedlings, inhibit the seedling growth, and even cause young roots to rot and seedlings to wilt and die in serious cases. Therefore, nitrate nitrogen should be selected as the main nitrogen source, and the proportion of ammonium nitrogen in total nitrogen should not exceed 30%.

2. Management of nutrient solution

The management of nutrient solution for soilless seedling culture will mainly adopt the solution supply method to scientifically control the time and amount of solution supply and the nutrient solution concentration. After the seedlings emerge from the soil, the solution should be properly supplied in advance during the transition stage from heterotrophic growth to autotrophic growth. Generally, after the seedlings emerge from the soil and are moved to the greening room, start pouring or spraying nutrient solution once a day or once every two days.

(1) Nutrient solution concentration: The seedlings of different crops have different requirements for nutrient solution concentration, and even those of the same crop also have different requirements for nutrient solution concentration in different growth stages. Generally, the nutrient solution concentration of young seedlings should be relatively low (1/2 or 1/3 of the standard concentration in the plant's adult stage), and the nutrient solution concentration should gradually increase with the growth of seedlings.

(2) Solution supply amount: When the solutions is supplied, it is necessary to prevent excessive liquid accumulation in the seedling culture container and keep a 0.5–1-cm-deep solution layer at the bottom of the seedbed after each solution supply. The results of previous studies showed that in the whole process of seedling culture, each of the tomato, eggplant, cucumber, and melon seedlings absorbed 800 mL, 1,000 mL, 500 mL, and 400 mL of standard concentration

nutrient solution, respectively. This standard can be used as a reference for small-scale seedling culture, and the nutrient solution can be applied a number of times, and the application rate for each seedbed should be controlled at about 10 L/m². In summer, the number of times of application of nutrient solution for seedling culture should be increased appropriately, and the seedbed should be often sprayed with water to keep it moist.

(3) Solution supply method: Nutrient solution supply should be combined with water supply. Applying the nutrient solution once or twice and then applying clear water once can prevent the excessive salt cumulative concentration in the substrate from inhibiting the growth of seedlings.

① Top solution supply is suitable for plug seedling culture or seedbed seedling culture. When the industrialized seedling culture scale or the seedling culture area is large, the double-arm suspension type or track type water-spraying fertilizing cart can be moved back and forth to spray the solution. During the high-temperature days in summer, spray water 2–3 times a day and spray fertilizer once every other day. In winter, as the temperature is low, spray water and fertilizer alternately once every 2–3 days.

② Bottom liquid supply is suitable for the seedling culture by rock wool block and foam cube. Store the water or nutrient solution in the seedbed. The seedbed is generally surrounded by plastic or foam boards in a trough shape, with a length of 10–20 m, a width of 1.2 –1.5 m, and a depth of about 10 cm. The bottom of the bed is flat and watertight; a black plastic film with a thickness of 0.2–0.5 mm is laid at the bottom as the liner, and the thickness of the thin layer of nutrient solution should be kept at about 2 cm. In some cases, the bottom of the bed is made into many small lattices with a depth of 2 mm, on which the seedling blocks are arranged. The bottom is supplied with the solution, and the excess nutrient solution is discharged from small holes arranged at certain intervals (Fig. 5-10).

In order to reduce the cost of seedling culture, it is better to adopt the circulating solution supply method to supply the solution and increase oxygen through the circulating flow of the nutrient solution, but the concentration and pH value of the nutrient solution should be adjusted in time.

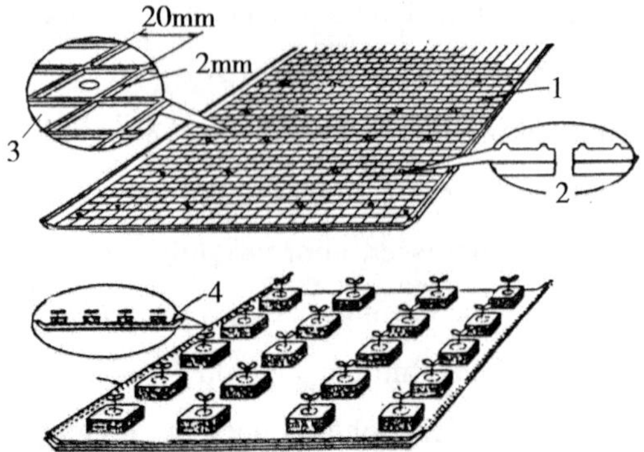

Fig. 5-10 Solution Supply System for Seedbed and Seedling Block

1. Seedbed 2. Drainage hole 3. Enlarged view of seedbed 4. Solution supply hole

II. Environmental Regulation in Soilless Seedling Culture Period

(I) Temperature

Temperature is an important environmental factor affecting seedling growth. Temperature not only directly affects the speed of seed germination and seedling growth but also affects the development process of seedlings. If the temperature is too low, the growth and development of seedlings will be delayed, and the growth vigor will become weak, which could easily produce weak seedlings or rigid seedlings. Under extreme conditions, a bed temperature that is too low will cause chilling injury or freezing injury. If the temperature is too high, the seedlings will grow too fast and become leggy seedlings.

1. Substrate temperature

Substrate temperature affects the root system growth and root hair generation, thus affecting the absorption of water and nutrients by seedlings. In a suitable temperature range, the root elongation rate increases with the rise in temperature. Beyond this range, although the root growth accelerates, the root system is thin and its life span will be shortened. The low temperature of the seedling culture substrate in early spring will lead to slow growth of the root system or cause physiological disorders, so an electric hotbed should be set up. In summer and autumn, in order to prevent high-temperature damage, low bed seedling culture method can be used.

2. Temperature difference between day and night

Keeping a certain temperature difference between day and night is very important in cultivating sound seedlings. Low night temperature is an effective measure to control excessive elongation of seedlings. It is important to maintain the proper temperature for seedling growth during the day, increase photosynthesis and material production, and reduce the temperature at night by 8–10℃ compared with that during the day, so as to promote the operation of photosynthetic products and reduce respiratory consumption. Alternating temperature management can be implemented for seedling culture in facilities with a high level of automatic control. On cloudy and rainy days, the temperature is relatively low during the day, and the temperature at night should be reduced accordingly.

3. Air temperature

Different crop types and growth periods have different requirements for temperature. Generally speaking, the temperature should be high after sowing, before seedling emergence, and after transplanting, before recovery; it should be low after seedling emergence, after recovery, and during the hardening stage. The temperature should be high in the early stage of growth and gradually decrease after the middle stage. The seedlings should be hardened at low temperatures 7–10 days before planting to enhance their adaptability to the environmental conditions after planting. A relatively high temperature should be maintained after grafting and before surviving.

Generally, the most suitable germination temperature for thermophilic solanaceous, legume, and melon vegetables is 25–30℃, and that for cold-resistant cabbage and root vegetables is 15–25℃. From seedling emergence to cotyledon flattening, the hypocotyl is sensitive to temperature. Especially when the night temperature is too high, it could easily grow excessively, so it is necessary to reduce the temperature. For solanaceous and melon vegetables, the daytime temperature should be controlled at 20–25℃ and the nighttime temperature should be controlled at 12–16℃, while those for cryophilic vegetables should be slightly lower. After the true leaves unfold, the temperature of thermophilic vegetables should be kept at 25–28℃ during the day and 13–18℃ at night; the temperature of hardy or semi-hardy vegetables should be kept at 18–22℃ during the day and 8–12℃ at night. For the vegetables that need to separate the seedlings, the temperature of the seedbed should be appropriately reduced to the lower limit of suitable temperature 2–3 days before the seedling separation, and the temperature should be increased as much as possible after the seedling separation. During the seedling stage, the temperature of thermophilic vegetables should be 23–30℃ during the day and 12–18℃ at night; the temperature of cryophilic vegetables should be 3–5℃ lower than that of thermophilic vegetables. The suitable temperature for the seedling culture of several vegetables is shown in Table 5-7.

Table 5-7 Suitable Temperature for the Seedling Culture of Several Vegetables

Vegetable species	Suitable air temperature (℃)		Suitable soil temperature (℃)
	Daytime temperature	Nighttime temperature	
Tomato	20–25	12–16	20–23
Eggplant	23–28	16–20	23–25
Pepper	23–28	17–20	23–25
Cucumber	22–28	15–18	20–25
Pumpkin	23–30	18–20	20–25
Watermelon	25–30	20	23–25
Melon	25–30	20	23–25
Kidney bean	18–26	13–18	18–23
Chinese cabbage	15–22	8–15	15–18
Kale	15–22	8–15	15–18
Strawberry	15–22	8–15	15–18
Lettuce	15–22	8–15	15–18
Celery	15–22	8–15	15–18

The suitable temperature for flower seed germination varies with the species and the place of origin and is generally 3–5℃ higher than the suitable temperature for its growth. The suitable temperature for germination of most kinds of flower seeds native to temperate zones is 20–25℃, and that of hardy perennial flower seeds and outdoor biennial flower seeds is generally 15–20℃. Some tropical flower seeds may only germinate at a relatively high temperature (32℃). During sowing, it is better to keep the substrate temperature relatively stable, with the variation range not exceeding 3–5℃. The temperature of flowers after seedling emergence should be gradually decreased with the growth of seedlings, generally 15–30℃ during the day and 10–18℃ at night. The temperature of the substrate or nutrient solution should be 15–22℃, which should be relatively lower for cold-resistant flowers and relatively higher for heat-resistant flowers.

4. Temperature control

During seedling culture in winter, as the temperature is obviously low, various measures should be taken to raise the temperature. The electric hotbed is the most effective way to raise and control the substrate temperature. If the air temperature cannot be raised to a suitable temperature for seedling growth after making full use of solar energy and heat preservation measures, heating equipment should be used to raise the temperature. The heating cost of the coal-fired stove is low and the management is simple, but the thermal efficiency is low and the pollution is serious. The heating boiler is clean and easy to control. It mainly includes two types: coal-fired boiler and oil-fired boiler. The heating is divided into hot water circulation heating and steam circulation heating. Besides, the hot-blast stove is also a commonly used heating device, which uses coal, kerosene, or liquefied petroleum gas as fuel. The air is first heated and then sent into the greenhouse by a blower. In addition, geothermal energy, solar energy, and factory waste heat can also be used for heating.

As the temperature is high in summer, the seedling culture facilities need to be cooled. When the outside air temperature is relatively low, the main cooling measure is through natural ventilation. Other cooling measures include forced ventilation, sunshade net, non-woven fabric, external sunshade of bamboo curtain, wet curtain fan, surface spraying of transparent covering, whitewashing, and indoor water spraying. The test shows that the wet-curtain fan cooling system can reduce the room temperature by 5–6°C. Spray cooling is only suitable for vegetable or flower crops that are resistant to high air humidity.

(II) Light

Not all vegetable and flower seeds require light to germinate, e.g. lettuce, celery, and primrose can germinate under certain light conditions, while leek, onion and *Amaranthus* are stunted in light. 90%–95% of dry matter of seedlings comes from photosynthesis, and the intensity of photosynthesis is mainly restricted by light conditions. Moreover, the lighting intensity also directly affects the ambient temperature and leaf temperature. Lighting intensity affects the growth rate and external appearance of seedlings. Strong light facilitates the development of flowers, while weak light could easily cause leggy seedlings. The lighting time has a great effect on the formation of plant organs, which can restrict the formation and differentiation of flower buds. Light quality also has a great influence on seedlings. Reddish-orange light can promote photosynthesis, and ultraviolet light can promote the sound growth of seedlings and prevent overgrowth.

The focus of seedling stage management is to improve the utilization rate of light energy, especially in winter and spring, when the light duration is short and the light intensity is weak, various measures should be taken to improve the light conditions of seedlings, which is one of the key prerequisites for cultivating sound seedlings. The main control measures are as follows: ① Increasing the daylighting amount of seedling culture facilities, increasing incident light and reducing shadows; ② Selecting a covering material with good light transmission to keep the surface clean and increase the light intensity; ③ Strengthening the covering/uncovering management of opaque covering material, uncovering as early as possible and covering as late as possible to extend the light duration; ④ Greening the seedlings in the light in time, and preventing the seedlings from shading each other.

If the lighting is insufficient, it can be artificially supplemented, which may be used as energy for photosynthesis or used to inhibit and promote flower bud differentiation and regulate the anthesis. The power density of supplementary lighting varies with the types of light sources, which is generally 50–150 W/m^2. In order to reduce the cost of seedling culture, fluorescent lamps are generally used.

In summer, in order to avoid strong light, reduce the temperature, and create a good environment for seedling growth, the shade seedling culture should be adopted. The commonly used materials include sunshade net, non-woven fabric, and straw curtain. Sunshade net is a kind of agricultural covering material with a lightweight, high strength, aging resistance, air permeability, and light transmittance, which is mainly made of polyolefin resin and woven by wire drawing. It has different sizes and colors. If full shade is required, a black plastic film can be used, which can block all the light, thus adjusting the sunshine hours. This method is mainly used to control the development time.

(III) Moisture Content

Moisture is an indispensable condition for the growth and development of seedlings. During the seedling-raising period, proper water supply is an effective way to increase the substance accumulation of seedlings and cultivate sound seedlings. The moisture content of the substrate suitable for the growth of most seedlings is generally 60%–80%. After sowing and before seedling emergence, the substrate should be kept at a high humidity, preferably 80%–90%. The moisture content should be properly controlled 7–10 days before planting. If the substrate contains too much water, the seedlings could easily overgrow under high temperatures and weak light or cause diseases or macerated roots under low temperatures and weak light. On the contrary, if the substrate moisture content is too low, the growth of seedlings will be inhibited, and the seedlings will become stiff when there is a long-time water shortage. The suitable air humidity at the seedling stage is generally about 60%–80% during the day and 90% at night, and the air humidity before seedling emergence and at the early stage of seedling separation should be appropriately increased. The moisture content of substrate in different growth stages of vegetables is shown in Table 5-8.

Table 5-8 Moisture Content of Substrate in Different Growth Stages[①]

Vegetable species	Sowing to seedling emergence(%)	Cotyledon unfolding 2 leaves and 1 heart(%)	3 leaves and 1 heart to seedling(%)
Eggplant	85–90	70–75	65–70
Sweet pepper	85–90	70–75	65–70
Tomato	75–85	65–70	60–65
Cucumber	85–90	75–80	75
Celery	85–90	75–80	70–75
Lettuce	85–90	75–80	70–75
Kale	75–85	70–75	55–60

Note: ①Equivalent to the percentage of maximum water-holding capacity.

The general requirement of water management at the seedling stage is to ensure the proper water content of the substrate and reduce the air temperature appropriately, which shall be controlled flexibly according to the crop type, seedling culture stage, seedling culture method, and seedbed facility conditions. For example, the watering amount of nutrition bowl seedling culture should be more than that of seedbed soil seedling culture. For industrialized seedling culture, it should not be watered with a sprinkler or hose, but a spraying device should be set up to realize mechanization and automation of watering. Nutrient solution or water should be applied on a sunny morning. For seedling culture at low temperatures, water or nutrient solution should be heated before being applied. Water spraying can increase the humidity of both substrate and air. The measures to reduce the humidity of seedbeds mainly include reasonable irrigation, ventilation, and temperature increase.

(IV) Gas

1. CO_2

CO_2 is the raw material of plant photosynthesis. The volume fraction of CO_2 in the outside atmosphere is about 330 μL/L, with a relatively small daily variation. However, in relatively closed greenhouses, polytunnels, and other seedling culture facilities, the variation in CO_2 concentration is much larger than that in the outside. The indoor CO_2 concentration is the highest before sunrise in the morning; after sunrise, with the improvement of light and temperature conditions, the photosynthesis of plants will be constantly enhanced, and the CO_2 concentration will rapidly decrease even to a level lower than the outside concentration, exhibiting a deficit. When cultivating seedlings in winter and spring, because of the low outside air temperature and little or no ventilation, the internal CO_2 content is even more insufficient, which restricts the photosynthesis and normal growth of seedlings. The lack of CO_2 in facilities will put seedlings in a state of carbon starvation, and CO_2 fertilization is the most effective at this time. According to previous research results, CO_2 fertilization should be carried out as early as possible at the seedling stage, preferably at the cotyledon stage. CO_2 fertilization for 3 hours every morning in winter can significantly promote the growth of seedlings, facilitate the formation of strong seedlings, and increase the early yield and total yield. The fertilization concentration should be controlled at about 1,000 μL/L. CO_2 fertilization at the seedling stage has become one of the characteristics of modern seedling culture techniques.The comparison of effects of CO_2 Fertilization on Cucumber and Tomato Seedlings at the Seedling Stage is shown in Table 5-9.

Table 5-9 Comparison of Effects of CO_2 Fertilization on Cucumber and Tomato Seedlings at the Seedling Stage

Vegetable	Fertilization concentration	Plant height (cm)	Stem diameter (cm)	Leaf area (cm^2)	Dry weight of whole plant (g/plant)	Net assimilation rate [$g/(m^2 \cdot d)$]	Sound seedling index	Water content (%)
Cucumber	1,100±100 μL/L	22.15	0.494	284.68	1.1945	3.292	0.1978	83.30
	700±100 μL/L	21.30	0.473	247.66	0.9171	2.867	0.1272	83.299
	No fertilization	17.04	0.433	186.82	0.6812	2.754	0.0902	83.741
Tomato	1,100±100 μL/L	40.25	0.556	296.33	1.5615	2.895	0.1836	83.016
	700±100 μL/L	37.25	0.531	249.99	1.2656	2.775	0.1327	83.172
	No fertilization	29.55	0.511	197.55	0.8723	2.410	0.1045	83.534

2. O_2

The O_2 content in the substrate is also important for seedling growth. If O_2 is sufficient, the root system will produce a lot of root hairs, forming a strong root system. If O_2 is deficient, it will cause the root system to suffer from hypoxia and asphyxia, and the aerial part will wilt and stop growing. Generally, the total porosity of the substrate is about 60%.

3. Toxic gases

Toxic gases harmful to seedlings mainly come from incomplete combustion of fuel, decomposition of organic fertilizer or chemical fertilizer, and release of plasticizer from plastic products during heating or CO_2 fertilization. Therefore, it is required to thoroughly check the plastic film and water pipes used for seedling culture and make sure that the fuel is fully burned, the chimney is well sealed, and no fermented organic fertilizer is accumulated in the greenhouse for seeding culture. Facility ventilation not only can reduce the temperature and humidity and supplement the internal CO_2 but also can discharge toxic gases. However, when the outside air temperature is too low, ventilation is not allowed.

III. Common Problems of Soilless Seedling Culture

1. Yellowing of seedlings

Because the nutrient solution prepared for soilless seedling culture is substantially or all nitrate nitrogen, compared with ammonium nitrogen applied in soil seedling culture, the color of seedlings is lighter, showing a yellowish-green color, which is caused by different nitrogen forms and does not affect the quality of seedlings. If the seedlings grow and develop normally and encounter no other growth obstacles, this is a normal situation.

2. Overgrowth of seedlings

Seedlings cultivated by soilless seedling culture techniques grow faster and are more prone to overgrowth, which should be properly controlled. The measures to control the overgrowth of seedlings are mainly to lower the temperature properly rather than overcontrol the supply of nutrient solution.

If the growth of seedlings is restrained as if they were raised in soil, and no nutrient solution is supplied for a long time, the quality of the seedlings will be reduced as a result of lack of nutrition and water even though the overgrowth of the seedlings can be controlled.

3. Rotten roots or stunted root system

This situation is generally caused by poor aeration of the substrate. If there is no problem in the selection and use of the substrate, it may be caused by the excessive solution supply. That is, most of the time, when the solution accumulates in the tray (bed) for a long time, it could easily cause the root system to rot or rust in the nutrient solution, resulting in stunting. This situation is more likely to occur, especially when using hygroscopic substrates for seedling culture, such as rock wool block seedling culture, and carbonized rice husk seedling culture. Therefore, the nutrient solution amount should be better controlled when using these substrates for seedling culture.

4. Seedling growth stagnation, small growth point, and shriveling of leaves

Under normal nutrient solution management, this situation may be caused by the following reasons:

① The proportion of ammonium nitrogen in nutrient solution is too high, resulting in the harm of ammonium ions. Therefore, the proportion of ammonium nitrogen in nutrient solution at the seedling stage had better not exceed 30% of total nitrogen.

② Saline injury: After spraying and applying nutrient solution continuously, the water in the substrate evaporates quickly, so salt accumulates in the substrate, and saline injury symptoms gradually appear. Once a saline injury is found, the solution supply should be stopped immediately and water should be applied instead so that the saline injury symptoms can be relieved. This kind of situation is very likely to happen, especially under high temperatures and strong light, so attention should be paid to it.

[Skill Training]

Skill Training 5-1 Soilless Seedling Culture of Vegetables

I. Purpose and Requirements

Select appropriate soilless seedling culture methods according to crop species and seedling age.

Master the operation process of soilless seedling culture.

Be able to accurately analyze and effectively solve the practical problems of soilless seedling culture.

II. Plan

1. Preparation of materials and tools

Lettuce, pepper, tomato and other crop seeds (100 g each), appropriate amount of vermiculite, perlite and peat, 3%–5% trisodium phosphate solution, 0.1% mercuric chloride solution, 0.3%–0.5% sodium hypochlorite (calcium hypochlorite) solution, 3 mmol/L NaOH or KOH dilute solution, 3 mmol/L sulfuric acid or phosphoric acid dilute solution, and agricultural or industrial fertilizer for the formulation of nutrient solution formula for leafy vegetables and fruit vegetables in South China Agricultural University.

2. Preparation of nutrient solution

(1) Nutrient solution formula for leafy vegetables in South China Agricultural University:

$Ca(NO_3)\cdot 4H_2O$: 472 mg; KNO_3: 202 mg; NH_4NO_3: 80mg; KH_2PO_4: 100 mg; K_2SO_4: 174 mg; $MgSO_4\cdot 7H_2O$: 246 mg; pH: 6.1–6.6.

(2) Nutrient solution formula for fruit vegetables in South China Agricultural University:

$Ca(NO_3)\cdot 4H_2O$: 472 mg; KNO_3: 404 mg; KH_2PO_4: 100 mg; $MgSO_4\cdot 7H_2O$: 246 mg;

pH: 6.4–7.8.

3. Instruments and tools

Germination accelerating room, seedling plug tray (72 holes, 128 holes), plastic seedling-raising pot (7 cm × 8 cm), watering can, dry hygrometer, tweezers, tray balance (or electronic analytical balance), steelyard, 50 mL measuring cylinder, acidimeter or pH indicator paper, conductivity meter, 500 mL beaker, plastic basin (barrel), and plastic label.

4. Implementation plans

(1) Seed selection: select full, neat, and pest-free seeds for future use.

(2) Disinfection of tools and hands: Apply 3%–5% trisodium phosphate solution for disinfection.

(3) Disinfection of seeds and substrates: Disinfect the selected seeds with 0.1% mercuric chloride solution for about five minutes, and rinse them with clear water 3–5 times to remove residual disinfectants. Soak the sand or peat substrate in a 0.3%–0.5% sodium hypochlorite (calcium hypochlorite) solution for 30 minutes, and then rinse it with clear water several times.

(4) Substrate loading: Mix the perlite and vermiculite in a ratio of 2 : 1, and load them evenly into a tray (1 cm away from the tray edge).

(5) Sowing: Carefully put the seeds into the plug tray with tweezers. Sow 1–2 seeds in each hole, cover them with a thin layer of vermiculite substrate after sowing, then scrape the soil flat, gently press it, and water it thoroughly.

(6) Transplanting seedlings: After the first true leaf is unfolded, move it into a seedling-raising pot filled with a single substrate such as rock wool, peat, sand, perlite, or vermiculite, and apply 1/3 dose of nutrient solution.

(7) Nutrient solution management: Before seeds germinate, apply only clear water instead of nutrient solution; apply 1/3 dose of nutrient solution in the early stage after seedling transplanting; apply 1/2 dose of nutrient solution in the middle and late stages. Water it every 2–3 days. In summer, water may be poured 1–2 times a day as appropriate to prevent the substrate from being too dry.

(8) Environment management: After sowing, move the seedling tray into the germination accelerating room in an overlapping way, and control the temperature

at 25–26℃. After seedling emergence, promptly move it into the greenhouse for greening in the light. Ensure ventilation, cooling, and shading at noon, and prevent excessive evaporation; widen the diurnal temperature difference at night (5–10℃ lower than the daytime temperature), preserve heat, and build a small Polytunnel when necessary. When the first true leaf unfolds, transplant the seedlings in time to avoid mutual influence. After the seedlings are moved to a plastic bowl, widen the planting spacing in time as the seedlings grow up. Carry out the planting after the physiological seedling age or calendar seedling age and seedling specifications required by different crops are met.

(9) Track record: During seedling culture, track and investigate the growth status and environmental changes of seedlings, and make records in time. Sort out and archive the seedling records in time.

III. Implementation

(1) Work in groups to develop a plan for plug seedling culture of vegetables.

(2) According to the characteristics of different vegetables, select the appropriate plug trays, and work in groups to carry out the plug seedling culture.

(3) Communicate and evaluate.

Module 6 Construction and Management of Soilless Culture Production Facilities

[Learning Objectives]

I. Knowledge Objectives

(1) Learn about the characteristics of deep flow technique, and be able to carry out production with deep flow hydroponic culture;

(2) Learn about the characteristics of nutrient film technique, and be able to carry out production with nutrient film hydroponic culture;

(3) Learn about the characteristics of aeroponics and be able to carry out production with aeroponics;

(4) Learn about the characteristics of trough culture, master the construction method of trough culture facilities, and be able to carry out production with trough culture;

(5) Learn about the characteristics of bag culture, master the construction method of bag culture facilities, and be able to carry out production with bag culture;

(6) Learn about the characteristics of rock wool culture, master the construction method of rock wool culture facilities, and be able to carry out production with rock wool culture;

(7) Learn about the characteristics of stereoscopic culture, master the construction method of stereoscopic culture facilities, and be able to carry out production with stereoscopic culture.

II. Skill Objectives

Learn to consult relevant materials, and get familiar with the concepts and characteristics of deep flow technique, nutrient film technique, aeroponics and solid substrate as well as their applications in production.

[Preparation for Learning]

I. Required Resources

(1) Learn about stereoscopic culture through the Internet;

(2) Standardized soilless culture production base.

II. Background Knowledge

(1) Master the basic knowledge of deep flow technique, nutrient film technique, aeroponics, and solid substrate culture facilities;

(2) Master the basic knowledge of literature review;

(3) Learn about certain basic principles of deep flow technique, nutrient film technique, aeroponics, and solid substrate culture facilities;

(4) Understand agricultural meteorology and agricultural regulations.

[Learning Tasks]

Task 1 Construction and Management of Deep Flow Hydroponic Facilities

I. Characteristics of Deep Flow Technique

Deep flow technique (DFT), also known as the deep flow circulation culture technique, refers to a cultivation technique in which the root system of plants grows in a deep (5–10 cm) and flowing nutrient solution layer. Most parts of the root system of plants are soaked in the nutrient solution, and the aeration of the root system is realized by adding oxygen to the nutrient solution. DFT is the earliest soilless culture technique developed for the commercial production of crops. Since the 1930s, it has been deemed as an effective, practical, and competitive hydroponic production type after continuous improvement. DFT has been widely applied in Japan. Besides, it has also been popularized to a certain extent in provinces and municipalities in China, such as Taiwan, Guangdong, Shandong, Fujian, Shanghai, Hubei, and Sichuan. Through this technique, many fruit vegetables such as tomato and cucumber and leafy vegetables such as lettuce and crown daisy have been successfully produced. Therefore, this type of hydroponic facility is suitable for China's current situation, especially for the hydroponic types with tropical and subtropical climate characteristics in southern China.

Deep flow hydroponic facilities are generally composed of four parts: nutrient solution plantation trough, planting plate (or planting net frame), reservoir, and nutrient solution circulating flow system and control system. Because of different building materials and designs, several types have been developed. Through practice and trial, the Japanese Kamizono hydroponic facilities (which are made of cement components and can be made by users themselves) are suitable for China's situation. The improved Kamizono deep flow hydroponic facilities are described below.

II. Commonly used Deep Flow Hydroponic Facilities

(I) Plantation Trough Construction

The plantation trough is generally 100–150 cm wide. The depth of the trough should be about 12–15 cm, with the deepest depth not exceeding 20 cm, and the length of the trough should be about 10–20 m. During the construction of the plantation trough, first level and tamp the ground, lay a layer of river sand or stone powder of about 3–5 cm at the position where the trough is built, then lay 5-cm-thick concrete on the river sand or stone powder layer as the bottom of the trough, lay bricks with cement mortar to form a trough frame on the trough bottom, then use high-grade cement mortar to plaster both inside and outside the plantation trough, and finally apply a layer of cement paste to smooth the surface so as to prevent the leakage of nutrient solution. The foundation of the plantation trough must be solid; otherwise, the uneven subsidence of the foundation may cause the plantation trough to break, thereby leading to the leakage of nutrient solution. When building a plantation trough in a place where the foundation is relatively soft, in order to prevent the subsidence of the foundation from causing the plantation trough to break, a ϕ 8 steel bar can be added to the concrete layer at the bottom of the trough every 20 cm. Waterproof paint may also be applied to prevent leakage when the plantation trough frame is built (Fig. 6-1).

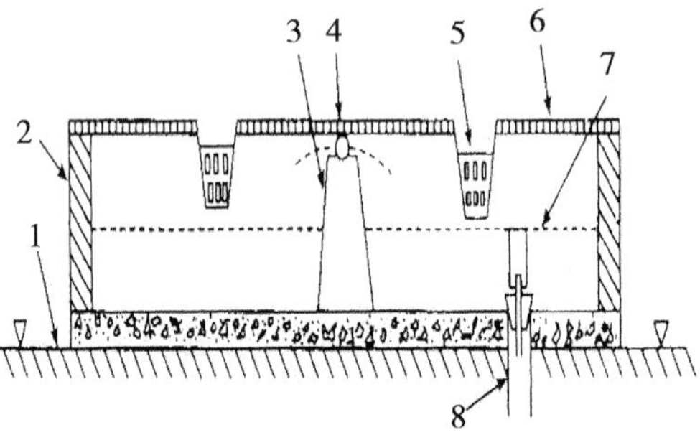

Fig. 6-1 Schematic Diagram of Cross Section of Plantation Trough

1. Ground 2. Plantation trough 3. Supporting pier 4. Liquid supply pipe 5. Planting cup 6. Planting plate 7. Liquid level 8. Return flow and liquid layer control device

This kind of trough does not need to be lined with plastic film but is directly filled with the nutrient solution for cultivation. However, the key to success lies in the selection of acid-resistant and corrosion-resistant cement materials. This kind of trough has the following advantages: It can be constructed by farmers themselves, it is easy to manage, and it has strong durability and a low cost. The disadvantage is that it cannot be removed or relocated. as it is a permanent building. The trough is quite heavy, so it must be built on a solid foundation; otherwise, it will break and cause leakage as a result of the subsidence of the foundation.

(II) Planting Plate Making

The planting plate (Fig. 6-2) is made of a hard foam polystyrene plate, with a thickness of about 2–3 cm. The plate has a number of planting holes with a diameter of 5–6 cm, which can be used for both fruit and leaf vegetables. A plastic planting cup (Fig. 6-3) is embedded in each planting hole, with a height of 7.5–8.0 cm. The diameter of the cup mouth is the same as that of the planting hole, and the outer edge of the cup mouth has a lip that is about 5 mm wide so as to be stuck on the planting hole and not fall into the bottom of the trough. There are many 3-mm holes in the lower half and bottom of the cup. The width of the planting plate is consistent with that of the outer edge of the plantation trough so that both sides of the planting plate can be supported on the wall of the plantation trough and thus the planting plate can be hung together with the planting cup embedded in the plate hole. The length of the planting plate is generally 150 cm, which can be expanded or contracted for the convenience of work. The planting plate covers the entire plantation trough piece by piece, so that the light cannot penetrate into the trough.

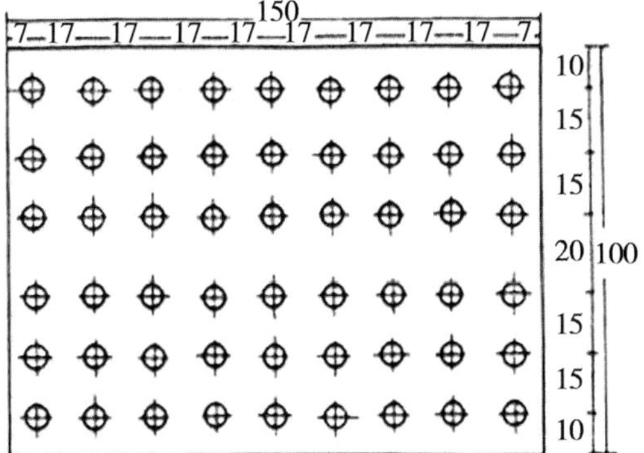

Fig. 6-2 Planting Plate Layout Plan (Unit: cm)

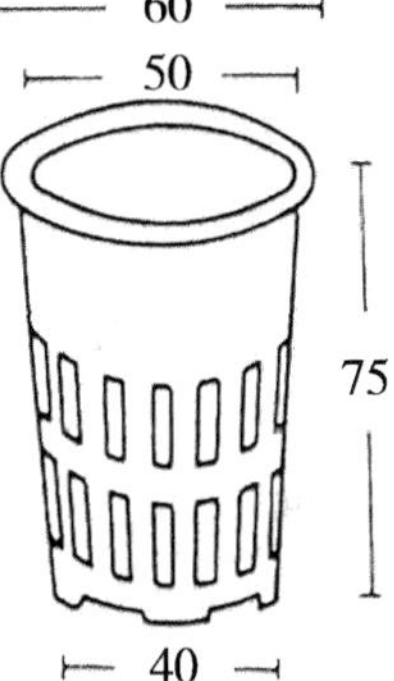

Fig. 6-3 Planting Cup (Unit: mm)

Planting mode of suspended cup planting plate: The weight of the plant is borne by the planting plate and the trough wall. When the liquid level in the trough is lower than the top of the trough wall, a space is formed between the bottom of the planting plate and the liquid level, which creates conditions for oxygen in the air to diffuse into the nutrient solution. When the width of the trough is 80–100 cm and the thickness of the planting plate is maintained at 2.0–2.5 mm, it is necessary to erect supports in the center of the width of the trough to support the weight of the planting plate, so that the planting plate will not bend down into an arc shape due to the growth and weight gain of plants. The supports can be made of truncated cone cement piers, each of which is set every 70 cm or so along the width center line of the trough. A hard plastic liquid supply pipe is placed on each pier to supply liquid and support the weight of the planting plate. The diameter of the truncated cone bottom of the cement pier is 10 cm, and the diameter of the top surface is 5 cm. The height of the pier plus the diameter of the liquid supply pipe should be equal to the height of the inner wall of the plantation trough. There should be a small pit on the top surface of the pier so that the liquid supply pipe will not slide when placed on it. The liquid supply pipe on the pier should be clung to the bottom of the planting plate to bear the gravity of the planting plate and maintain its horizontal state. Under the condition that the top surface of the trough wall is horizontal, all points between the bottom of the planting plate and the bottom of the planting cup and the liquid surface should be equidistant, so that every plant has an equal chance to contact the liquid surface. It is necessary to avoid the root systems of some plants touching the nutrient solution while others are still hanging in the space, resulting in uneven growth.

(III) Construction of Underground Reservoir

The underground reservoir is designed to increase the buffer capacity of nutrient solution and create a stable living environment for the root system. Some types of deep flow hydroponic facilities are not provided with underground reservoirs, and they directly pump nutrient solution from the bottom of the plantation trough for circulation, such as the Japanese M-type hydroponic facilities. This can undoubtedly save land and cost, but it also loses many advantages of the underground reservoir.

The underground reservoir is constructed on the general principle of no leakage. When building the reservoir bottom, 10–15 cm concrete should be used together with ϕ 8 steel bar. The wall of the reservoir should be made of 18–24 cm bricks, and #100 cement mortar should be used for plastering. The cement used for building the reservoir should be of high grade and resistant to plastering. Besides, the surface of the underground reservoir should be 10–20 cm higher than the ground and covered to prevent rain or other debris from falling into the reservoir. The inside of the reservoir should be kept dark to prevent algae from growing.

Advantages of underground reservoir: ① As a place for nutrient solution regulation, the adjustment of the pH value of the nutrient solution and the supplement of nutrients and water are all carried out in the reservoir. ② It can increase the total amount of nutrient solution in the planting system so that the amount of nutrient solution occupied by each plant will increase, so that the concentration, pH value, dissolved oxygen content, and temperature of the nutrient solution will not change drastically.

(IV) Nutrient Solution Circulating Supply System

The nutrient solution circulating supply system consists of two parts: the liquid supply system and the return flow system.

1. The liquid supply system

It comprises a liquid supply pipe, a water pump, and a valve for regulating the flow rate. After pumping up the nutrient solution from the reservoir by a water pump, the liquid supply pipe is divided into two branch pipes, each of which is controlled by a valve. One branch pipe is turned back to the upper part of the reservoir to spray part of the nutrient solution back into the reservoir for oxygenation. If the entire planting system is to be cleaned, this pipe can be used for thorough drainage. The other branch pipe is connected to the main liquid supply pipe, and the main liquid supply pipe branches off to the edge of each plantation trough and is then connected to the liquid supply pipe in the trough. The liquid supply pipe in the trough is a long plastic pipe running through the whole trough, and there are small holes for spraying liquid at regular intervals on the pipe so that the nutrient solution can be evenly distributed throughout the trough.

The liquid supply pipe in the plantation trough with a trough width of 80–90 cm is made of 25 mm polyethylene rigid pipe, and a pair of small holes with a diameter of 2 mm are arranged every 45 cm, which are located on both sides below the horizontal diameter line of the pipe, and the angle between the small hole to the pipe center line and the horizontal diameter is 45°. The liquid supply pipe of each plantation trough is provided with a control valve before it enters the trough, so as to regulate the flow rate.

2. The return flow system

It includes a return pipe, a liquid level regulator in the plantation trough, and liquid level regulators in the return pipe and plantation trough (Fig. 6-4 and Fig. 6-5). A return pipe is arranged at the bottom of one end of the plantation trough, with the pipe orifice being flush with the bottom of the trough, and the lower section of the pipe is buried underground and externally connected to the main return pipe. If the return pipe orifice in the trough is not plugged, the nutrient solution entering the trough can completely flow back into the reservoir. In order to keep a certain amount of nutrient solution in the trough, a section of liquid level control pipe with a rubber plug should be used to plug the return pipe orifice. When the liquid level rises due to the continuous supply of liquid from the liquid supply pipe and exceeds the liquid level control pipe orifice, it will flow back through the orifice. In addition, a movable rubber hose may be arranged on the upper section of the liquid level control pipe so that the liquid level can be raised (lowered) when the rubber hose is raised (lowered). A loose cofferdam cylinder (made of plastic; the inner diameter of the cylinder only needs to be twice as large as that of the liquid level control pipe) can be arranged outside the liquid level control pipe. The height of the cylinder should exceed the liquid level control pipe orifice, and there are jagged notches at the foot of the cylinder so that the nutrient solution cannot flow from the liquid surface into the return pipe during return flow, thereby forcing the nutrient solution to pass through the notches at the foot of the cofferdam before turning to the return pipe. In this way, the oxygen-enriched nutrient solution sprayed from the liquid supply pipe can drive the original oxygen-deficient nutrient solution from the bottom of the trough to flow back, and the cofferdam can prevent the root system from growing into the return pipe orifice. If the liquid level control pipe with a rubber plug is removed as a whole, the nutrient solution in the trough can be completely drained.

The diameters of the return pipe of each trough and the main return pipe should be determined according to the amount of liquid supply. The diameter of the return pipe should be large enough to discharge the amount of liquid that needs to be flowed back in time, so as to avoid flooding resulting from the greater volume of

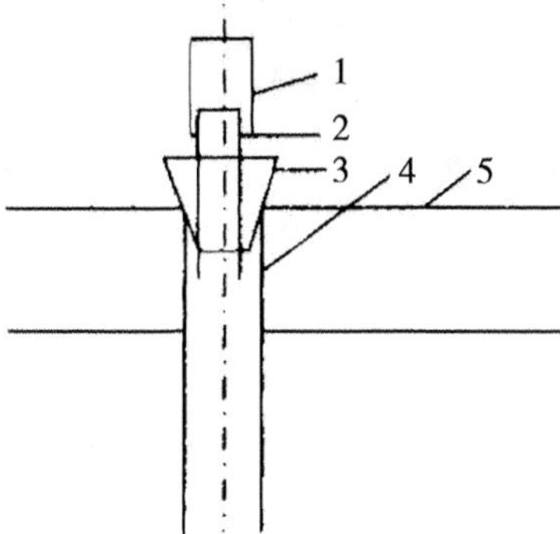

Fig. 6-4 Liquid Layer Control Device

1. Liftable rubber pipe arranged outside the hard plastic pipe 2. Hard plastic pipe 3. Rubber plug 4. Return pipe 5. Bottom of plantation trough

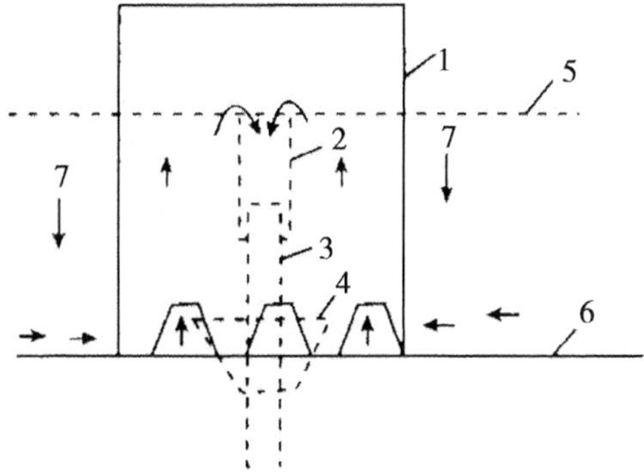

Fig. 6-5 Plastic Pipe Covering Liquid Level Regulator

1. Hard plastic pipe with notch 2. Liquid level regulating pipe 3. PVC hard pipe 4. Rubber plug 5. Liquid level 6. Trough bottom 7. Nutrient solution and its direction (indicated by an arrow)

liquid supply in the trough than the returned liquid volume.

3. Water pump and timer

The water pump is equipped with a timing controller to control the working time of the water pump as required. During large-scale cultivation, all the plantation troughs in the greenhouse can be divided into four groups, with each group equipped with a liquid supply control valve to supply the liquid by turns, so as to ensure that the small liquid flow ejected from the small holes has enough pressure during liquid supply, thereby improving the oxygenation effect.

(V) Treatment of DFT components

1. Treatment of new plantation troughs

New plantation troughs and reservoirs with a cement structure will ooze alkaline substances, which need to be neutralized by impregnation with dilute sulfuric or phosphoric acid and may not only be used until they are not alkaline. The specific treatment method is as follows: Soak the planting facilities in clear water for 2–3 days, wash the soaked alkaline substances, pump out the soaking solution, then soak them in clear water for 2–3 days, and repeat this process several times until the pH value is stable between 6.5 and 7.5 after adding clear water. In order to accelerate the processing speed of the new planting system and shorten the processing time, at the beginning, wash away most alkaline substances by soaking them in water for several days and then soaking them in acid. At the beginning, the pH value of the acid solution is adjusted to about 2, and the pH value will rise again during soaking. The acid solution should be added continuously until the pH value is stable between 6.5 and 7.5. Then, drain the soaking solution and rinse them with clear water for 2–3 times.

2. Cleaning and disinfection in the crop rotation stage

The next-stubble crops may not be planted until the facility system is disinfected during crop rotation.

(1) Cleaning and disinfection of the planting cup: Take the planting cup, together with the residual roots left in the cup, out of the planting plate, pour out the residual roots and small gravel in the cup, wash the gravel and the planting cup with water, wash away the broken root system and other impurities as far as possible, and then place the gravel and the planting cup in separate containers for disinfection. During disinfection, soak it in the hypochlorous acid solution containing 0.3%–0.5% available chlorine or formalin solution containing 0.4% formaldehyde for one day or soak it in the potassium permanganate solution with a concentration of 1/5,000 for 30 minutes, then pour out the disinfectant and rinse it with clear water.

(2) Cleaning and disinfection of the planting plate: Use a brush to brush away the residual roots attached to the plate in water, then soak the planting plate in

the sodium hypochlorite or calcium hypochlorite solution containing 0.3%–0.5% available chlorine, pick it up when it is soaked all over, stack it up piece by piece, cover it with plastic film, keep it moist for more than 30 minutes, and then rinse it with clear water for later use.

(3) Disinfection of plantation trough, reservoir, and circulating pipe: Spray all parts inside and outside the trough and reservoir with sodium hypochlorite or calcium hypochlorite solution containing 0.3%–0.5% available chlorine to make them wet (about 250 mL/m^2), then cover them with the planting plate and reservoir cover plate to keep them moist for more than 30 minutes, and then wash off the disinfectant with clear water for later use. The inner parts of all circulating pipes are to be circularly flushed with the sodium hypochlorite or calcium hypochlorite solution containing 0.3%–0.5% available chlorine for 30 minutes. There is no need to leave a liquid layer in the trough during circulation so that the solution can flow back completely after ejection, and this can be carried out in groups to save the amount of liquid used.

(VI) Common Deep Flow Culture Types

1. Box culture

Box culture is a complex substrate culture method that uses a polystyrene foam plastic box as a cultivation container. It has a good overall effect and is convenient to move. When an individual vegetable is ill, the foam box, together with the plant, can be easily replaced (Fig. 6-6).

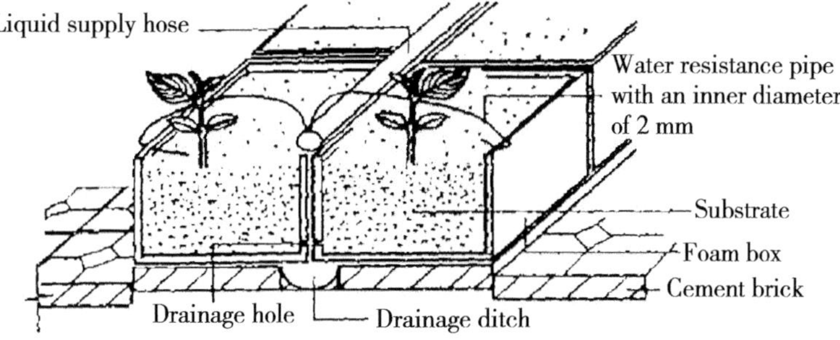

Fig. 6-6 Schematic Diagram of Complex Substrate Box Culture Facilities

(1) Composition of box culture facilities.

① Ground: Level the ground and make a slope of 1∶100–1∶200, and spread the whole ground with cement bricks. At the position where the foam box will be placed, lay two rows of cement bricks and make sure that they are slightly higher than the walkway, leave a gap of 10–15 cm between the two rows of bricks, and plaster the drainage ditch with cement mortar. Make sure that the lower end of the drainage ditch is connected with the drainage trough at one end of the greenhouse so that the excess nutrient solution can be discharged outdoors.

Later, a row of foam boxes will be placed on each row of square bricks. There are also those who only lay cement bricks on the ground of the greenhouse without making drainage ditches, and appropriately reduce the amount of liquid supply. A small amount of excess nutrient solution can infiltrate into the ground through the gaps between the cement bricks.

② Foam plastic box: The basic facilities of box culture are similar to those of trough culture and bag culture, except that the culture trough and bag are replaced with foam plastic box (white). When cultivating large vegetables such as melon and solanaceous vegetables, foam boxes with a height above 20 cm should be selected. When cultivating small vegetables such as cabbage and green leafy vegetables, foam boxes with a height of about 10 cm can be selected. The foam box should have a certain strength, generally up to 20 kg/m^3, so as to prolong the service life of the box. Before use, drill 2–3 holes on the bottom of the foam box or on the side wall 2–3 cm away from the bottom to prevent the water in the box from macerating the root.

③ Liquid supply system: The open liquid supply method is adopted, and the nutrient solution is not to be recycled. The PE pipe is used as the liquid supply pipe, and the main pipe is equipped with a filter. Each foam box is provided with two water-resistance pipes with an inner diameter of 2 mm. Two ends of each water-resistante pipe are sharpened, with one end inserted into the liquid supply branch pipe between the cultivation rows and the other end passing through and fixed at the upper edge of the foam box, extending to the substrate surface in the foam box. The water outlet is kept at a distance of 1–2 cm from the substrate surface, so as to prevent the substrate from being sucked into the water-resistance pipes and causing blockage when the submersible pump stops and the nutrient solution flows back.

(2) Box culture management technique.

During the box filling, the box should not be filled full with the substrate, and it should be ensured that the surface of the substrate is 1–2 cm away from the box opening after crop planting, so as to avoid the overflow of nutrient solution. During the culture of small vegetables, because the plants are relatively dense, it is not necessary to use the box cover. During the culture of large vegetables, planting holes of ϕ 10 cm or above can be punched on the foam box cover; after planting the vegetables, the box cover can be put on tightly, and then the dropper can be inserted through the planting hole. For other management techniques, please refer to the substrate trough culture.

2. Pot culture

Pot culture is a kind of substrate culture, and it is also the simplest way of soilless culture. The pot is relatively large and can hold relatively more substrate soil. It is suitable for planting plants with a large root system and climbing plants, such as tomatoes, pumpkins, and sweet potatoes.

Pot culture, also known as basin culture, mainly adopts miniaturized cultivation containers such as basins and pots. The culture management techniques are similar to those of box culture. The cultivation method is to drill holes of ϕ 5 mm at the height of 1/3 from the bottom of the plastic flowerpot. The bottom of the plastic bucket is filled with water-washed slag, and the substrate, such as rock wool, is placed in it. The nutrient solution is applied every 3–5 days until the overflow outlet drains liquid, and clear water is poured once a month, so that various flowers, fruit vegetables, and green leafy vegetables can be cultivated (Fig. 6-7 and Fig. 6-8).

The small-scale drip irrigation culture may also be used in families. Hang a plastic bucket over the windowsill to hold the nutrient solution, install a drip tube used by the hospital, and insert it into the surface layer of the cultivation pot below. Adopt a plastic bucket or pot as the cultivation pot, drill a hole at a distance of 3 cm from the bottom of the pot, and install a waste infusion tube. Fill the cultivation pot with the substrate, and perform drip irrigation three times a day (30 minutes each time). This facility can be used to cultivate fruit vegetables and medium-sized flowers.

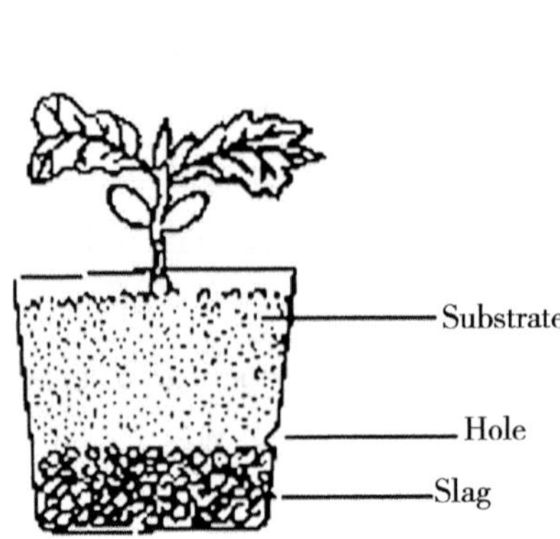

Fig. 6-7 Substrate Pot Culture

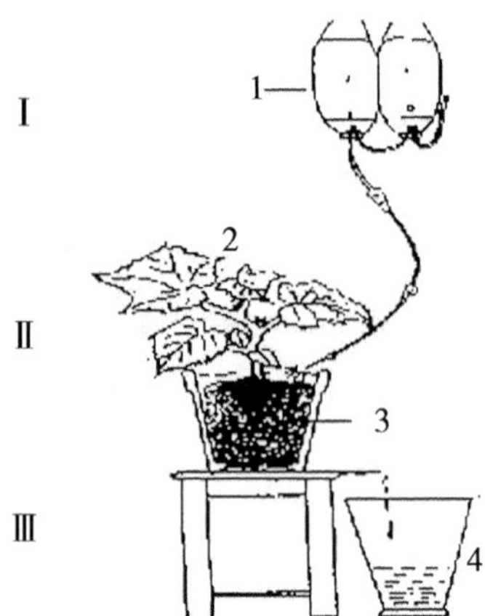

Fig. 6-8 Small-scale Drip Irrigation Culture

Ⅰ. High part Ⅱ. Middle part Ⅲ. Low part
1. Nutrient solution 2. Cucumber seedling
3. Sand and gravel 4. Recovery liquid barrel

III. Management of Deep Flow Technique

(I) Selection of Crop Species for Hydroponic Culture

Select crop varieties with high economic value, and make use of greenhouse conditions to carry out out-of-season or off-season production. During the initial hydroponic production, the crops that are suitable for hydroponic culture, such as tomato, squash, straight-leaf lettuce, water spinach, *Cryptotaenia japonica*, and chrysanthemum, are selected for planting, so as to achieve success in hydroponic culture. When planting in a polytunnel without temperature control, select crops that are suitable for growing in the current season, and avoid blindly planting out-of-season crops. Do not misunderstand the out-of-season function of the soilless culture technique.

(II) Seedling Raising and Planting

(1) Seedling raising: Apply the plug seedling culture method to raise seedlings (The hole diameter of the plug tray should be slightly smaller than that of the

planting cup).

(2) Transplanting seedlings into the planting cup: Prepare non-calcareous gravel for stabilizing the seedlings (the particle size should be larger than the small hole in the lower part of the planting cup), place 1–2 cm gravel in the bottom of the planting cup to prevent the root collar of seedlings from pressing directly to the bottom of the cup, then pull out the seedlings with the substrate from the plug tray and move them into the planting cup (without removing the substrate attached to the roots), and then cover a layer of small gravel on the root mass of seedlings to stabilize the seedlings. Small gravel must be used to stabilize the seedlings. As gravel has no capillary effect, it can prevent the nutrient solution from rising and forming salt rust (salt rust could cause stem base necrosis). Materials with a strong capillary effect (very fine peat, plant residues, etc.) shall not be used to stabilize the seedlings because such materials are easy to form salt rust.

(3) Centralized transitional planting in the transition trough: After the seedlings are transplanted into the planting cup, they can be immediately moved into the holes of the planting plate on the plantation trough for formal planting. However, the hole spacing of the planting plate is determined according to the space that the plants need to occupy when they grow up. When the seedlings are too thin, it will take a long time to fill up the space. In order to improve the utilization rate of greenhouse and hydroponic facilities, the very small seedlings that have been moved into the planting cup are densely placed in a transition trough, and directly placed at the bottom of the trough without using the planting plate for transitional planting. By putting the nutrient solution into the bottom of the trough to a depth of 1–2 cm to soak the cup foot, the seedlings can absorb water and nutrients and grow up quickly and some roots will stick out of the cup. When they grow to a large enough plant, they will be formally transplanted onto the planting plate of the plantation trough. Soon after being transplanted, the space will be filled (closed) up to the extent that it can be harvested, which greatly shortens the time of occupying the plantation trough. This method of centralized transitional planting is very useful for leafy vegetables with a short growth period, but not useful for fruit vegetables with a long growth period.

(4) Control of liquid level in the trough after formal planting: After the planting cup with seedlings is moved onto the planting plate on the plantation trough, it is deemed as formal planting. When the root system does not stick out of the cup or only a few roots stick out at the initial stage of planting, it is required that the liquid level can soak the bottom of the cup for a depth of 1–2 cm so that each seedling has an equal chance to absorb water and nutrients in time. This is the key measure to ensure the uniform growth of plants and avoid the emergence of both large and small seedlings. However, the liquid level should not be adjusted too high to stick to the bottom of the planting plate, which will hinder the diffusion of oxygen into the liquid, soak the root collar of the plant, and suffocate it. In production, the liquid level regulator should be used to gradually reduce the liquid level of the nutrient solution along with the growth of plants and the increase of the root system, so that some root segments are exposed to the air. Once the liquid level is reduced and more root hairs are generated in the root system, the nutrient solution layer with reduced liquid level should not be raised again; otherwise, it may cause damage to the root hairs or even the entire root system, or cause death in serious cases. However, the liquid layer in the plantation trough should not be too shallow. Generally, it should be ensured that the depth of the liquid layer can be maintained to the extent that the plants can normally grow for 1–2 days when there is no power supply and the water pump cannot work normally.

(III) Formulation and Management of Nutrient Solution

The regulation of the liquid level in the plantation trough is a very important technical part of the suspended cup deep flow technique. If it is not done well, it may cause damage to the root system, so it must be carried out very carefully.

At the beginning of planting, the liquid should soak the bottom of the planting cup for a depth of 1–2 cm. When the root system is deeply immersed in the nutrient solution, the liquid level should be lowered accordingly, so that more roots are exposed to the air, which is conducive to breathing and reduces the energy consumption of circulating oxygenation. In this case, the root segments exposed to the humid air will reproduce many root hairs. These root segments with many root hairs should not be submerged in the nutrient solution for too long; otherwise, they will die and cause damage to the entire root system, so the liquid level should not be raised or lowered arbitrarily. In principle, after the liquid level is lowered, if a large number of root hairs have been produced in the upper root segment, the liquid level should be stable at this level. In addition, it should be noted that the amount of liquid maintained at the bottom of the trough should be enough for the plants to absorb water for 2–3 days, and it should not be lowered to a level that cannot maintain the water absorption by the plants for 1 day. In production, attention should also be paid to the problem of the water pump being out of order or the liquid not being supplied as a result of power supply interruption.

(IV) Establishment of a Scientific and Efficient Management System

This is necessary for socialized mass production. Every technical department and every technical measure should have a dedicated person in charge. Clarify the post responsibilities, establish management files, list items that need to be recorded, make sheets and work diaries, and register them item by item. Only in this way can the problems identified in production be scientifically analyzed and effectively solved. A scientific management system is a necessary guarantee for advanced science and technology to come into play. Without a scientific management system, advanced science and technology can hardly play a role in improving productivity. Under the influence of the ideology formed on the basis of China's long-term self-supporting economy, the scientific management system is often ignored. Therefore, we must solve this problem while learning and introducing advanced science and technology to improve productivity.

(V) Summary

The deep flow technique has become a practical and efficient soilless culture technique because of its easy management, durable facilities, and low investment in subsequent production materials. It has also been widely promoted and applied in South China. The construction of deep flow hydroponic facility includes the construction of four parts: nutrient solution plantation trough, planting plate (or planting net frame), reservoir, and nutrient solution circulation system and control system. During the deep flow hydroponic culture, crops that are suitable for hydroponics should be selected for planting in the initial production so as to achieve the success of hydroponics.

Task 2 Construction and Management of Nutrient Film Hydroponic Facilities

I. Characteristics of Nutrient Film Technique

Nutrient film technique (NFT) is a hydroponic method of cultivating the plants in nutrient solution that flows in the shallow layer. It was invented by A.J. Cooper from Glasshouse Crops Research Institute in 1973. After 1979, this technique was rapidly promoted and applied in the world. According to the data recorded in 1980, 68 countries were studying and applying this technique for soilless culture. In 1984, China began to study and apply this soilless culture technique and achieved good results.

II. Commonly Used Nutrient Film Hydroponic Facilities

The facilities applying the nutrient film technique are mainly composed of three parts: plantation trough, reservoir, and nutrient solution circulation device. In addition, according to the actual production and the possibility of funding, some other auxiliary facilities can be selected, such as concentrated nutrient solution storage tanks and automatic delivery devices, and nutrient solution heating and cooling devices.

(I) Plantation Trough Construction

The plantation trough of NFT can be divided into two types according to the types of planted crops: the trough for cultivating large crops and the trough for cultivating small crops.

(1) Plantation trough for cultivating large crops: This kind of trough is an isosceles triangular trough temporarily enclosed by 0.1–0.2-mm-thick polyethylene film with a white surface and black bottom. The trough is 20–25 m long, 25–30 cm wide at the bottom, and 20 cm high. That is, take the above film with a width of 75–80 cm

and a length of 21–26 m, and lay it on the pre-leveled and compacted ground with a certain slope (about 1 : 75), with the long side parallel to the slope direction. When planting, place the seedlings with seedling pots in a row in the center of the film width, and then pull up the two sides of the film so that the width of 20–30 cm in the center of the film is closely attached to the ground, and the pulled up two sides are folded together and clamped with clips to form an isosceles triangular trough with a height of 20 cm. The stems and leaves of the plants extend out of the trough from the crevices at the top of the trough, and the root is placed at the opaque bottom of the trough.

The nutrient solution should flow from the high end to the low end, so there should be no potholes on the ground at the bottom of the trough to avoid water accumulation in the trough. The slope can be adjusted by using a hard board (wood or plastic) to pad the trough. The slope should not be too small or too large so that the nutrient solution can flow smoothly in the trough. The nutrient solution should flow in a shallow layer in the trough, and the depth of the liquid layer should not exceed 1–2 cm. It is appropriate to inject 2–4 L of nutrient solution per minute in the trough with a bottom width of 20–30 cm and a trough length of no more than 25 m.

In order to improve the water absorption and aeration of crops, a layer of non-woven fabric can be laid at the bottom of the trough, which can absorb water and make the water diffuse, and the root system cannot pass through it, so the plants can then be planted on the non-woven fabric. Its main functions are as follows: ① Shallow layer: When the nutrient solution flows directly on the plastic film, it will cause turbulence. When the plant is young, the nutrient solution cannot flow into the root system, resulting in a water shortage. Non-woven fabric can make nutrient solution spread to the bottom of the trough, thereby ensuring that plants can absorb water. ② The root system grows directly against the plastic film. When the plant grows big enough, it will have a large root volume with a heavy weight, forming a thick root pad that sticks tightly with the plastic film, and the nutrient solution cannot flow smoothly at the bottom of the root pad, resulting in hypoxia at the bottom of the root pad, which is prone to necrosis. With a layer of non-woven fabric that cannot be penetrated by the root system, the roots can only grow on the non-woven fabric. There is a layer of non-woven fabric between the roots and the plastic film, and the nutrient solution can flow through it, thus solving the problem of hypoxia at the bottom of the root pad. ③ Non-woven fabrics can absorb a lot of water, and when the power is cut off, it can alleviate the risk of rapid crop wilting due to water shortage.

(2) Plantation trough for cultivating small plants: This kind of trough is made of glass fiber-reinforced plastic or cement, with corrugated tiles as the trough bottom. The depth of corrugated tile is 2.5–5.0 cm, the peak distance varies depending on the size of the plant type, and the width is 100–120 cm. It can be planted in 6–8 rows, so the peak distance can be calculated accordingly. The whole trough is about 20 m long with a slope of 1 : 75. When the corrugated tiles are connected, the overlap should be deep enough to coincide, so as to prevent the nutrient solution from leaking. Generally, the trough is set up on a wooden frame or a metal frame, with a height that is convenient for operation. A plate cover is required on the corrugated tile to cover it and make it opaque. The plate cover is made of rigid foam plastic board, with planting holes drilled on it, and the spacing of holes is determined according to the row spacing of plants. The length and width of the plate cover are matched with the corrugated tile trough bottom, and the thickness is about 2 cm (Fig. 6-9).

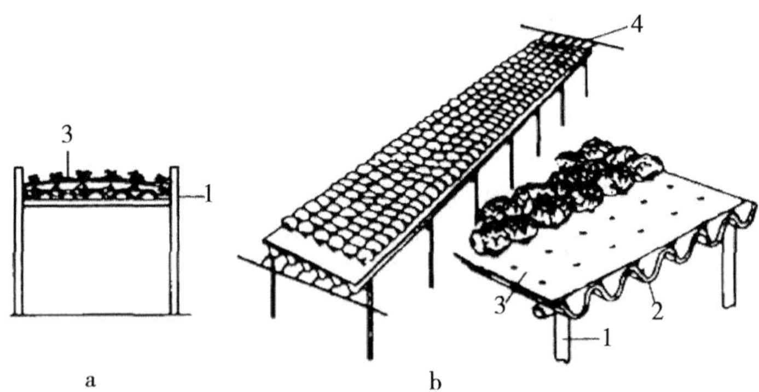

Fig. 6-9 NFT Plantation Trough for Small Crops

a. Cross section b. Side view

1. Support 2. Plastic corrugated tile 3. Planting plate cover 4. Liquid supply pipe

(II) Reservoir Construction

Generally, the reservoir is located below the ground so that the nutrient solution flowing out of the plantation trough can flow back into the reservoir. If it is a plantation trough with support for planting small crops, the reservoir can be constructed on the ground, and the volume of the reservoir should be determined on the premise of ensuring adequate water supply and circulation of the plants.

(III) Circulating System Assembly

The circulating system assembly is mainly composed of a water pump, pipes, flow-regulating valves, etc.

1. Water pump

It shall follow the principle of durability and sufficient water output. A self-priming pump or submersible pump with acid and alkali resistance and corrosion resistance may be adopted.

2. Pipes

Plastic pipes should be adopted to prevent corrosion. Strict sealing is required during pipe installation, and it is better to adopt thread connection instead of socket connection. Besides, it is necessary to bury the pipe as far below ground as possible, which is convenient for work and avoids accelerated aging caused by sunlight exposure. There are two kinds of pipes: The liquid supply pipe and the return pipe. The liquid supply pipe connects to the main pipe from the water pump, and the branch pipes are led out from the main pipe. One of the branch pipes is led back to the reservoir, so that a part of the pumped nutrient solution flows back into the reservoir. On the one hand, it plays a role in stirring the nutrient solution to make it more uniform and increase the dissolved oxygen in the solution; on the other hand, the flow can be adjusted through the valve on it. A number of capillary tubes are connected to the branch pipes and delivered to the high end of each plantation trough. The capillary tubes of each trough are provided with a flow-regulating valve, and then small infusion tubes are connected to the capillary tubes and introduced into the plantation trough. Several small infusion tubes of 2–3 mm are set in each large plantation trough, and the number of tubes should be controlled to the extent that the flow rate is controlled within 2–4 L/min in each trough. The purpose of setting up several small infusion tubes is to make sure that at least one or two of them remain unblocked to ensure that there is no water shortage when another one or two of them are blocked. Two small infusion tubes are arranged in each small plantation trough valley to ensure that there is liquid flow in each valley, with the flow rate being 2 L/min per valley. The return pipe is connected to the liquid discharge port, which is located at the lower end of the plantation trough. The collected liquid flows back to the reservoir through the connection of the return pipe and the liquid collecting and return main trough. The liquid collecting and return main trough should have a large enough diameter to avoid stagnation and overflow.

(IV) Treatment of NFT Components

1. Treatment of plantation trough

For a new plantation trough, check whether all components meet the requirements, especially whether the bottom of the trough is smooth and whether the plastic film is damaged or leaking. For a trough reused after crop rotation, check whether there is any leakage before use, and thoroughly clean and disinfect it.

2. Seedling raising and planting

(1) Seedling raising and planting in a plantation trough for large plants: Because the nutrient solution layer of NFT is very shallow and the root systems of crops are placed at the bottom of the trough during planting, all the planted seedlings need a solid substrate or a porous plastic bowl to anchor the plants. Therefore, a solid substrate block (usually rock wool block) or porous plastic bowl should be used to raise seedlings. When planting, the solid substrate block or plastic bowl should not be removed, and the seedlings together with the bowl (block) should be placed at the bottom of the trough.

The triangular plantation trough for cultivating large plants is closed at a relatively high position, so the seedlings raised should have an adequate height before being planted so that the stems and leaves of the seedlings can protrude through the crevices at the top of the triangular trough when placed in the trough.

(2) Seedling raising and planting in a plantation trough for small plants: Rock wool block or sponge block may be used to raise the seedlings. The rock wool block should have an appropriate size so that it can be rotated into the planting hole

without lying on the bottom of the trough. Non-woven fabric may also be rolled or rock wool can be cut into square strips to raise the seedlings. Cut a small slit at the upper end of the seedling raising strip, and put the germinated seeds in it, so that the seedlings with 2–3 leaves can be densely raised. Then, move them into the planting holes of the plate cover. After planting, make sure that the seedling-raising strips touch the bottom of the trough and that the young leaves protrude above the plate surface.

3. Other auxiliary management facilities

Because of the small amount of nutrient solution to be used in NFT, the nutrient solution changes rapidly, and thus it must be regulated frequently. In order to reduce the labor intensity and make a timely regulation, some automatic control auxiliary facilities can be selected for automatic regulation. Even without these auxiliary facilities, normal production can also be carried out by manual regulation, which is relatively complex. Auxiliary facilities include the timer, electrical conductivity (EC) automatic control device, pH automatic control device, nutrient solution temperature regulating device, safety device, etc. (Fig. 6-10).

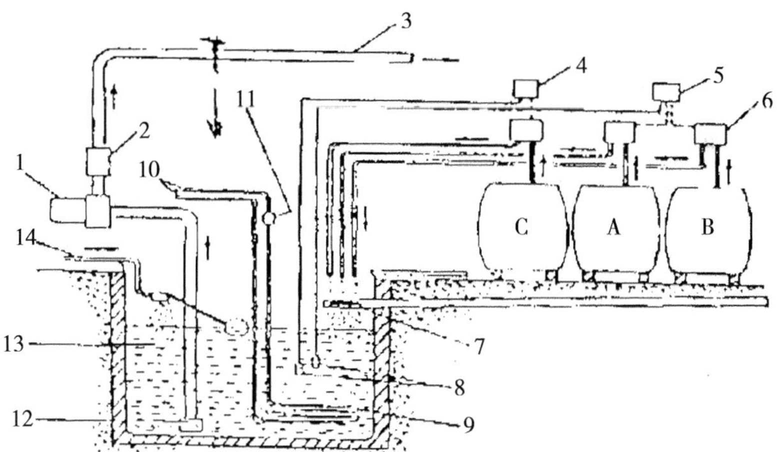

Fig. 6-10 Schematic Diagram of Nutrient Solution Automatic Control Device

A & B: Concentrated nutrient solution storage tank C: Concentrated acid (alkali) storage tank
1. Pump 2. Timer 3. Liquid supply pipe 4. pH control instrument 5. EC control instrument
6. Injection pump 7. Nutrient solution return pipe 8. EC and pH sensors 9. Heating or cooling pipe 10. Warm air (cold water) return pipe 11. Warm air (cold water) control valve
12. Water pump strainer 13. Reservoir 14. Water source and floating ball

(1) Timer: Intermittent liquid supply is a unique management measure of NFT hydroponics. By installing a timer on the water pump, it can realize the accurate control of intermittent liquid supply. However, the setting of intermittent time should be in line with the actual crop growth.

(2) Electrical conductivity (EC) automatic control device: This device consists of the EC sensor and control instrument, concentrated nutrient solution tanks A and B, and an injection pump. When the EC sensor senses that the concentration of nutrient solution has decreased to the set limit, the control instrument will instruct the injection pump to inject the concentrated nutrient solution into the reservoir, so that the concentration of the nutrient solution can be restored to the original concentration. On the contrary, if the concentration of the nutrient solution is too high, it will instruct the water source valve to open and add water to dilute the nutrient solution so as to reach the specified concentration.

(3) PH automatic control device: This device consists of a pH sensor and control instrument, and a concentrated acid (alkali) storage tank with an injection pump. Its working principle is similar to that of the EC automatic control device. (Generally, the concentrated acid is nitric acid or phosphoric acid, or the concentrated alkali is sodium hydroxide or sometimes potassium hydroxide).

(4) Nutrient solution temperature regulating device: Too-high or too-low solution temperature will inhibit the growth of crops. Improving the growth conditions of crops by adjusting the solution temperature is much more economical than heating or cooling the whole polytunnel or greenhouse. The nutrient solution temperature control device is mainly composed of a heating device or cooling device and an automatic temperature controller.

(5) Safety device: The characteristics of NFT determine that the liquid layer in the plantation trough is very thin. Once the power is cut off or the water pump fails, the liquid supply cannot be circulated in time, and in this case, crops could easily wilt because of water shortage. Tomatoes with water-absorbent non-woven fabric as the bottom liner of the trough will wilt in summer after stopping the liquid supply for two hours. Leafy vegetables planted in the plantation trough without non-woven fabric liner will dry up and die after stopping the liquid supply for more than 30

minutes. Therefore, the NFT system must be equipped with a backup motor and a water pump. Besides, it is also necessary to install an alarm device in the circulation system, so as to give an alarm when the water pump fails and remedy it in time. The quality of automatic regulating devices for electrical conductivity, pH value, temperature, etc. should be sensitive and stable, and it is necessary to regularly monitor these devices every day to see whether they fail, so as to ensure that these devices will not go wrong and endanger crops.

(V) Other Hydroponic Facilities

1. Floating capillary hydroponic facility

Floating capillary hydroponic(FCH) facility is a new soilless culture system developed jointly by Zhejiang Academy of Agricultural Sciences and Nanjing Agricultural University on the basis of Japan's root floating technique (Fig. 6-11). This kind of deep flow hydroponic facility can effectively solve the contradiction between water and vapor in the nutrient solution. It can provide a stable rhizospheric environment and liquid level, eliminating concerns of power

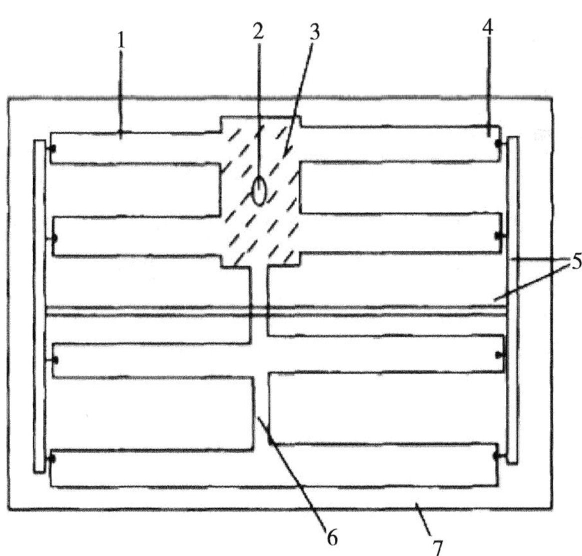

Fig. 6-11 Layout Plan of FCH Facility

1. Plantation trough 2. Water pump 3. Reservoir 4. Air mixer 5. Liquid supply pipe 6. Liquid discharge pipe 7. 6 m × 30 m polytunnel

and water outages halfway. It has the characteristics of low cost, less investment, convenient management, energy saving, practicality, and wide adaptability, and it is suitable for various climatic conditions and ecological types in the north and south of China.

The floating capillary hydroponic facilities include the plantation trough, reservoir, circulation system, and control system. Except for the plantation trough and the air mixer installed in the circulation system, the other three facilities are basically the same as the NFT facilities. The plantation trough is a concave trough made of molding polystyrene board with a length of 1 m, and then connected to form a long trough with a length of 15–20 m, a width of 40–50 cm and a height of 10 cm. The 0.1-cm-thick plastic film is laid in the trough. The slope of the plantation trough is 1 : 100. The depth of the nutrient solution is 5–7 cm. A polystyrene foam board with a thickness of 1.25 cm and a width of 10–20 cm floats on the liquid surface. The board is covered with a layer of non-woven fabric, and its two sides extend into the nutrient solution. Through capillary action, the floating board is kept moist all the time. The planting plate is a 2.5-cm-thick and 40–50-cm-wide polystyrene foam board, which covers the plantation trough. Two rows of planting holes are drilled on the planting plate. The hole diameter is consistent with the outside diameter of the planting cup, and the hole spacing is 40 cm×20 cm. The seedlings are planted in the planting cup and then hung in the planting hole of the planting plate, which clamps the floating plate in the trough right in the middle. After the root system extends out of the hole of the planting cup, part of the roots climb and grow onto the floating plate, producing root hairs to absorb oxygen, and part of the roots extend into the nutrient solution to absorb water and nutrients. The upper end of the plantation trough is provided with a liquid inlet pipe, the lower end is provided with a drainage device, and the liquid inlet pipe is equipped with an air mixer to increase the dissolved oxygen content of the nutrient solution. The drain pipe is connected to the reservoir. The depth of the nutrient solution in the trough is regulated through the base plate or liquid layer control device (Fig. 6-12).

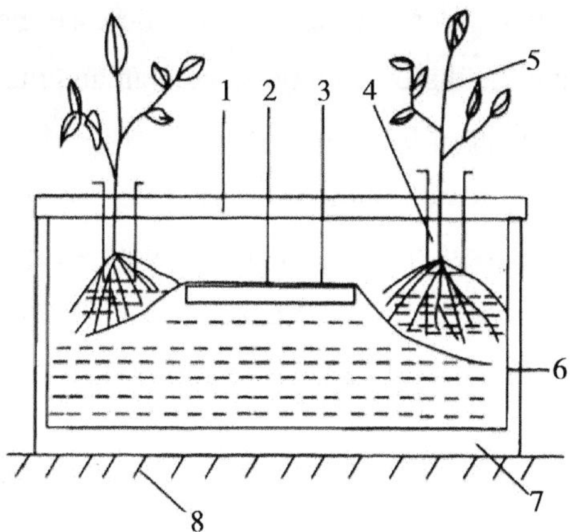

Fig. 6-12 Schematic Diagram of Cross Section of FCH Plantation Trough

1. Planting plate 2. Floating plate 3. Non-woven fabric 4. Planting cup 5. Plants 6. Nutrient solution 7. Molding polystyrene plantation trough 8. Ground

The circulating liquid supply system consists of the water pump, valve, pipe, and air mixer. The circulation route of nutrient solution is as follows: reservoir → valve → pipe → air mixer → culture bed → liquid outlet → reservoir. The control system consists of a timer and a temperature control instrument, which are mainly used to start and stop the water pump and automatically control the liquid temperature.

2. Dynamic root floating system

Dynamic root floating system(Fig. 6-13) is a deep flow technique that was developed and applied in Taiwan, China. The crop root system is placed in the nutrient solution of the culture bed, which can fluctuate up, down, left, and right depending on the change of the liquid level of the nutrient solution. The culture bed is formed by pressing the foam plastic plate, with a length of 180 cm, a width of 90 cm, and a depth of 8 cm. There is a 2 cm bulge in the middle to make the culture bed more firm. The slope of the culture bed is 1%–2%. The culture bed is covered with a 90 cm × 60 cm × 3 cm foam board, and a ϕ 2.5 cm hole is dug at regular intervals on the board for planting leafy vegetables. The hole spacing depends on the needs of the crops. This kind of culture bed also needs a bracket. The air mixer is installed at the inlet where the nutrient solution flows into the culture bed, and its internal structure is two groups of overlapping plastic gates in a cross shape, which will generate eight water streams and can increase the air mixing by about 30%, thereby increasing the dissolved oxygen content and maintaining it at 3–6 μg/mL. A drainer with a height of 0–8 cm is installed at the drain outlet of the culture bed, which works in conjunction with the timer. One underground reservoir with a capacity of about 6 t or a large liquid storage tank with a capacity of more than 100 L can be set up per mu of hydroponic area, and one 750 W high-speed water pump can be installed. The nutrient solution layer is controlled by a floating switch.

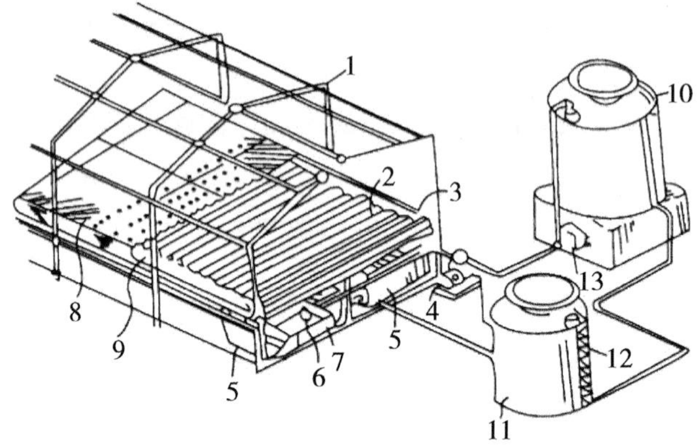

Fig. 6-13 Main Components of Dynamic Root Floating System

1. Tubular greenhouse 2. Culture bed 3. Air mixer 4. Water pump 5. Pool 6. Nutrient solution level regulator 7. Nutrient solution exchange box 8. Slats 9. Nutrient solution outlet plug 10. High-level nutrient solution tank 11. Low-level nutrient solution tank 12. Floating switch 13. Automatic power controller

3. Floating hydroponic technique facility

Floating hydroponic technique (FHT) refers to a deep flow technique in which plants are planted on the floating plate and the floating plate floats naturally in the culture bed. The depth of nutrient solution in the culture bed is generally 10–100 cm. According to the depth of nutrient solution in the culture bed, the floating hydroponics can be divided into deep water floating hydroponics and shallow pool floating hydroponics. The deep water floating hydroponics is suitable for the cultivation of various leafy vegetables. Its planting plate floats on the nutrient solution, so it is easy to move. One of the characteristics of deep water floating hydroponics is that the planting plate can be changed several times according to the size of plants to save greenhouse space. Although this culture method involves large investment and high production cost, it has the characteristics of large scale and modernization, which realizes the industrialized production of leafy vegetables and can significantly improve the utilization rate of greenhouse and the yield per unit area. The main difference between shallow pool floating hydroponics and traditional DFT hydroponics is that both the planting plate and the plant root system float up and down with the change of the liquid level of the nutrient solution in the plantation trough.

The deep water floating hydroponics includes culture bed, planting plate, circulating liquid supply system, automatic control system, and nutrient solution disinfection device. Generally, the culture bed is a pool made of bricks and cement. In the greenhouse, one or several pools are built, each with a depth of 80–100 cm, a width of 4–10 m, and a length of 10 m, except that a small space is reserved at both ends to serve as a working channel and for placing the conveying devices for transplanting seedlings and planting. In large multi-span greenhouses, several culture beds are often arranged in parallel and separated by aisles in the middle. The bottom of the culture bed is provided with an air outlet connected with a compressed air pump and a liquid outlet connected with a concentrated solution distribution pump. The 80–90-cm-deep nutrient solution is put into the cultivation culture bed. The nutrient solution in the reservoir is connected to another water pump through a return pipe, and the self-circulation of the nutrient solution in the reservoir is realized through the water pump.

The planting plate is a white polystyrene foam plastic plate with many planting holes, and the hole spacing varies depending on the crop species and growth stages. The planting plate floats on the nutrient solution by buoyancy without other supports. The nutrient solution circulation system comprises a reservoir, water pump, liquid-adding system, liquid return system, and oxygen supplementing device. The automatic control system comprises a conductivity meter, pH meter, hygrothermograph, illumination measuring device, and alarm device connected to a computer. It can monitor the concentration, pH value, and temperature of the nutrient solution as well as the temperature, humidity, and illumination of the greenhouse at any time, and can automatically adjust the pH value of the nutrient solution according to the set program.

4. Small hydroponic facilities

They are mainly used for small-scale research experiments in home hydroponics and scientific research institutes or as teaching aids in primary and secondary schools. Most of them are simple in structure and generally need a plastic container for liquid storage and a small number of PVC pipe fittings, and some need to add a small oxygen pump. The types of facilities mainly include the simple NFT device (Fig. 6-14), vegetable wall (Fig. 6-15), and simple static hydroponic box (Fig. 6-16).

Fig. 6-14 Simple NFT Device

Fig. 6-15 Vegetable Wall

Fig. 6-16 Simple Static Hydroponic Box

III. Nutrient Film Culture Management Technique

1. Selection of plantation crops

It is advisable to first select crops that are easy to cultivate successfully, such as leaf lettuces and tomatoes and then plant other crops with experience.

2. Seedling raising and planting

Refer to the DFT hydroponics.

3. Selection of nutrient solution formulas

Because the concentration and composition of nutrient solution in the NFT system change rapidly, the nutrient solution formulas with better stability should be selected.

4. Liquid supply method

The liquid supply method of NFT is relatively more particular because its liquid layer is very shallow (no greater than 1.0–2.0 cm). In such a shallow liquid layer, the nutrients and oxygen contained in it could easily be consumed to a very low level. After the nutrient solution is input from one end of the trough head and flows over a long distance (not exceeding the limit of 25 m), many plants will absorb its nutrients and oxygen so that there is not much oxygen and nutrients left in the nutrient solution from the trough head to the trough tail, resulting in a great difference in plant growth between the trough head and the trough tail. When the liquid supply reaches a certain limit, it will impact the output. This shows that the liquid supply of NFT is related to multiple factors.

When the length of the NFT trough exceeds 30 m and the plants are dense, the intermittent liquid supply method should be adopted to meet the oxygen requirement of the root system. Thus, the liquid supply method of NFT is derived into two types: continuous liquid supply method and intermittent liquid supply method.

(1) Continuous liquid supply method: The case of absorption of oxygen by the root system under NFT can be divided into two stages. The first stage is from the planting to the formation of the root pad, during which the root system is immersed in the nutrient solution, thus mainly absorbing dissolved oxygen from the nutrient solution. In the second stage, with the increase of root volume, some roots are exposed to the air after the root pad is formed, thus absorbing oxygen from

the nutrient solution and air. The appearance of the second stage is related to the amount of liquid supply. If the liquid supply is large, the root pad needs to reach a relatively thick level before it can be exposed to the air, thus entering the second stage relatively late; If the liquid supply is small, it will enter the second stage very soon. As the second stage is the stage when the root system obtains sufficient oxygen source, we should promote the early appearance of the second stage. The amount of continuous liquid supply can be in the range of 2–4 L/min, which varies depending on the growth vigor of crops. In principle, liquid supply is required in both daytime and nighttime. If the liquid supply is stopped at night, the absorption of nutrients and water by crops will be inhibited (the absorption will be reduced by 15%–30%), which will lead to crop yield reduction.

(2) Intermittent liquid supply method: This is an effective method to solve the problem of root hypoxia caused by a too-long trough and too-many plants in the NFT system. In addition, under the condition of normal trough length and normal plant number, the yield and fruit weight of the intermittent liquid supply method are higher than those of the continuous liquid supply method. In the case of intermittent liquid supply, when the liquid supply stops, the nutrient solution flows out from the large pores in the root pad, and the air is injected so that the oxygen in the air can be absorbed by the root pad through to the root bottom, thus increasing the oxygen absorption of the entire root system.

The intermittent liquid supply should start at the early stage of the root pad formation. When the root pad is not formed (that is, the root system is small in volume and has not accumulated into a thick layer), the intermittent liquid supply has little effect. Taking the planting of tomatoes with non-woven fabric at the bottom of the trough as an example, the degree of intermittent liquid supply is as follows: supplying the nutrient solution for 15 minutes and stopping the supply for 45 minutes every 1 hour in summer; supplying the nutrient solution for 15 minutes and stop supplying liquid for 105 minutes every 2 hours in winter; repeating the above process every day. These parameters should be adjusted according to the specific growth vigor of the crops and the climate conditions. The time for stopping the liquid supply should not be too short. If the time is shorter than 35 minutes, the effect of supplementing oxygen will not be achieved. However, the liquid supply should not be stopped for too long; otherwise, it will make crops wilt due to the lack of water.

5. Management of liquid temperature

Because of the poor heat insulation performance of the NFT plantation trough (especially the triangular trough made of plastic film) and the small amount of liquid used, the stability of liquid temperature is poor, which could easily cause a significant difference in the liquid temperature between the head and the tail of the same trough. Especially in winter and spring, the temperature difference between the liquid inlet and the liquid outlet of the trough could be up to 6℃, which makes the liquid temperature that is already adjusted to meet the requirements of crops become significantly lower at the end of the trough. Thus, particular care should be taken with regard to the management of liquid temperature for NFT. Different crops have different requirements for liquid temperature. It is advisable to keep the temperature below 28℃ in summer and above 12℃ in winter.

Task 3 Construction and Management of Spray Culture Facilities

I. Types and Facilities of Spray Culture

Spray culture refers to a kind of soilless culture technique in which the root system of crops hangs and grows in a closed, opaque container (trough, box, or bed), and the nutrient solution is sprayed onto the root system of crops intermittently by special equipment to provide the water and nutrients needed for crop growth. Spray culture is also known as aeroponic culture. The root system grows in the air with a relative humidity of 100% rather than in the nutrient solution, and the growth of crop stems and leaves is the same as that of general cultivation.

Spray culture can meet the needs of the crop root system for water, nutrients and oxygen at the same time with the atomized nutrient solution. The crop root system can absorb oxygen more easily when growing in humid air than growing in nutrient solution or solid substrate. It is a better way to solve the contradiction between water and air of the root system in all soilless culture techniques, which is the physiological basis for the success of spray culture. Meanwhile, spray culture is easy to control automatically and suitable for stereoscopic culture, which can improve the utilization rate of greenhouse space.

Spray culture was first developed in Italy and was used to grow lettuce, cucumber, melon, tomato, and other vegetables. Researchers at the Environmental Research Laboratory of the University of Arizona have developed and improved spray culture, and displayed this advanced culture technique in Disneyland, California, USA, for tourists to visit. Japan has applied the spray culture technique to produce leaf lettuce on a large scale. In China, the spray culture technique is also applied and displayed in modern agricultural parks in big cities.

1. A-shaped spray culture

The facility structure of this culture technique is shown in Fig. 6-17. A-shaped cultivation frame is a typical structure of this type of spray culture, where the crops grow on the side plates, with their root systems hanging laterally inside the A-shaped container and intermittently bathed in the atomized nutrient solution. If the angle between the side edge and the bottom edge of the frame is 60°, the area of cultivated crops is twice the area of land occupied. Therefore, A-shaped spray culture can save the greenhouse area and improve the land utilization rate. This culture technique is suitable for occasions with narrow spaces, such as spaceships. The main facilities of A-shaped spray culture include the culture bed (Fig. 6-18), spraying device, nutrient solution circulation system, and automatic control system.

2. Column aeroponics

The crops are planted around a vertical column container, and the root system grows inside the container. The top of the column is equipped with a spraying device, which can spray the atomized nutrient solution onto the root system, and the excess nutrient solution is recovered through the drain pipe at the bottom of the column for recycling (Fig. 6-19). The column aeroponics is characterized in

Fig. 6-17 Facility Structure of A-shaped Spray Culture

1. Foam plastic plate 2. Plastic film 3. Aerial part of head lettuce 4. Root system 5. Liquid supply pipe 6. Sprayer

Fig. 6-18 Culture Bed

that it can make full use of space and save the area of land occupied. This culture technique was originally developed by researchers at the University of Pisa, Italy. The main facilities of column aeroponics include the column, spraying device, nutrient solution circulation system, and automatic control system. The column height is 1.8–2.0 m, the diameter is 25–35 cm, and the column spacing is 80–100 cm. Generally, the column is made of white opaque rigid plastic, and there are many planting holes around the column for planting crops. There is a spray nozzle at the top of each column to spray atomized nutrient solution and air into the column for the root system growth.

Fig. 6-19 Column Culture

3. Semi-spray culture

Semi-spray culture means that most of the crop root system grows in the air for most of the time, and only a small part of the root system grows in the nutrient solution for a short time. The atomized nutrient solution is sprayed into the culture bed, which can be divided into various forms according to the spraying speed and the depth of the liquid layer at the bottom of the culture trough. When the spraying amount is large, the culture bed is quickly filled with nutrient solution after the liquid is added each time and the root system is completely or partially soaked in the nutrient solution. After the spraying is stopped, the nutrient solution in the culture bed flows out from the drain pipe at the bottom of the bed at a certain speed, and the root system is exposed to the humid air again (Fig. 6-20 and Fig. 6-21). This cultivation method was originally designed by the Japanese to mainly solve the contradiction between liquid supply and oxygen supply in hydroponics and save energy consumption.

The main facilities of semi-spray culture include culture bed, spraying device, nutrient solution circulation system, and automatic control system. The culture bed should have a width of 40 cm, a height of 30 cm, and a flexible length according to needs. The upper part of the culture bed should be covered with 2–3-cm-thick polystyrene foam planting plate. The spraying device should be arranged at the upper part of the side wall of the culture bed and should have a spray nozzle arranged every

1–1.5 m. A large amount of liquid can be added to the spraying device, and the culture bed can be filled with the nutrient solution quickly after the liquid is added each time.

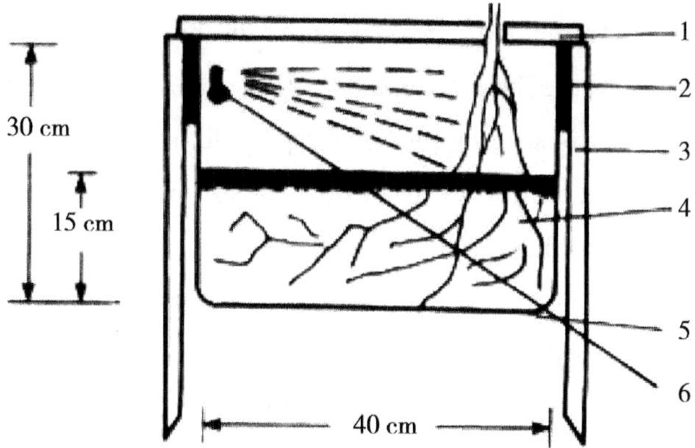

Fig. 6-20 Sectional View of Semi-spray Culture Bed

1. Polystyrene planting plate 2. Slats for fixing the shape of the culture bed
3. Stake 4. Nutrient solution 5. Plastic film (0.3 mm thick, 100 cm wide)
6. Spray nozzle

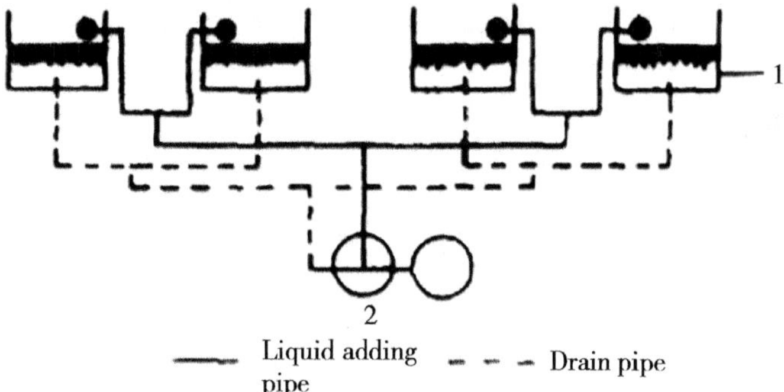

Fig. 6-21 Pipeline Diagram of Liquid Adding and Drainage in Semi-spray Culture

1. Culture bed 2. Pump

II. Spray Culture Management

1. A-shaped spray culture

(1) Selection of crop species: The A-shaped spray culture is used to cultivate leafy vegetables and other dwarf crops such as lettuce. Besides, it can also be used

to cultivate large plant crops such as tomatoes, which need to be cultivated by hanging them.

(2) Seedling raising and planting: The A-shaped spray culture adopts the seedling transplantation method. Seeds are sown in the seedling-raising pots with holes (one plant per pot). There are many holes around and at the bottom of the pot, and the root system can grow out of the pot. Before planting, it is necessary to check whether the facilities are in normal operation and prepare the nutrient solution. When the seedlings grow to a suitable size (the root system grows out about 3 cm from the bottom of the seedling pot), put the seedling-raising pot into the planting hole on the side of the culture bed.

(3) Temperature and humidity management: As the root system of aeroponic crops grows in the air, the root temperature could easily be affected by the air temperature, which is not as stable as growing in the nutrient solution. A relatively stable and suitable air temperature is the basic requirement of all spray culture techniques. Therefore, spray culture is generally carried out in a greenhouse in good condition. By controlling the air temperature in the greenhouse, we can ensure that the crop root system grows in a suitable temperature range.

Because the culture bed for A-shaped spray culture is closed to some extent, the root system will grow in a relatively isolated space, and there is no special requirement for the humidity of the greenhouse environment, but the relative humidity inside the culture bed must be maintained at 100%.

(4) Nutrient solution management: The formula and formulation method of nutrient solution for spray culture are the same as those for other hydroponic culture methods. The daily management of nutrient solution for A-shaped spray culture mainly includes the management of the starting time and ending time of spray, the spraying frequency, the duration of each spray, and the spraying rate of the spray nozzle. As the water demand of crops is affected by environmental factors such as light, temperature, and humidity, as well as the growth stage of crops, the above management indicators need to be adjusted according to seasons, weather, and crop growth stages. In seasons with strong sunlight and relatively high

outside temperatures, it is necessary to start spraying at an earlier time, delay the ending time of spraying, and appropriately increase the spraying frequency (including the nighttime). When the plant grows big, the duration of each spray should be slightly longer than when the plant is small. In particular, it should be emphasized that the spray supplies not only water and nutrients but also air to the root system.

The nutrient solution management of mobile spray culture is mainly realized by the design of facilities and the adjustment of the rotating speed of the support.

2. Column aeroponics

Column aeroponics is suitable for cultivating leafy vegetables, small fruit vegetables, and ornamental plants. These plants include lettuce, parsley, strawberry, petunia, coleus, and other horticultural plants. The root system of crops grows in a relatively closed container, so the cultivation management is similar to that of A-shaped spray culture.

3. Semi-spray culture

Semi-spray culture is suitable for cultivating all kinds of horticultural crops, such as vegetables and flowers. The requirements for environmental conditions of crops are the same as those for A-shaped spray culture. The key to nutrient solution management lies in the number of times the pump works every day (i.e. the number of times of liquid adding), the length of working each time, the height and speed of nutrient solution rising in the culture bed, and the speed of its discharge from the culture bed. The management varies according to the cultivation seasons, weather, and crop growth stages. In the seedling stage, every time the nutrient solution is added, the liquid level can reach the lower edge of the planting plate, submerging the entire root system. After the plant grows up, the liquid level is appropriately reduced, and only the lower part of the root system is submerged. For young plants, add the solution once a day; for large plants, add the solution 2–4 times a day, with an interval of 4 hours after adding the solution every 2 times. Generally, the solution does not need to be added at night. It is necessary to add the nutrient solution in the highest temperature period of the day (12:00–14:00). The ideal liquid-adding effect is that the nutrient solution is sprayed through the spray nozzle and then quickly rises to the required height in the culture bed. The ideal drainage is that the nutrient solution can be discharged at a relatively slow rate so that the root system can fully absorb water and nutrients. Every time the liquid adding stops (the pump stops working), the excess nutrient solution should be completely discharged from the culture bed for recycling.

III. Construction and Management of Spray Culture Facilities

(I) Plantation Trough Construction

The plantation trough for spray culture can be made of rigid plastic board, foamed plastic board, wooden board, or cement concrete in a variety of shapes. The shape and size of the plantation trough should take into account the following: After the root system of the plant extends into the trough, there should be sufficient space for the nutrient solution to be uniformly sprayed onto the root system of each plant by the spray nozzle installed in the trough. Therefore, the plantation trough should be made neither too narrow nor too wide to spray the atomized nutrient solution; otherwise, the spray nozzle may not spray the nutrient solution onto all the root systems.

Fig. 6-22 and Fig. 6-23 are trapezoidal and A-shaped plantation troughs, respectively. The bottom of the trough can be made into a trough with a depth of

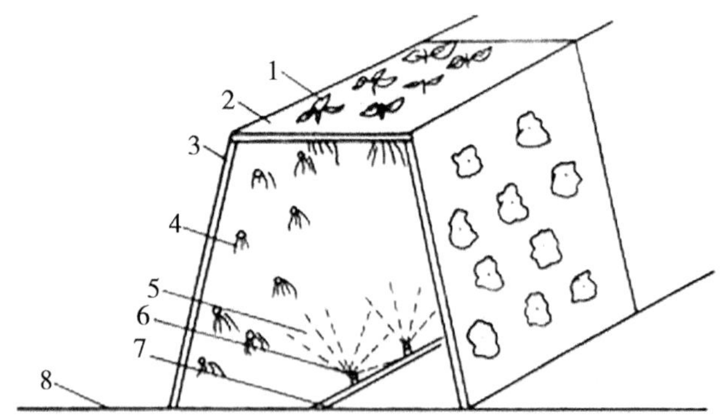

Fig. 6-22 Schematic Diagram of Plantation Trough for Trapezoidal Spray Culture

1. Plant 2 & 3. Foam plastic board 4. Root system 5. Misty nutrient solution 6. Spray nozzle 7. Liquid supply pipe 8. Ground

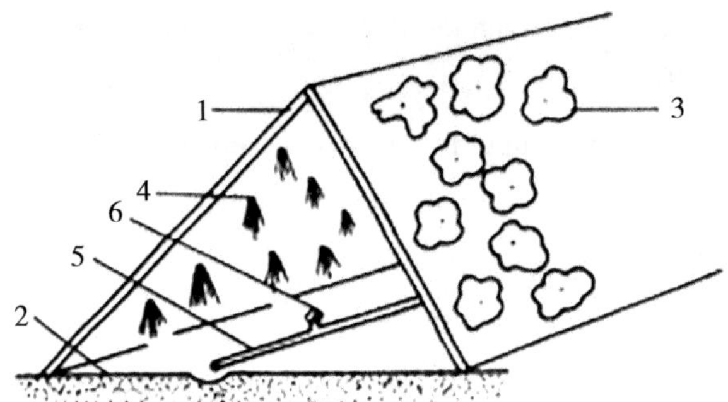

Fig. 6-23 Schematic Diagram of Plantation Trough for A-shaped Spray Culture

1. Foam plastic plate 2. Plastic film 3. Aerial part of head lettuce 4. Root system 5. Liquid supply pipe 6. Sprayer

about 10 cm for holding excess nutrient solution. The upper part of the trough can be made into an A-shaped or trapezoidal frame with iron bars, and then a foam plastic planting cup with planting holes can be placed above the frame so that crops can be planted.

(II) Assembly of the Liquid Supply System

The liquid supply system mainly comprises the nutrient solution reservoir, water pump, pipe, filter, and spray nozzle. Some spray culture techniques apply an ultrasonic atomization machine instead of a spray nozzle to atomize the nutrient solution.

(1) Nutrient solution reservoir: A large-volume nutrient solution reservoir built with cement bricks can be used for large-scale spray culture. For small-scale spray culture, a large plastic bucket or box can be used instead. The volume of the reservoir should be large enough to ensure that the water pump has a certain period of liquid supply so that the nutrient solution in the reservoir will not be drained quickly. If conditions permit, the volume of the nutrient solution reservoir can be made larger, but in all cases, it should at least meet the 1–2 days of water consumption requirement of plants.

(2) Water pump: The power of the water pump should be determined by taking comprehensive consideration of the planting area, pipe layout, selected spray nozzle and its required working pressure. Corrosion-resistant water pumps should be

selected. Generally, for a polytunnel of 1 mu, the power of the water pump should be about 1,000–1,500 W.

(3) Pipe: Plastic pipes should be selected, and the sizes of pipes at all levels should be determined according to the working pressure of the spray nozzle on the selected spraying device.

(4) Filter: The water or raw materials for formulating the nutrient solution contain some impurities, which may block the spray nozzle. Therefore, a filter with a sound filtering effect should be selected.

(5) Spray nozzle: Different spray nozzles can be selected according to the spray culture form and the installation position of the spray nozzle. Some spray nozzles have a plane fan-shaped spray surface, while others are of all-round spray. The selection of nozzles is based on the principle that nutrient solution can be sprayed to all the root systems in the facility, and the droplets are relatively small.

(6) Ultrasonic atomization machine: The ultrasonic atomization machine uses the ultrasonic wave generated by the ultrasonic generator to atomize the nutrient solution into a mist stream of fine droplets, which will spread all over the root system growth area (in the plantation trough), thereby superseding the above-mentioned liquid supply system. Ultrasonic atomization of the nutrient solution can help kill pathogenic bacteria that may exist in the nutrient solution, which is beneficial to crop growth. The water outlet of the nutrient solution reservoir or tank should be located at a position higher than the water inlet of the ultrasonic atomization machine. The nutrient solution reservoir or tank is connected to the water inlet of the ultrasonic atomization machine through pipe fittings so that the nutrient solution flows into the ultrasonic atomization machine under the action of gravity. Because the power of the built-in blower in the ultrasonic atomization machine is limited, the planting bed should not be too long and generally not exceed 8 m.

(III) Cultivation Management

The planting method of spray culture can be similar to that of deep flow hydroponics. However, if the planting plate is inclined, the plants cannot be fixed with small gravel. A small amount of rock wool fiber, polyurethane fiber, or sponge block should be used to wrap the root collar of the seedlings, and then the seedlings should be put in the planting cup before being put into the planting hole of the planting plate. Alternatively, the seedlings wrapped with rock wool, polyurethane fiber, or sponge can be directly inserted into the planting hole without using the planting cup. At this time, the amount of rock wool, polyurethane fiber, or sponge used to wrap the seedlings should be such that the seedlings will not fall off from the planting hole after being inserted into the planting hole, but it should not be inserted too tightly so as not to prevent the growth of crops.

(IV) Nutrient Solution Management

The nutrient solution concentration for spray culture can be higher (generally about 20%–30% higher) than that for other hydroponics. This is mainly because when the nutrient solution is supplied in the form of spray, the nutrient solution attached to the root system surface is only a thin water film, so the total amount is relatively small. In order to enable the plants to absorb enough nutrients when the liquid supply is stopped, the concentration of the nutrient solution should be slightly increased. However, in the case of semi-spray culture, the concentration of nutrient solution does not need to be increased, and it can be the same as that for deep flow hydroponics. Spray culture is subject to intermittent liquid supply. The liquid supply and intermittent time should depend on the plant size and different climatic conditions. When the plant is large, the sunshine is abundant, and the air humidity is low, the liquid supply time should be relatively longer, and the intermittent time can be relatively shorter. In the case of semi-spray culture, the intermittent time of liquid supply can be slightly extended, but the liquid supply time can be relatively shorter. The liquid supply time during the day should be longer than that at night, and the intermittent time should be relatively shorter. Some people just cannot be bothered to adjust the liquid supply time every day, and instead, they choose to shorten both the liquid supply time and the intermittent time, i.e. pausing for 5–10 minutes after supplying the liquid for

5–10 minutes each time, so that the frequency of the liquid supply is increased, thus solving the problem of untimely supply of nutrient solution. However, in this case, the water pump needs to be started frequently, and its service life will be shortened.

Task 4 Construction and Management of Solid Substrate Culture Facilities

I. Sand Culture

Sand is one of the earliest substrate materials used in soilless culture, which can be obtained from a variety of sources and is cheap. The physical properties of sand with different particle sizes vary greatly, and the cultivation effects are quite different. Coarse sand has good air permeability and weak water-holding capacity, while fine sand and mealy sand are the opposite. The Shive research results show that the water-holding capacity of sand particles is 26.8% for a particle size of 1.5–1.0 mm, 30.2% for a particle size of 1.0–0.5 mm, 32.4% for a particle size of 0.5–0.32 mm, and 37.6% for a particle size of 0.23–0.25 mm.

From the perspective of the chemical properties of sand, the pH value and trace element content of sand vary greatly with the types and sources of sand. In view of the above, when sand is used as the substrate for soilless culture, attention should be paid to the following aspects.

(1) Sand particles should not be too fine, with 0.6–2.0 mm as the best particle size. Sand particles should be uniform, and soil or fine sand particles should not be added to large sand particles. According to the research, sand with a particle size less than 0.6 mm should account for about 50%, and that with a particle size greater than 0.6 mm should account for about 50%. The particle size composition of sand culture is as follows: 1.1% for the sand particle size greater than 2 mm, 6.9% for 2–1 mm, 19.7% for 1–0.5 mm, and 72.3% for less than 0.5 mm.

(2) Before use, the sand should be sieved to remove large gravel and then washed with water to remove soil and silt.

(3) Before use, the sand should be chemically analyzed to determine the content of relevant components, so as to maintain the reasonable application rate and effectiveness of nutrients.

(4) Reasonable liquid supply amount and time should be determined to prevent water shortage caused by insufficient liquid supply.

(I) Facility Structure of Sand Culture

The schematic diagram of sand culture on the greenhouse ground is shown in Fig 6-24.

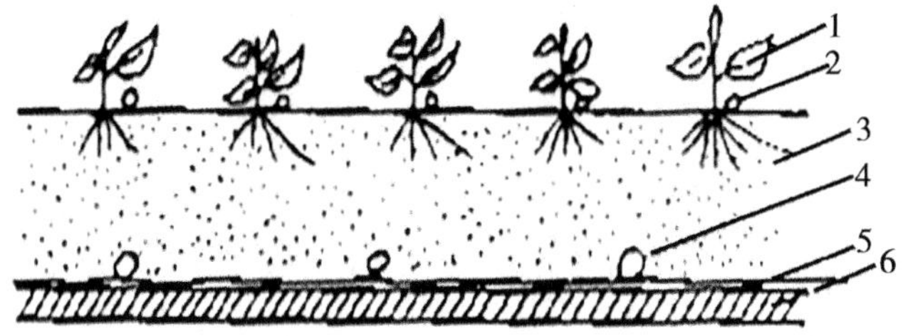

Fig. 6-24 Schematic Diagram of Sand Culture on the Greenhouse Ground

1. Plant 2. Thin-walled drip irrigation tape 3. Sand 4. Drain pipe 5. Black plastic film 6. Greenhouse ground

1. Plantation trough

According to different shapes, the plantation can be divided into the following types.

(1)V-shaped plantation trough (Fig. 6-25).

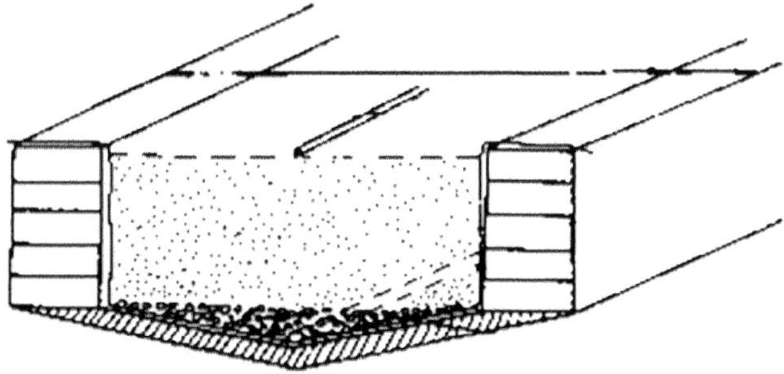

Fig. 6-25 V-shaped Plantation Trough

(2) Λ-shaped plantation trough (Fig. 6-26).

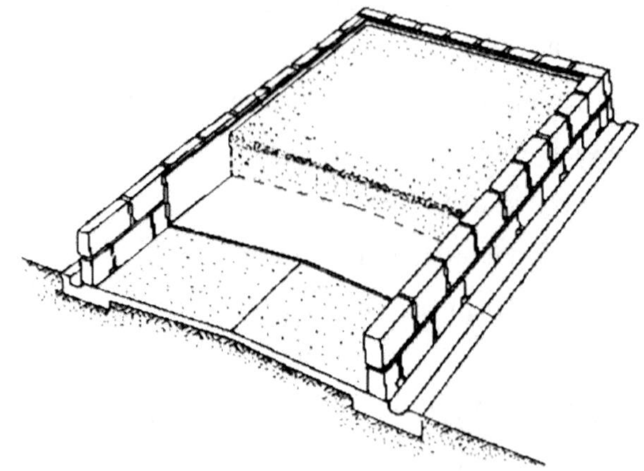

Fig. 6-26　Λ-shaped Plantation Trough

(3) Flat-bottom plantation trough (Fig. 6-27).

The bottom of the trough is horizontal and can be paved with 3–4 layers of red bricks on the ground. The trough is lined with a layer of black plastic film.

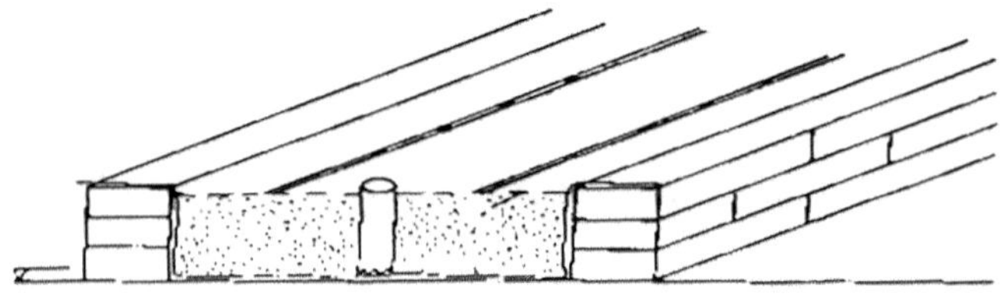

Fig. 6-27　Flat-bottom Plantation Trough

Thin-walled drip irrigation tape may also be used for liquid supply. One or two small holes are punched on the thin-walled drip irrigation tape every 25–30 cm. The thin-walled drip irrigation tape is flat before use and directly laid on the surface of the substrate. During liquid supply, the drip irrigation tape will be inflated and the nutrient solution will be sprayed from the small holes. Once the liquid supply is stopped, the drip irrigation tape will deflate. Such kind of drip irrigation tape has a low cost and is very convenient to use.

2. Reservoir

If a water pump is used for liquid supply, the reservoir can be built underground.

If the gravity-flow liquid supply method is used, the reservoir should be built high above the ground according to the pressure requirements of the drip irrigation system (Fig. 6-28).

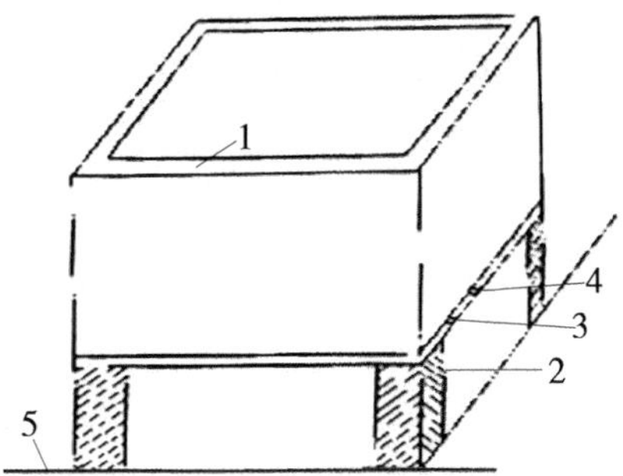

Fig. 6-28 Schematic Diagram of Reservoir

1. Reservoir body 2. Support column 3. Drain outlet for cleaning 4. Liquid supply outlet 5. Ground

3. Liquid supply system

Sand culture usually uses the drip irrigation method for liquid supply, and the liquid supply system consists of main liquid supply pipe (ϕ 32–50 mm), branch pipe (ϕ 20–25 mm), capillary pipe (ϕ 13 mm), drip tube and dropper. The drip tube and dropper are connected to the capillary pipe. Each plant is provided with a dropper to ensure the same amount of liquid is dropped per plant. The length of the capillary pipe on the horizontal bed surface should not exceed 15 m. If it is too long, the liquid supply for the plant at the end will be less than that at the liquid inlet end, resulting in inconsistent crop growth.

The more economical and convenient way is to choose a porous micro-irrigation hose instead of the above-mentioned drip irrigation system so that the capillary pipe, drip tube, and dropper can be integrated. The water outlet is located above the hose axis. The wall thickness of the hose is generally 0.1–0.2 mm, the aperture of the outlet hole is 0.7–1.0 mm, and the hole spacing is 250–400 mm. The requirements for water sources have also been lowered significantly. The hose is directly laid between rows, and the nutrient solution flows out from the micropores, moistening the substrate. The outlet hole of the micro-irrigation tape is formed by using a special machining method so that the flow rate is uniform. Soft irrigation tape is low in cost and convenient to use, but its service life is short.

The nutrient solution for the irrigation system should pass through a filter with a 100-mesh-filter screen to prevent impurities from blocking the dropper.

(II) Key Points of Sand Culture Technique

1. Nutrient solution management

(1) Formula and concentration: According to the chemical properties of sand, the pH value is generally neutral or slightly acidic. Except for high Ca content, the contents of other major elements are relatively low. All kinds of trace elements have a certain content in sand. The content of Fe is high in many kinds of sand, which can be used by plants. The contents of Mn and B are second only to that of Fe, which can sometimes meet the needs of crops.

The buffer capacity of the sand substrate is relatively low, and the open liquid supply leads to less liquid stored in the substrate, so the composition, concentration, and pH value of the nutrient solution in the substrate vary greatly.

Therefore, when selecting the nutrient solution formula, the formula should be adjusted according to the contents of various elements in the sand used to ensure the balance of various nutrients. In addition, the physiological acidity and alkalinity of the nutrient solution should be relatively stable, and a low-dose formula should be used. If the dose of the original formula is high, 1/2 of its dose can be adopted.

(2) Liquid supply amount and method: Normally, the liquid supply frequency can be determined according to the water requirement of crops. Drip irrigation can be carried out 2–5 times a day, with an adequate amount of water to be irrigated each time, allowing 8%–10% of water to be discharged. The liquid filling volume should be determined based on this.

The total amount of soluble salt in drainage water should be determined twice a week (with a conductivity meter). If the total amount of soluble salt exceeds 2,000 mg/L, drip irrigation with clear water should be used for several days to dissolve the salt and reduce the concentration. When the concentration of the nutrient solution is reduced to a level below that for drip irrigation, drip irrigation with the nutrient solution should be used again.

In the case of continuous low temperatures and rainy weather, it may not be necessary to carry out drip irrigation several times a day from the perspective of water requirement, but it may be necessary to carry out drip irrigation from the perspective of nutrient requirement. At this point, the nutrient solution can be continuously dripped, so that the new nutrient solution can replace the old nutrient solution whose nutrients have been consumed by crops in the sand, thereby ensuring the crops' requirement for nutrients. If a lot of water is discharged when the drop volume is not large, the concentration of the nutrient solution can be increased (the total nutrient salt concentration should not exceed 2.5 g/L) before drip irrigation.

2. Substrate disinfection

Generally, the substrate can be disinfected once a year or once per crop to eliminate soil-borne diseases and pests, including nematodes. Commonly used disinfectants include 1% formalin solution, 0.3%–1% calcium hypochlorite, or sodium hypochlorite solution. After the disinfectant is applied in bed for 24 hours, wash it with water for 3–4 times until the disinfectant is completely washed away. Besides, agents such as methyl bromide and other methods may also be used for disinfection. For the specific methods, please refer to the substrate disinfection part of this book.

II. Rock Wool Culture

The soilless culture with rock wool as the substrate is called rock wool culture. In 1840, Americans manufactured rock wool for the first time. In 1968, Grodan Company of Denmark developed rock wool culture. In 1970, the Netherlands began to adopt rock wool culture and succeeded in cultivating crops. After 1980, rock wool culture was rapidly popularized in European countries, with the Netherlands as the center. At present, many countries are experimenting and applying rock wool culture, among which the Netherlands has the largest application area of over 3,000 hm^2.

In China, rock wool culture technique is still in its infancy. In 1987, the Vegetable Research Institute of Jiangsu Academy of Agricultural Sciences worked with Nanjing Fiberglass Research Institute to develop domestic agricultural rock wool suitable for soilless culture. In recent years, Sunqiao Vegetable Base in Shanghai has carried out 3 hm^2 of rock wool culture in the imported modern greenhouse, and successfully cultivated cucumbers and tomatoes. China's rock wool raw material resources are extremely rich, and the domestic rock wool production lines spread almost all over the country. With the continuous updating of rock wool production techniques, the production cost of rock wool will also decrease.

(I) Characteristics of Rock Wool Culture

Rock wool is a kind of loose, porous, and moldable solid substrate formed by melting a variety of rocks together, spraying them into filaments, and then cooling and bonding them. The bulk density of agricultural rock wool is generally 80–90 k/m^3, and the total porosity is 97.2%.

The three-phase proportion of soaked rock wool is: solid phase accounts for 4.6%, liquid phase accounts for 45.2%, and gas phase accounts for 50.2%. It has multiple buffering effects of soil culture, such as water absorption, water retention and aeration. Its texture is soft and uniform, which is conducive to the growth of the crop root system.

The rock wool culture is to cultivate plants in a rock wool block with a certain volume so that crops can take root and anchor in it and absorb water and fertilizer. The basic model is to cut the rock wool into modular blocks and use plastic film to wrap them into a pillow bag, which is called the rock wool plantation pad (Fig. 6-29). When planting, cut a small hole in the film on the surface of the rock wool plantation pad, plant seedlings with seedling blocks, and drip nutrient solution into it, so that the plants can take root in it and absorb water and nutrients to grow.

Compared with other solid substrate culture and hydroponic culture methods, rock wool culture has the following advantages:

(1) Rock wool culture can properly resolve the problems associated with the supply of water, nutrients, and oxygen. Hydroponics mainly relies on the configuration of an oxygen cascading device, water spraying on the water surface, installation of the bubbler, circulation of nutrient solution, intermittent supply of liquid in a thin layer, and other methods to raise the dissolved oxygen content in the

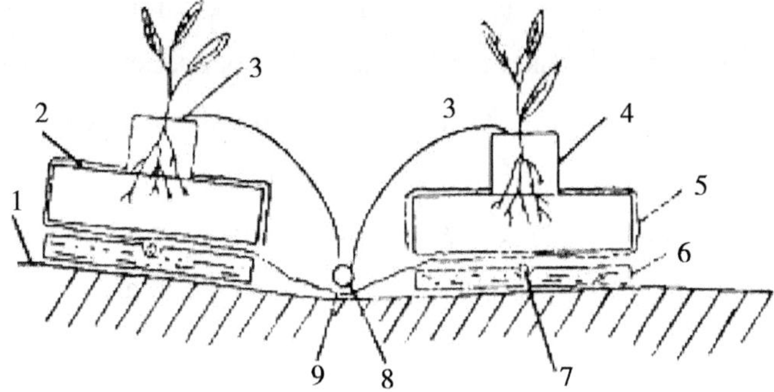

Fig. 6-29 Cross Sections of Open Rock Wool Plantation Bed and Rock Wool Plantation Pad

1. Bed surface plastic film 2. Rock wool plantation pad 3. Drip irrigation pipe 4. Rock wool seedling block 5. Black and white plastic film 6. Foam plastic block 7. Heating pipe 8. Drip irrigation capillary pipe 9. Plastic film ditch

nutrient solution, or expose part of the root system to the air to supplement oxygen. Rock wool culture makes use of the water retention and aeration characteristics of rock wool to coordinate the relationship among fertilizer, water, and air, without adding other devices.

(2) Rock wool culture has multiple buffering effects. It can create a stable growth environment for the root system of crops by making use of its functions of water absorption, water retention, fertilizer retention, aeration, and root mass fixation, and it is less affected by the outside world. Besides, due to the uniform texture of rock wool, the nutrient solution and oxygen supply at different positions in the culture bed are similar, which will not cause much difference among plants and is conducive to a balanced yield increase.

(3) The device of rock wool culture is simple in structure and easy to install and use. Its culture bed only requires rock wool felt, black plastic film, non-woven fabric, and a drip irrigation device. If thin wool felt is used as the culture bed, or rock wool block is used instead of rock wool felt, it can save on materials. Because drip irrigation is adopted for liquid supply in rock wool culture, the requirement on the ground slope is not as strict as that for NFT. The nutrient solution supply frequency can be greatly reduced, and it is not limited by power failure and water cut-offs, which can save water and electricity.

(4) Rock wool does not spread diseases, insects, and weeds. In cultivation management, soil-borne diseases rarely occur. Without serious diseases, rock wool can be used continuously for 1–2 years or reused after disinfection.

(II) Facility Structure of Rock Wool Culture

Rock wool culture can be divided into open rock wool culture and recycling rock wool culture according to different liquid supply methods.

The main feature of open rock wool culture is that the nutrient solution supplied to crops is not recycled. The excess part of the nutrient solution dropped into the rock wool plantation pad through drip irrigation flows out from the pad bottom to the outside. The main advantages include simple facility structure, easy construction, low cost, convenient management, and the fact that there is no danger of disease spreading caused by nutrient solution circulation. In areas prone to soil-borne diseases, open rock wool culture is a very effective cultivation method. The main disadvantages include cosuming a large amount of nutrient solution, and the discarded excess nutrient solution can cause pollution to the external environment (leading to eutrophication of nitrogen and phosphorus in the external environment).

Recycling rock wool culture is designed to overcome the shortcomings of open rock wool culture. The so-called recycling means that after the nutrient solution is dripped into the rock wool, the excess nutrient solution is not discarded but flows back to the underground liquid collecting reservoir through the return pipe for recycling. Its advantage is that it will neither cause waste of nutrient solution nor pollute the environment. The disadvantages are that the design is relatively open and complicated, the capital construction investment is relatively high, and it could easily spread rhizosphere diseases. It should be selected according to local conditions.

In order to avoid soil pollution caused by the discharge of nutrient solution and protect the environment, rock wool culture is developing towards a closed cycle. A few years ago, the Dutch government stipulated that rock wool culture in the Netherlands should all be switched to a closed cycle by the year 2000.

The basic devices of rock wool culture include the culture bed, liquid supply device, and liquid discharge device. If the circulating liquid supply is adopted, the liquid discharge device will not be required.

1. Culture bed

The culture bed consists of beds and rock wool pads connected together. The rock wool pads are covered with black or black-and-white polyethylene plastic film bags. Before planting, holes are made on the film for planting seedlings with rock wool blocks.

The rock wool culture bed has strict construction technology requirements. The ground of the culture bed must be flat; otherwise, it will cause uneven liquid supply or even cause salt accumulation and pH value rise, which will affect the cultivation effect.

The two types of rock wool culture have different requirements for drainage, so the culture beds also have certain differences, which are described below.

(1) Culture bed structure of open rock wool culture.

① Bed building: After leveling the ground in the polytunnel, build a turtle-back-shaped soil bed according to the specifications and compact it. The specifications of the bed depend on the crop type. Taking tomato cultivation as an example, the bed width (between bed ditches) is 150 cm, and the bed height is about 10 cm (from the bottom of the bed ditch to the highest point of the bed surface). It begins to gently incline at the places 30 cm, on both sides, from the midpoint of the bed width, thereby forming a ditch between two beds. The bed length is about 30 m, and the bed ditch has a 1 ∶ 100 slope along the long side to facilitate drainage. After compacted beds are built on the ground of the polytunnel, a 0.2-mm-thick milky white plastic film will be laid to cover all beds and ditches. The film should be tightly attached to the beds and ditches so that the beds and ditches still show their shapes after the film is laid. The milky white film has the following functions: preventing the infestation of diseases, pests, and weeds in the soil; preventing the excess nutrient solution from infiltrating into the soil and causing salinization; increasing the light reflectivity so that the lighting intensity of the lower leaves of tall crops planted in the greenhouse can be increased, which is beneficial to their growth.

② Arrangement of rock wool plantation pads on the bed: Two rows of rock wool plantation pads are placed on the bed back one after another, and the long sides of the pads should be in the same direction as the bed length. Each row is placed on the slope of the bed so that the pad tilts to one side of the bed ditch to facilitate future drainage. The distance from the rock wool plantation pad to the bed ditch is shorter than that to the center of the bed, resulting in a relatively larger distance between the two rows on the bed back and a relatively smaller distance between the two rows separated by the bed ditch (Fig. 6-30). Unlike field planting, the open rock wool plantation bed uses the bed back as the pedestrian or working channel, and the bed ditch is only used for placing the drip irrigation capillary pipe and discharging excess nutrient solution, not for the pedestrian channel.

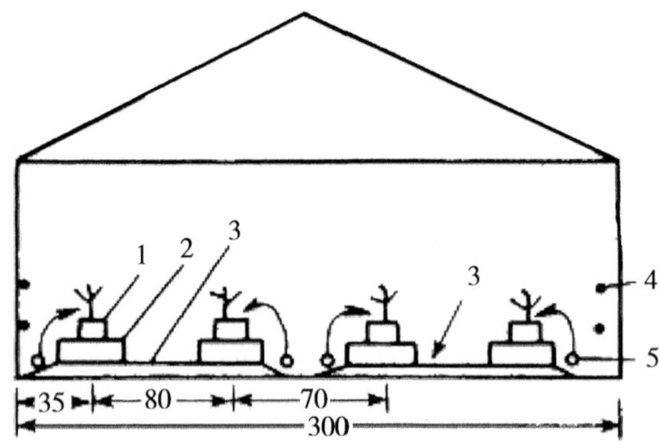

Fig. 6-30 Arrangement of Rock Wool Plantation Pads (Unit: cm)

1. Seedling pot 2. Rock wool plantation pad 3. Bed surface 4. Heating pipe 5. Drip irrigation pipe

In cold areas in winter, a root heating device can be arranged. The method is to place a rigid foam plastic board under all the rock wool plantation pads connected in rows, with a width consistent with the rock wool plantation pad, a thickness of about 3 cm, and the length depending on the convenience of work. A small concave groove is provided in the central extension direction of the board for placing the heating pipe. The foam plastic board and the rock wool plantation pad should be separated by a plastic film (0.1 mm thick) with a white surface and black bottom.

The plastic film should be wide enough to span the bed ditch so that the bed ditch, together with two rows of foam plastic boards on both sides, can be covered, and it can be bent to the bottom of the bed ditch to keep the bed ditch still showing a ditch shape. The two sides of the film width are turned up, exposing the black bottom and covering the rock wool plantation pad. It should cover the entire pad surface, and only one hole should be cut in the position where the crop is planted, so that the pad surface shows black, thereby making it easy to absorb the heat of sunlight and increase the pad temperature.

(2) Culture bed structure of closed rock wool culture.

① Bed frame building: First, build a bed frame on the ground with wooden boards or rigid foam plastic boards, with a height of about 15 cm, a width of about 32 cm (to the extent that it can hold the rock wool block with a width of 30 cm and its envelope), and a length of 20–30 m. The ground in the frame is built into a small ditch, and the ditch inclines to the direction of the liquid collecting reservoir at a slope of 1 : 200. The ground should be compacted. Second, a milky white plastic film with a thickness of 0.2 mm is laid, and the film should be tightly attached to the bottom of the ditch on the ground to show the ditch shape and wrap the objects placed on the film.

② Placement of rock wool planting pads and drainage and irrigation pipes: After the bed frame is built and covered with plastic film, the rock wool plantation pads are to be placed on it. The rock wool plantation pad should be 30 cm wide, 91 cm long, and 10 cm high, and the bottom and both sides should be covered with non-woven fabrics to prevent the roots from reaching into the ditch. A hard PVC drainage pipe with a diameter of 20 mm is placed in the small ditch at the bottom of the bed frame, and this pipe is connected to the liquid collecting reservoir outside the bed. A hard foam plastic strip with a height of 5 cm, a width of 5 cm, and a length the same as that of the above-mentioned rock wool plantation pad is placed on both sides of the ditch to support the rock wool plantation pad so as to keep the plantation pad off the plastic film at the bottom and prevent the stagnant nutrient solution from soaking to the pad bottom. The rock wool plantation pads are placed on the hard foam plastic strips, one by one, to fill the whole bed, with a small gap left at the joint between the pads for nutrient solution discharge. A soft drip irrigation pipe with a diameter of 20 mm is placed on the rock wool plantation pad, and a hole with a diameter of 0.5 mm is drilled every 40 cm on the pipe so that the nutrient solution can drip from the hole. The flow rate of each hole is about 30 mL/min. The drip irrigation pipe is connected to the outdoor liquid supply reservoir.

③ Determination of specifications of rock wool plantation pad: The size of the plantation pad relates to the nutrient area occupied by each plant or the amount of nutrient solution held per unit of time. The amount of nutrient solution held by rock wool directly affects the cultivation effect, which is worthy of further study. Some ranges have been put forward so far. It is generally believed that it is better to adopt the flat rectangle shape, with a thickness of 7–10 cm, a width of 25–30 cm, and a length of about 90 cm. Taking the cultivation of tomatoes and cucumbers as an example, according to the relevant research data, the maximum daily transpiration rate of tomatoes and cucumbers is 3 L/plant, or 4 L/plant with the 1/3 liquid supply guarantee coefficient. Generally, the porosity of rock wool is 95%, and its maximum water-holding capacity conducive to crop growth should not exceed 60% of its volume. Based on the above two values, it can be calculated that the volume of rock wool to be occupied by each tomato is 6.7 L. If one rock wool plantation pad is suitable for cultivating two plants, its volume should be 13.4 L. The 13.4 L rock wool can be made into a flat rectangle of 90 cm (length) × 20 cm (width) × 7.5 cm (thickness). Then, the whole rock wool is tightly wrapped with milky white plastic film, thereby turning into the rock wool plantation pad suitable for planting crops such as tomatoes and cucumbers.

2. Liquid supply device

(1) Liquid supply device for open rock wool culture: All open rock wool culture techniques adopt the drip irrigation system to supply liquid. Drip irrigation is a very water-saving irrigation method that supplies water in the form of small drops to crops slowly through droppers (the volume of water dropped by a dropper per hour is controlled at 2–8 L). The drip irrigation system consists of a liquid source, filter, and its control parts, plastic main pipe and branch pipe, capillary

pipe, and dropper pipe. The various parts of the drip irrigation system and their arrangement are shown in Fig. 6-31.

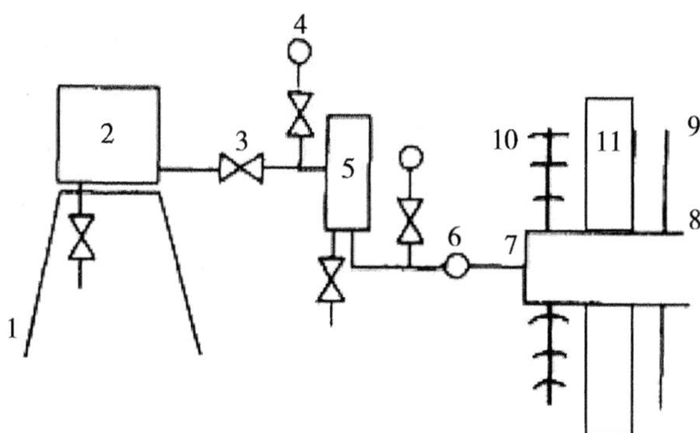

Fig. 6-31 Gravity Drip Irrigation System for Open Rock Wool Culture

1. Iron bracket 2. High-level nutrient solution tank 3. Valve 4. Pressure gauge 5. Filter 6. Water meter 7. Main pipe 8. Branch pipe 9. Capillary pipe 10. Dropper pipe 11.Plantation bed

① Liquid source, filter, and control parts: There are two methods to provide the liquid source. The first method is to set up a large-capacity nutrient solution reservoir, in which the working nutrient solution that can be directly supplied to crops for absorption is prepared. Its capacity should be able to meet the requirement of liquid supply for crop growth in a certain period and a certain area. This kind of reservoir (built at a high place) can supply nutrient solution by gravity. The nutrient solution can be pressed into a filter with a filter screen of more than 100 meshes. After impurities such as sediment are filtered out, the nutrient solution will enter the main pipe (the filter is an essential part of the drip irrigation system) and then be diverted to each branch pipe and the irrigation area. There are pressure gauges and flow control valves arranged at the front and rear of the filter. This method of liquid supply by gravity is relatively simple, with a low requirement of power (as long as tap water is available), and is easy to manage. This kind of large-capacity nutrient solution reservoir can also be built underground, so it is necessary to add a water pump with a certain power to pump the nutrient solution from the reservoir to the filter and then distribute it to the irrigation area.

The second method is to set only a storage tank for concentrated nutrient solution (divided into concentrated solutions A and B) without setting a large-capacity nutrient solution reservoir (Fig. 6-32). When liquid supply is required, the concentrated nutrient solution in tanks A and B will be delivered to the water source pipe by piston quantitative pump, and then the nutrient solution will enter the fertilizer-water mixer together with the water source and be mixed into a working nutrient solution with any set concentration. Then, it will be filtered by a filter as in the first method and then delivered to the irrigation area through the delivery pipe. The key to this liquid source supply method lies in the quantitative pump, the water source flow control valve, and the fertilizer-water mixer. These devices must be well-designed automatic control systems, which can accurately input the concentrated solution and water volume according to the instructions and mix them evenly to form a working nutrient solution with the specified concentration. Generally, they are manufactured in sets by a specialized factory for purchase and can hardly be manufactured by a non-specialized factory. This is a highly automated system that requires managers with a relatively high level of technical proficiency.

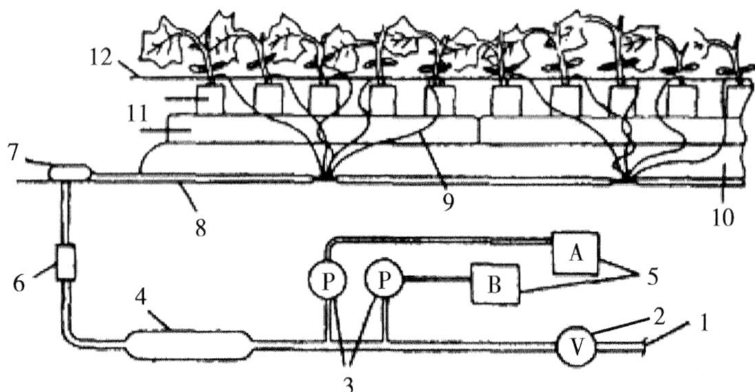

Fig. 6-32 Schematic Diagram of Drip Irrigation System without Large-capacity Nutrient Solution Reservoir in Open Rock Wool Culture

1. Water source 2. Electromagnetic valve 3. Quantitative injection pumps of concentrated nutrient solution 4. Nutrient solution mixer 5. Concentrated nutrient solution tanks 6. Filter 7. Flow control valve 8. Liquid supply pipe 9. Dropper pipe 10. Bed 11. Rock wool seedling block and rock wool plantation pad 12. Support wire

② After the liquid source of the main pipe and branch pipe passes through the filter, it is distributed to the first-level and second-level pipes before each planting row. All of the pipes are made of hard plastic pipes with the appropriate pipe diameter for the required liquid supply, and the length depending on the liquid delivery distance.

③ The capillary pipe is the pipe that goes into the planting row, and the last-level dropper pipe that drops liquid directly into the plant is connected to the capillary pipe. The diameter of the capillary pipe is usually 12–16 mm, and it is made of flexible plastic because it is to be tightly embedded into the dropper pipe. There is a capillary pipe arranged between every two rows of plants, the length of which is the same as that of the planting row, and it is placed in the bed ditch. A capillary pipe is used to connect the dropper pipe required by the two rows of plants.

④ The dropper pipe is the last-level pipe that drops liquid directly into the plant and is made of flexible hard plastic. One end of the dropper pipe is embedded in the capillary pipe. The method is to drill a small hole (with a diameter slightly smaller than the outer diameter of the dropper pipe) on the capillary pipe, then tightly embed the dropper pipe into the hole, and make sure that it is not easy to loosen and leak. The other end of the dropper pipe is supported by a small plastic rod and inserted into the planting hole of each plant. The drop outlet is 2–3 cm away from the substrate surface so that the nutrient solution drips out and falls into the planting hole at a very slow speed. The most commonly used dropper flow rate is 2–4 L per hour.

There are two forms of dropper pipe. One is called the hair pipe. The hair pipe is very thin, with a standard inner diameter of 0.5–0.875 mm. When water passes through it, water will drip out in the form of droplets, so the hair pipe itself is a dropper. Its flow rate is affected by the length of the pipe. The longer the pipe, the smaller the flow rate, which is the earliest form of the dropper. Its disadvantage is that the diameter of the entire pipe is too thin, which could be easily blocked and hard to dredge when it is used for nutrient solution drip irrigation. The other one is called the water-resistance pipe, which is a plastic pipe with a relatively large aperture (about 4 mm) tightly covering a small section of pipe (i.e. the dropper) with a relatively small aperture (0.5–1.0 mm). One end of the water resistance pipe is embedded in the capillary pipe, and the other end (used as the dropper) is mounted on the planting hole. This kind of dropper with a water resistance pipe can avoid the blockage.

(2) Liquid supply device for recycling rock wool culture: The schematic diagram of recycling rock wool culture is shown in Fig. 6-33.

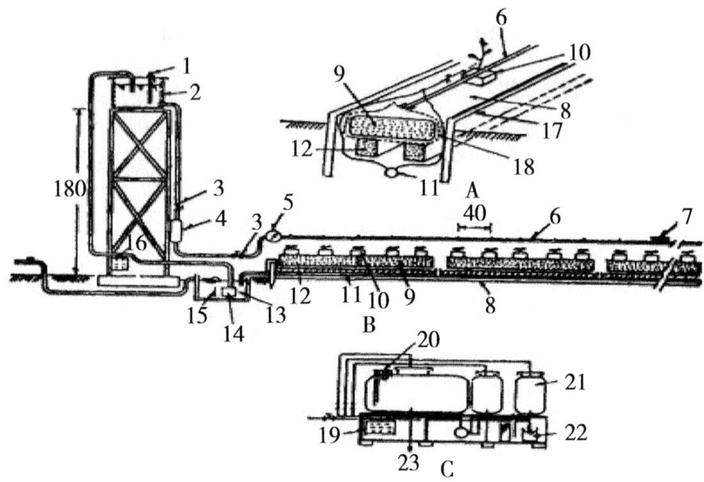

Fig. 6-33 Schematic Diagram of Recycling Rock Wool Culture (Unit: cm)

1. Liquid level inductor 2. Elevated liquid supply trough 3. Valve 4. Filter 5. Flowmeter 6. Liquid supply pipe 7. Regulating valve 8. Polyethylene film 9. Rock wool plantation pad 10. Rock wool seedling block 11. Return pipe 12. Foam plastic block 13. Liquid collecting reservoir 14. Water pump 15. Ball valve 16. Control panel 17. Bed frame 18. Non-woven fabric 19. Control panel 20. Liquid level inductor 21. Mother solution tank 22. Fertilizer dissolving tank 23. Mixing tank and reserve nutrient solution

① The liquid supply reservoir is arranged on a shelf with a height of 1.8 m, and the nutrient solution is delivered by gravity to each planting bed. A liquid level inductor is arranged to control the liquid level in the reservoir, and an electromagnetic valve and a timer are arranged on the output pipe to control it.

② The nutrient solution from the liquid supply reservoir should be filtered before flowing into each bed.

③ The nutrient solution drips from the drip hole of the drip irrigation pipe in the bed at a slow speed, passes through the rock wool plantation pad, flows into the drainage pipe at the bottom of the bed, and then flows back into the liquid collecting reservoir.

④ The liquid collecting reservoir is located underground at one end of the bed and collects the returned nutrient solution for recycling. Besides, a liquid level inductor is also arranged.

⑤ The water pump is arranged in the liquid collecting reservoir and connected with the liquid level inductor to control the startup and shutdown of the water pump.

3. Drainage facility

The nutrient solution for recycling rock wool culture is recycled, so there is no drainage facility. The drainage facility for open rock wool culture is very simple. It only needs to punch several holes in the plastic packaging bag at the bottom of the rock wool plantation pad to let the excess nutrient solution flow out. Under the action of the slope of the bed surface, the nutrient solution will flow into the bed ditch, then flow into the drainage ditch at the transverse head of the bed, and finally be led outdoors. A liquid collecting pit shall be placed outdoors to collect the nutrient solution that flows out. If there is a large amount, we should try to deliver it back to the field for fertilization of crops and prevent it from being scattered and polluting the environment.

III. Seedling Raising and Planting of Rock Wool Culture

(I) Seedling Culture by Rock Wool Block

The shape and size of the rock wool block used for seedling raising can be determined according to the types of crops. Generally, the square blocks have the following specifications: 3 cm × 3 cm × 3 cm, 4 cm × 4 cm × 4 cm, 5 cm × 5 cm × 5 cm, 7.5 cm × 7.5 cm × 7.5 cm, 10 cm × 10 cm × 5 cm, etc. A small square hole is made in the center of a larger square block for embedding a small square block. The size of the small square hole just matches the embedded small square block, which is called "pot in pot". Except for the upper and lower sides, the big rock wool block should be wrapped with black or milky white opaque plastic film to prevent water evaporation, salt accumulation, and algae breeding around it. First, select a seedling raising box or seedbed of a certain size, and for the latter, a film should be laid at the bottom of the seedbed to prevent the leakage of nutrient solution. Then lay the rock wool block flat in it and soak the block in clear water for 24 hours before use. Seeds can be directly sown in the rock wool blocks, or in the seedling tray or smaller rock wool blocks. When the first true leaf of the seedlings appears, the seedlings will be moved into the big rock wool blocks (Fig. 6-34). When sowing seeds, poke a small hole in rock wool with a bamboo pole or tweezers and then put the seeds in the hole. Sow 1–2 seeds per block, and make sure that the sowing should be shallow rather than deep. After watering, cover them with old newspaper or non-woven fabric, and wait until seedling emerges. After the emergence of true leaves, apply 1/3–1/2 of the nutrient solution of standard concentration until the seedlings are grown. In the process of seedling raising, the spacing between rock wool blocks should be kept at all times to prevent the seedlings from overgrowing.

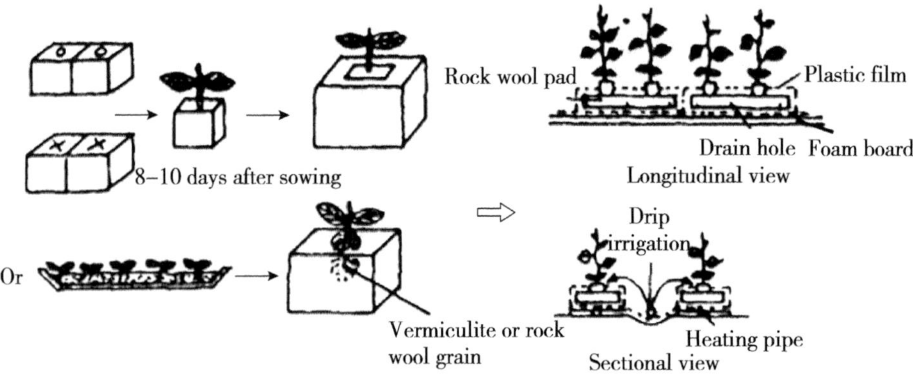

Fig. 6-34 Seedling Raising by Rock Wool

(II) Planting

Seedlings raised by rock wool blocks will be planted on rock wool plantation pads arranged according to specifications. That is, first cut the envelope on the rock wool plantation pad to make a planting hole that matches the bottom area of the seedling block, then arrange the dropper pipe of the drip irrigation system above it, drop the nutrient solution into the hole to make the entire rock wool plantation pad absorb enough nutrient solution, and then poke several small holes at the bottom of both ends of the rock wool plantation pad by the side of the bed ditch to make the excess nutrient solution flow out. After that, place the seedling block with seedlings on the planting hole of the rock wool plantation pad, and then set up the dropper of the dropper pipe on the seedling block so that the nutrient solution is dripped into the seedling block before flowing into the plantation pad. After the roots stretch into the plantation pad, move the dropper to a position above the plantation pad so that the nutrient solution can be dripped directly onto the plantation pad. In this way, the planting process is completed. After that, supply the nutrient solution as needed.

(III) Nutrient Solution Management

1. Determination of solution supply amount

The solution supply amount for soilless culture is different from that for field irrigation. Water supply and nutrition supply are carried out at the same time, and it needs to take into account the requirements from both aspects. Here, we will focus on water supply, while nutrition supply will be described separately.

The determination of the water supply amount is restricted by three factors: the allowable water-holding capacity of the substrate, the volume of substrate per plant,and the crop water requirement. The crop water requirement is affected by crop species, growth period, light, and temperature. Current data are mostly empirical or within a safe range, so we have to use them flexibly in practice. Some basic data are listed here for reference, and we will illustrate the ideas for using these data.

(1) Allowable water holding capacity of rock wool substrate.A Japanese researcher Yasui Hideo proposed that the maximum water-holding capacity of rock wool should not exceed 80% of the rock wool volume. However, considering that water is divided into upper, middle, and lower layers by gravity, if the overall water supply is 80%, the water-holding capacity of the lower layer will exceed 80%. To keep the water-holding capacity of the lower layer at a level no higher than 80%, the total water supply should be set at 60% of the total volume, so that there will be a situation of 80% in the lower layer, 60% in the middle layer and 40% in the upper layer, which can coordinate the water-air contradiction in rock wool. The practice

has proven that it is feasible to cultivate tomatoes.

(2) Crop water requirement. Table 6-1 shows the crop water requirement under soilless culture. Tanaka investigated the successful cases of tomato cultivation by 52 Japanese farmers using the open rock wool culture technique. The application amount of drip irrigation nutrient solution is shown in Table 6-2.

Table 6-1 Water Absorption Capacity of Several Crops at Different Growth Stages

Crop type	Early stage of planting [L/(plant·d)]	After first flowering date [L/(plant·d)]	Peak harvest stage [L/(plant·d)]
Tomato	0.1–0.2	0.8–1.0	around 1.5
Cucumber	0.2–0.3	around 1.0	around 1.6
Melon	0.1–0.2	around 0.5	around 1.0
Strawberry	around 0.02	around 0.04	around 0.15

Table 6-2 Drip Irrigation Solution Supply Amount by Month for Open Rock Wool Culture of Tomato

Month	1	2	3	4	5	6	7	8	9	10	11	12
Average value [L/(plant·d)]	0.79	0.74	0.84	1.14	1.52	1.53	1.64	1.85	1.48	1.05	0.81	0.67
Standard deviation [L/(plant·d)]	0.28	0.25	0.25	0.27	0.46	0.38	0.41	0.33	0.14	0.23	0.22	0.23
Sample size	13	11	13	17	18	20	13	8	7	11	16	20

These data are all rough parameters, which may only be used as a reference. They must be adjusted according to the actual conditions (plant size, weather conditions, etc.), and the adjustment range may be very large. For example, according to Yamazaki's data, the daily water absorption per tomato plant is 1.5 L at the peak harvest stage, but this peak stage is a rather long period, and the plant type cannot remain unchanged, and there will also be differences between sunny and rainy days. Therefore, 1.5 L is only an average value, and it will inevitably change. According to Yasui Hideo's data, the maximum water consumption of tomato can reach 3 L/(plant·d), and Tanaka's data also indicates such a value [1.64 + 3 × 0.41 = 2.87 L/(plant·d), calculated according to statistical rules] will occur. So, under what circumstances should this value be used depends on how the manager adapts to the change. It can be seen that the data currently available is very rough, but at least it can be used as a reference.

(3) The first method for determining the water supply amount—measurement of water-holding capacity of substrate. Taking tomatoes as an example, it has been specified above that each tomato should have 6.7 L of rock wool pad, and the safe allowable water-holding capacity is 60% of the rock wool volume; that is, each tomato has 4 L of basic nutrient solution, which is enough to meet one day's requirement when the water absorption of tomatoes is at its peak. As long as this water-holding condition is maintained, the tomato's water requirement can be guaranteed. In principle, we should replenish the same amount of water as that absorbed by the tomato. The amount of water absorbed by the tomato can be determined by measuring the water-holding capacity of the rock wool plantation pad with a moisture tensiometer. The method is to select some rock wool plantation pads at different positions within the planting range, place a tensiometer on the upper, middle, and lower layers of each pad, observe its scale regularly, and calculate its average value. When the value shows that the water-holding capacity of the substrate is 10% lower than the original level (i.e. decreasing from 60% to 50%), it is necessary to replenish water. The specific amount of water to be replenished is 6.7 L × 10% = 0.67 L. If the flow rate of each dropper is 2 L/h, it is necessary to put the dropper to work for 20 minutes. We should observe it regularly every day and replenish water as soon as this limit is reached. This kind of water supply method can save the amount of liquid used and avoid the environmental pollution caused by excessive liquid supply. This method must have a reliable moisture tensiometer. With the tensiometer, the water replenishment procedure can be carried out manually or connected in series to an automation device controlled by a computer instead of manual operation.

(4) The second method for determining the water supply amount— estimation of crop water consumption. This is an empirical water supply method. Taking

tomatoes as an example, according to Yamazaki's data, the water consumption of tomatoes after the first flowering date is 0.8–1.0 L/(plant·d). Now, the object is the tomato that has been growing for many days after the first flowering date, and its plant is relatively large. On sunny days, the water consumption tends to be 1.0 L/(plant·d). In this way, if the manager increases the assurance coefficient by 30% according to experience, the water consumption should be 1.33 L/(plant·d). This estimate may be too large, but it is not dangerous as the excess liquid just flows away. After the total daily water supply is determined, it will be completed in several periods, usually 4–5 times, from 5:00 to 15:00. This method should be mastered by experienced people, and the reaction of crops should be observed frequently, so as to increase or decrease the water supply in time.

2. Determination of the concentration of the liquid supplied

According to Yamazaki's theory, water and fertilizer are absorbed in proportion by crops. For the nutrient solution formula based on the n/W value, in the process of being absorbed by crops, water and fertilizer are absorbed at the same time (we might as well call it the formula of simultaneous absorption of water and fertilizer). The nutrient solution prepared by this formula is supplied to the corresponding crops. When crops absorb of nutrient solution, they not only absorb 1 L of water but also absorb 1 dose of nutrients from the 1 L of water. Therefore, the Yamazaki formula can be used as long as the nutrient concentration is controlled at one dose. If the concentration of the Japanese garden test formula is twice as high as that of the Yamazaki formula, it is appropriate to use 1/2 dose of the Japanese garden formula for tomatoes. Tanaka also believes that this concentration is relatively safe according to experience.

3. Operation of circulating liquid supply system

The method of intermittent liquid supply within 24 hours is adopted. That is, under the condition that the rock wool plantation pad has absorbed enough nutrient solution, it drips at the rate of 2 L of nutrient solution per plant per hour and stops dripping 1 hour later when enough solution is dripped. All the dripped liquid is returned to the liquid collecting reservoir and pumped into the liquid supply reservoir before being dripped again (since the rock wool planting pad is in its maximum water-holding state, the liquid-holding capacity of each plant can reach 22 L according to the volume of the plantation pad; as each plant only absorbs 1–2 L per day, most of the dripped nutrient solution will return to the liquid collecting reservoir). During the automatic control operation, enough nutrient solution is stored in the liquid supply reservoir, and the sensor instructs to open the liquid supply magnetic valve so that the nutrient solution is delivered to the bed. One hour later, the timer will instruct the solenoid valve to close and stop the liquid supply. The nutrient solution dripped into the bed is collected in the liquid collecting reservoir through the drain pipe. When the liquid level in the liquid collection reservoir is high enough and touches the liquid level inductor, the water pump will be instructed to start pumping liquid into the liquid supply reservoir. When the nutrient solution in the liquid supply reservoir is high enough, the water pump will be instructed to shut down and stop pumping so as to prevent the nutrient solution in the liquid supply reservoir from overflowing. At the same time, the liquid supply solenoid valve will be instructed to open and drip the solution into the plantation bed again.

4. Elimination of excessive salt accumulated in the rock wool plantation pad

Due to the long planting time, the by-components in nutrient solution remaining in the substrate or the large dosage of the formula used may cause the accumulation of salt in the rock wool plantation pad. When a lot of salt is accumulated in the pad, it will cause an additional increase in the concentration of the nutrient solution in the pad, which will endanger the growth of crops. Therefore, salt should be washed with clear water in rock wool culture at a certain time. The method is to detect the electrical conductivity change of nutrient solution in the pad. Generally, the sample of the solution flowing out from the bottom of the rock wool plantation pad is measured 2–3 times a week. If it turns out to be higher than 3.5 mS/cm, the nutrient solution supply should be stopped. Drop more clear water to wash away the excess salt in a short time. When the electrical conductivity of the washing solution that comes out drops to a level close to that of the clear water, drop the nutrient solution again. Because the process of salt washing with clear water will keep the substrate in the state of being filled with clear water for a long time, which will lead to "starvation" of plants, it is better to wash the salt with a dilute nutrient solution (1/4–1/2 of the original concentration). When the electrical conductivity of the washing solution that comes out is close to that of the dilute nutrient solution, drop the original nutrient solution again.

5. Reuse of rock wool plantation pad

After one crop rotation, the rock wool plantation pad can be reused to plant the second crop. In commercial production in the Netherlands, it is proved that there is little difference in the yield of cucumbers planted on old and new rock wool. The British test also proves that rock wool can be used for at least two years, and the yield will decrease after two years. Because the rock wool has become compacted and disintegrated, the air permeability will decrease.

In principle, the rock wool pad should be disinfected before being reused. Specific practices can be combined with crop rotation to avoid diseases and reduce the disinfection work. For example, after the pad is used for planting tomatoes, it must be disinfected before being used for planting tomatoes again; if it is reused to plant cucumbers, disinfection is not required. It depends on the possibility of disease transmission between specific crops.

Runia (1986) studied the disinfection method of rock wool plantation pad in detail. That is, the rock wool plantation pad is placed in a basket for steam disinfection. The height of the rock wool stack for disinfection should not exceed 1.5 m. The disinfection takes 2 hours for bare rock wool and 5 hours for wrapped rock wool. The disinfection temperature is 70℃ for most germs. For cucumber viruses, it is required to be as high as 100℃ to kill them. Due to the high cost of disinfection, a cheap, low-density, and disposable rock wool has been developed for scientific research and production in recent years.

IV. Eco-organic Soilless Culture

(I) Characteristics of Eco-organic Soilless Culture Technique

Traditional organic substrate culture formulates a nutrient solution with a certain concentration from various inorganic fertilizers for crops to absorb and utilize. The eco-organic soilless culture technique is a form of organic substrate culture, in which various organic fertilizers in their solid forms are directly mixed and applied to the substrate, as the basis for supplying nutrients for the crops to be cultivated. Throughout the growth period of crops, solid organic fertilizers can be directly applied to the substrate surface several times every few days to maintain the nutrient supply intensity. Compared with the traditional organic substrate culture, it has the following advantages.

1. Simpleness

For the nutrient solution for traditional soilless culture, it is necessary to maintain a certain concentration of various nutrient elements and the balance among the elements, especially the availability of trace elements. As the eco-organic soilless culture adopts substrate culture and applies organic fertilizer, it not only has all kinds of nutrient elements but also has more than enough trace elements. Therefore, the total supply of nitrogen, phosphorus, and potassium and their balance are mainly considered in management, which greatly simplifies the operation and management process.

2. Significant reduction of one-time investment in soilless culture facility system

Since the eco-organic soilless culture does not apply nutrient solution, the equipment, testing system, timer, circulating pump, and other facilities required for formulating the nutrient solution can be canceled.

3. Great saving of production costs

The eco-organic soilless culture mainly applies disinfected organic solid fertilizer. Compared with nutrient solution, its fertilizer cost is reduced by 60%–80%, thus greatly saving the production costs of soilless culture.

4. No pollution to the environment

Under the condition of soilless culture, it is normal for about 20% of water or nutrient solution to be discharged to the outside of the system during irrigation. If the salt concentration in the drained liquid is too high, it will pollute the environment. The nitrate content in the drained liquid from the eco-organic soilless culture system is only 1–4 mg/L, which has no pollution to groundwater. Therefore, the application of eco-organic soilless culture technique to cultivate vegetables is not only sanitary and clean but also pollution-free to the environment.

5. Excellent product quality that meets the "green food" standard

From the culture substrate to the fertilizer applied, organic substances are the main components. After certain processing (such as high temperature and anaerobic fermentation), the organic fertilizer applied will not produce excessive harmful inorganic salts in the process of decomposing and releasing nutrients, and there will be no pollution from other harmful chemicals in the cultivation process, so that the product can meet the A-level or AA-level green food standard.

In short, eco-organic soilless culture has the obvious characteristics of low investment, low cost, less labor, easy operation, high yields, and high quality of products. It introduces organic agriculture into the soilless culture, which is a simple soilless culture technique combining organic and inorganic, with high efficiency and low cost, and is suitable for China's current situation.

(II) Structure and Management of Facilities

The facilities consist of the nutrient solution reservoir (tank), culture bed, liquid-adding system, and drainage circulation system.

1. Nutrient solution reservoir (tank)

There are two kinds of liquid feeding methods for substrate culture: circulating liquid feeding and non-circulating liquid feeding. The circulating liquid feeding is to be described here. No matter which circulation method is used to supply liquid, the volume of the nutrient solution reservoir is determined by the cultivation area and crop types. For example, 600 melons can be planted in a 200 m^2 polytunnel, and the maximum daily liquid consumption of each melon is 2 L, and the daily liquid consumption of 600 melons is 1,200 L, so the minimum design capacity of the reservoir should be no less than 1.5–2 t. In order to reduce the work of solution preparation every day and lower the labor intensity, the capacity of the nutrient solution reservoir can be designed as 4.5 t, that is, designing the reservoir to be 2 m long, 1.5 m wide, and 1.5 m deep. The nutrient solution reservoir is made of bricks and cement, and linoleum is laid at the bottom and four sides of the reservoir to prevent leakage. To facilitate the cleaning of the nutrient solution pool, a 20 cm^2 water tank is built at the corner of the nutrient solution reservoir below the pump.

2. Culture trough

The substrate culture trough consists of a trough body, substrate, and seepage layer. The size and shape of the culture trough depend on the ease of field operation of different crops. For example, tomatoes, cucumbers, and other high seedling crops are usually planted in two rows per trough to facilitate pruning, tendril binding, harvesting, and other field operations, and the trough width is generally 0.48 cm (inner diameter width). Some dwarf plants can be planted in multiple rows by setting a wide culture trough, as long as the hand can easily reach the middle of the trough for field management. The depth of the culture trough is preferably 15 cm. A shallower culture trough may also be used to reduce the cost, but special care is required for irrigation with a shallower culture trough. The length of the trough is determined by factors such as irrigation capacity (the irrigation system must be able to provide the same amount of nutrient solution for each plant), greenhouse structure, and walkways required for field operation. The slope of the trough should be at least 0.4% in order to obtain good drainage performance. If conditions permit, a porous drain pipe may also be laid at the bottom of the trough.

The trough body can be made of polystyrene board, glass fiber-reinforced plastic, silver-gray film, and cement. The trough body for simple substrate culture is made of bricks. When making the trough body, first level the ground surface in the facility to make an inclined surface with a slope of 1 : 100 from north to south; that is, the south side is 20 cm lower than the north side. Then, dig the culture trough with a length of 20 m. Dig a soil trough with an upper width of 48 cm, a lower width of 30 cm, and a depth of 20 cm from the ground, and then build two layers of bricks along the ground beside the trough, forming the trough body. In order to be isolated from the soil, lay a film along the bottom of the bed, make a hole on the film below the south side of the trough, and use a plastic pipe as a drain outlet leading to the drainage ditch. The drain outlet should be located at the lowest position of the culture trough. The plastic pipe should have a diameter of 0–25 mm and a length of about 20 mm. Then, lay a layer of walnut-sized broken bricks or stones on the film as the seepage layer. To prevent the substrate from being mixed in, lay a layer of window screen on the broken bricks, and then lay a 20-cm-thick substrate on the window screen. Usually, a complex substrate is adopted, e.g. turf and slag mixed at the ratio of 1 ∶ 1. A soft sprinkler irrigation pipe is laid in the middle of the substrate, where one end of the pipe is tied with iron wire, and the other end is connected to the branch pipe of the liquid adding pipe.

3. Liquid adding, drainage, and circulation system

This system can be either open or closed, depending on whether the excess nutrient solution is recycled and reused. In an open system, the nutrient solution is not recycled, while in a closed system, the nutrient solution is recycled.

The main pipe of the soft liquid feeding pipe from the nutrient solution reservoir to the culture trough is a ϕ 30mm iron pipe, and a water filter is arranged on the main pipe. The nutrient solution is pumped out from the cultivation tank, passes through a filter, and enters a sprinkler irrigation hose, which adds it to the culture bed by sprinkler irrigation so as to be absorbed by crops, and the remaining part seeps into a seepage layer composed of bricks. Because a film is laid under the seepage layer, the nutrient solution will not seep into the ground but will flow along the 1/100 slope to the liquid outlet on the south side of the culture trough, through which the nutrient solution will flow into the drainage ditch. The drainage ditch is located on the south side of the trough and is made of bricks and cement, and it is placed underground as a whole. The nutrient solution flows back to the nutrient solution reservoir through the return pipe that connects the drainage ditch to the nutrient solution reservoir (Fig. 6-35 and Fig. 6-36).

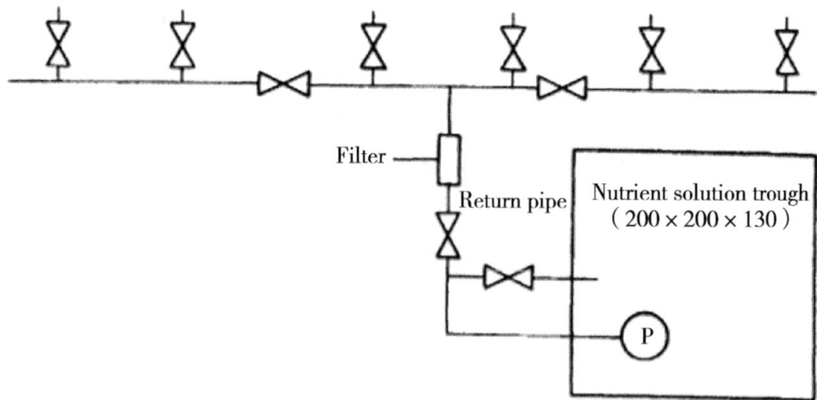

Fig. 6-35 Nutrient Solution Circulation System of Organic Substrate Culture (Unit: cm)

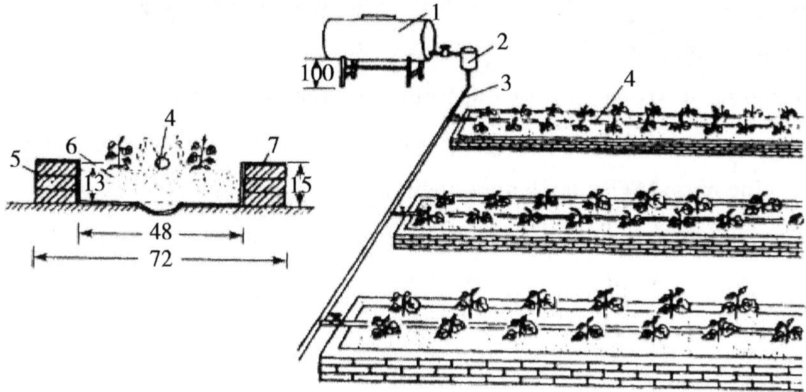

Fig. 6-36 Organic Substrate Culture System (Unit: cm)

1. Liquid storage tank 2. Filter 3. Liquid supply pipe 4. Drip irrigation tape 5. Brick 6. Organic substrate 7. Plastic film

(III) Operation Management of Eco-organic Soilless Culture Technique

1. Nutrition management

The supply of fertilizer adopts nitrogen, phosphorus, and potassium as the main indicators; 1 m^3 of substrate should contain 1.5–2.0 kg of total nitrogen (N), 0.5–0.8 kg of total phosphorus (P_2O_5), and 0.8–2.4 kg of total potassium (K_2O). This level of fertilizer supply is enough to meet the nutrient requirement of a mu of tomatoes to achieve a yield of 8,000–10,000 kg. In order to keep the best fertilizer supply state throughout the growth period of crops, fertilizers are usually applied in stages according to the types of crops and the fertilizers applied. A certain amount of fertilizer should be mixed into the substrate as basal fertilizer before filling the substrate into the culture trough or after harvesting the previous crop and before planting the next crop. In this way, topdressing is not required for fruit vegetables such as tomatoes and cucumbers within 20 days after planting, and it only needs to be watered. After 20 days, topdressing should be carried out every 10–15 days, and the fertilizer should be evenly spread around 5 cm away from the roots. The ratio of basal fertilizer to topdressing is 25 : 75 to 60 : 40, and the topdressing amount for 1 m^3 of substrate each time is as follows: 80–150 g of total nitrogen (N), 30–50 g of total phosphorus (P_2O_5), and 50–180 g of total potassium (K_2O). The topdressing frequency depends on the length of the growth period of each crop.

2. Water management

Determine irrigation quota according to the types of cultivated crops, and adjust the irrigation amount each time according to the water content of substrate in the growing period. The day before planting, the irrigation amount should reach the saturation water content of the substrate; that is, the substrate should be thoroughly watered. After crop planting, the irrigation frequency is not fixed every day, e.g. once or 2–3 times a day, so as to keep the water content of the substrate up to 60%–85% (of the dry substrate). Generally, 1.5–2 L of water should be applied to cucumber each day in the adult-plant stage, 0.8–1.2 L for tomato and 0.7–0.9 L for sweet pepper. The amount of irrigation water must be adjusted according to climate change and plant size. Irrigation should be stopped on rainy days and carried out once every other day in winter.

V. Complex Substrate Culture

The complex substrate is a mixture of two or more single substrates in a certain proportion. It is advisable to use two or three single substrates to prepare the complex substrate. The prepared complex substrate should have an appropriate bulk density, higher porosity, and increased water and air contents.

The complex substrate culture includes trough culture, bag culture, and pot culture.

1. Trough culture

Trough culture uses bricks, cement, or wooden board to make a relatively fixed culture trough on the greenhouse ground, put a solid substrate into the culture trough, and cultivate crops through facilities such as liquid supply and drainage system and reservoir that are matched with the culture trough (Fig. 6-37).

The plantation trough frame is made of brick cement mortar, cement precast slab, plastic board, or wooden board. The specifications of plantation troughs vary depending on the crop types. Generally, the length × width × height is (15–30) m × (50–120) cm × (25–35) cm. The shape of a plantation trough can be giant, V-shaped, inverted-V-shaped, etc. The giant trough is also called a flat-bottom trough; that is, the trough bottom is horizontal. The bottom of the V-shaped trough is deeper than the trough frame, forming a V-shape. The bottom of the inverted-V-shaped trough has a higher center and two lower inclined sides.

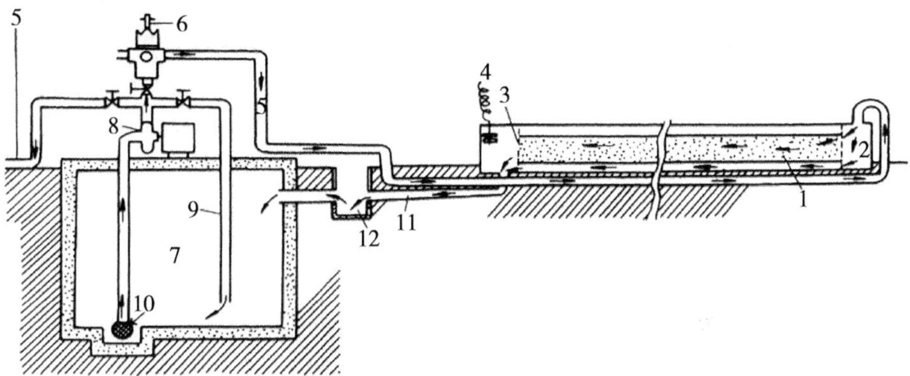

Fig. 6-37 Schematic Diagram of Trough Culture Facilities

1. Gravel layer 2. Liquid supply buffer room 3. Drainage buffer room 4. Liquid level sensor 5. Liquid supply pipe. 6. Switching liquid supply valve 7. Reservoir 8. Water pump 9. Liquid separation pipe 10. Water pump strainer 11. Drain pipe 12. Sedimentation tank

The liquid supply and drainage system matched with the plantation trough comprises the water pump, liquid supply pipe, solenoid valve, timer, automatic switching rotation irrigation valve, and liquid level sensor for controlling the water level and pump operation. The irrigation and drainage process is as follows: During liquid supply, use a water pump to pump up the nutrient solution, transport it to the higher end of the plantation trough through the liquid supply pipe, then inject it into the liquid supply buffer room, and then inject it into the irrigation and drainage pipe at the bottom of the trough. At this time, the drain valve of the drainage buffer room is closed, and the nutrient solution level rises from bottom to top in the substrate layer until it reaches a level about 3–5 cm below the substrate surface. When the liquid level contacts with the floating ball switch in the upper part of the drainage buffer room, immediately cut off the power supply of the water pump and open the drain valve to drain the liquid until all the nutrient solution in the trough is drained into the reservoir. At this time, a certain amount of nutrient solution can be adsorbed at the substrate surface, the contact position between substrate particles, as well as the parts between the crop root system surface and roots for crop growth.

In recent years, drip irrigation facilities have been widely applied in this culture mode. In particular, drip irrigation facilities (including infusion pipes and dripping parts) are used in the liquid supply process. The infusion pipe consists of the main pipe, branch pipe, and capillary pipe, where the main pipe is the main pipe in the facility, the branch pipe connects the four sides of the plantation trough in a polytunnel or greenhouse, the capillary pipe enters the planting row pipe through the branch pipe in the plantation trough, and the dropper pipe that drops liquid to the plants is embedded on the capillary pipe, with a diameter of 12–16 mm. The drip part consists of a dropper pipe and a small plastic rod, where one end of the dropper pipe is embedded in the capillary pipe, and the other end is supported by a small plastic rod and inserted into the planting hole of the plant, so as to drop liquid directly onto the plant. The liquid outlet of the dropper pipe should be 2–3 cm away from the substrate surface. At present, in order to save the cost, the drip irrigation tape may also be made by integrating the capillary pipe and the drip part. The drip irrigation tape is placed above the surface of the substrate in the plantation trough and under the plastic film mulch.

Generally, the construction position of the reservoir is determined according to the liquid supply method. If a water pump is used for liquid supply, the reservoir can be built underground; if the gravity-flow liquid supply method is adopted, the reservoir should be elevated above the ground according to the pressure requirements of the drip irrigation system. The reservoir can be built of reinforced concrete, and the inner wall of the reservoir should be coated with high-grade acid-resistant cement mortar, and anti-seepage treatment should be done properly. The volume of the reservoir should be determined according to the controlled area of liquid supply and the types of crops planted. At least one day’s liquid supply should be assured, which is generally about 75% of the total volume of the substrate.

2. Bag culture

Bag culture is a crop cultivation method that puts solid substrate into a plastic bag and supplies nutrient solution to it. The bag is usually made of UV-resistant polyethylene film, which can be used for at least 2 years. In the high-temperature season or southern China, the surface of the plastic bag should be white so as to reflect sunlight and prevent the substrate from being heated up. In the low-temperature season or region, the surface of the bag should be black so as to absorb heat and keep the temperature of the substrate in the bag.

Bag culture can be divided into ground bag culture and stereoscopic bag culture.

(1) The ground bag culture can be further divided into barrel culture and pillow culture. The barrel culture is to put about 10–15 L of substrate into a barrel-shaped plastic bag with a diameter of 30–35 cm and a height of 35 cm for cultivating a large plant. The pillow culture is to put 20–30 L of substrate into a pillow-shaped plastic bag with a length of 70 cm and a diameter of 30–35 cm, where two planting holes with a diameter of 10 cm are made on the bag, and the distance between the two holes is 40 cm, so as to plant two large crops. Before putting the cultivation bag in the greenhouse or polytunnel, milky white or black-and-white double-sided plastic film (with white facing out) should be laid on the ground to isolate the cultivation bag from the soil. On the one hand, it can prevent the invasion of pests and diseases in the soil, and on the other hand, it can help increase the indoor lighting intensity. After planting bags are arranged, lay out the drip irrigation pipe with one dropper per plant. It is worth noting that 2–3 small holes with a diameter of 0.5–1.0 cm should be made at the bottom or both sides of the substrate bag for bag culture so that the excess nutrient solution can flow out of the holes to prevent accumulated solution from macerating the roots (Fig. 6-38).

(2) Stereoscopic bag culture is to make use of polyethylene plastic barrel film with a diameter of 15 cm and a thickness of 0.15 mm, where its bottom end is tied tightly, and the substrate is filled in from the upper end to form a sausage shape, and then it is cut off at a length of about 2 m. After the upper end is tied, it is hung in the greenhouse; some holes with a diameter of 2.5–5.0 cm are made around the bag to cultivate plants. The supply of nutrients and water is completed by

the drip irrigation system installed at the top of the substrate bag. The nutrient solution is poured from the top and then permeates downward through the cultivation bag, and the excess nutrient solution is drained from the drain hole (Fig. 6-39). Clean water should be used to wash the salt once a month to prevent salt accumulation.

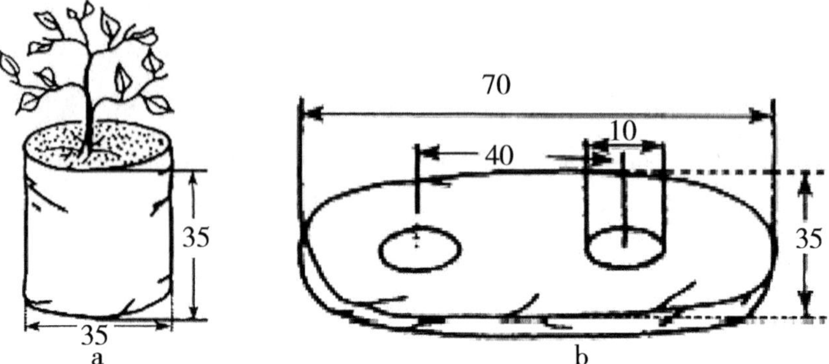

Fig. 6-38 Ground Bag Culture (Unit: cm)

a. Barrel culture b. Pillow culture

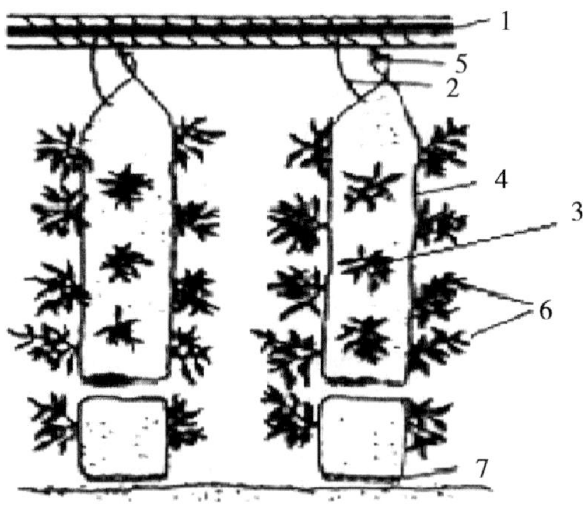

Fig. 6-39 Long Bag Stereoscopic Culture

1. Liquid supply pipe 2. Drip irrigation pipe 3. Planting hole 4. Film bag 5. Hanger 6. Crop 7. Drain hole

3. Pot culture

Pot culture is a kind of substrate culture, and it is also the simplest way of soilless culture. The pot is relatively large and can hold relatively more substrate soil. It is suitable for planting plants with a large root system and climbing plants, such as tomatoes, pumpkins, and sweet potatoes.

Pot culture, also known as basin culture, mainly adopts miniaturized cultivation containers such as basins and pots. The culture management techniques are similar to those of box culture. The cultivation method is to drill holes with a diameter of 5 mm at the height of 1/3 from the bottom of the plastic flowerpot. The bottom of the plastic bucket is filled with water-washed slag, and the substrate, such as rock wool, is placed in it. The nutrient solution is applied every 3–5 days until the overflow outlet drains liquid, and clear water is poured once a month, so that various flowers, fruit vegetables, and green leafy vegetables can be cultivated.

The small-scale drip irrigation culture may also be used in families. Hang a plastic bucket over the windowsill to hold the nutrient solution, install a drip tube used by the hospital, and insert it into the surface layer of the cultivation pot below. Adopt a plastic bucket or pot as the cultivation pot, drill a hole 3 cm from the bottom of the pot, and install a waste infusion tube. Fill the cultivation pot with the substrate, and perform drip irrigation 3 times a day (30 minutes each time). This facility can be used to cultivate fruit vegetables and medium-sized flowers.

VI. Stereoscopic Culture

Stereoscopic culture, also known as vertical culture, is a kind of stereoscopic soilless culture. Under the condition of not affecting flat culture, stereoscopic culture develops into space through the culture columns erected on all sides, and it makes full use of greenhouse space and solar energy to increase the land utilization rate by 3–5 times, which can raise the yield per unit area by 2–3 times.

In the 1960s, stereoscopic culture was first developed in developed countries. The United States, Japan, Spain, Italy, and other developed countries developed different forms of stereoscopic culture, such as multi-layer, dangling, sausage, and unit superposition forms. China began to research and popularize this technique in the

1990s. Column culture has become the first choice for leisure agriculture because of its high-tech, novel, and aesthetic features. In recent years, it has been adopted in Beijing, Shanghai, Hebei, Jiangsu, and other regions.

(I) Types of Stereoscopic Culture

1. Column culture

The culture column adopts asbestos cement pipe or rigid plastic pipe, where holes are made around the pipe at the helical position, and plants are cultivated in the substrate in the holes. A special soilless culture column may also be adopted, where the culture column consists of several short model pipes, and each model pipe has several protruding cups for cultivating crops (Fig. 6-40).

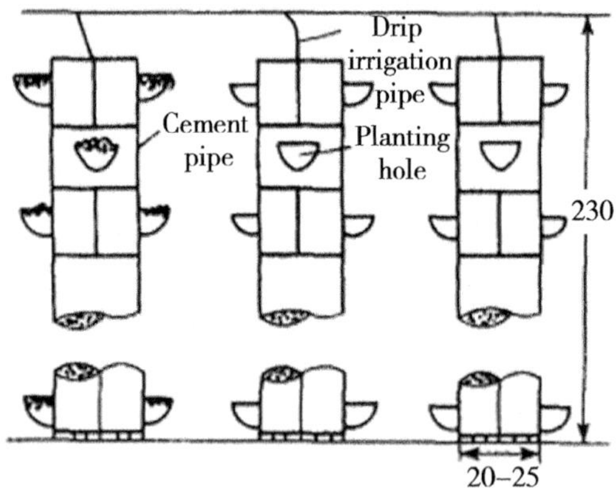

Fig. 6-40 Schematic Diagram of Column Culture(Unit: cm)

2. Long bag culture

The cultivation bag adopts a polyethylene barrel film with a diameter of 15 cm, a thickness of 0.15 mm, and a length of generally 2 m, where its bottom end is tied tightly to prevent the substrate from falling down, the substrate is filled into from the upper end to form a sausage shape, and the upper end is ligated, and then the bag is hung in the greenhouse; some 2.5–5 cm holes are made on all sides of the bag for planting crops.

In view of factors such as facility cost, cultivation effect, and requirements for the greenhouse environment, column soilless culture has a certain ornamental value, low investment cost, and high benefit, so it is widely applied in various parts of China. Next, we will use the column culture as an example to describe its facility structure and management.

(II) Column Soilless Culture Facility Structure

The column soilless culture facilities are composed of a nutrient solution reservoir, DFT system, culture column, column culture pot, and liquid adding and returning system for column culture.

(1) The volume of the nutrient solution reservoir is designed according to the standard that the hydroponic area of 1 mu requires 15–20 t of nutrient solution. The specific construction method is the same as above.

(2) For the DFT system, refer to the relevant content in the DFT hydroponics section.

(3) The culture column is the carrier used to support and fix the culture pots and drip boxes. The column integrates the culture pots into one, and the column leading to the air consists of two parts: the cement pier and the iron pipe. The cement pier is 15 cm^2, and there is a round hole with a diameter of 30 mm and a depth of 10 cm in the middle. It is buried underground on both sides of the hydroponic bed to fix the iron pipe of the column, and the pier spacing is 90 cm. The iron pipe has a diameter of 25–30 mm and a length of about 2 m. It can be made of thin-walled iron pipe or rigid plastic pipe, and the lower end of the pipe is inserted into the hole of the cement pier.

(4) The culture pot is a device for cultivating crops on a column. It is a hollow, six-petalled plastic pot with a height of 20 cm, a diameter of 20 cm, and a petal spacing of 10 cm, where granular rock wool or coconut shell fiber is put in the pot. Six plants are planted in the petals, and the culture column is composed of 8–9 culture pots and drip boxes according to the height of the greenhouse. The culture pots are stacked on the column with staggered petal positions and strung into a column shape.

(5) The liquid-adding system for column culture consists of water pump liquid-adding main pipe and branch pipe. The liquid-adding main pipe is a rigid dropping pipe with a diameter of 40–50 mm, and the liquid-adding branch pipe is a non-porous rigid dropping pipe with a diameter of 16 mm. The drip box is a

round plastic box. Two hollow short handles are arranged at both ends of the box to connect the liquid-adding branch pipe, and there are six small holes around the bottom of the box so that the nutrient solution can flow down. The bottom center of the drip box is fixed above the column.

During the liquid supply, the nutrient solution is pumped out of the reservoir by a water pump, enters the drip box through the liquid-adding main pipe and branch pipe, flows from the drip box into the culture pot, then flows into the second culture pot through the small hole at the bottom of the culture pot, flows down to the lowest culture pot, flows into the plane hydroponic bed, and then flows back into the nutrient solution reservoir to complete a cycle.

(III) Precautions for Column Soilless Culture

1. Selection of vegetables

Not all vegetables are suitable for column culture. Only by taking advantage of strengths and avoiding weaknesses can we play the role of column culture. Generally, dwarf leafy vegetables are suitable for column culture, and their upward growth height should not exceed 45 cm. So far, the varieties that have been successfully tested include small-plant leafy vegetables such as *Gynura bicolor*, strawberry, *Chrysanthemum coronarium*, lettuce, rape, and celery. Stems of vegetables of higher plant type will fall down due to space limitations and gravity, which will affect their growth. Fruit vegetables have higher requirements for light conditions, so they are generally not suitable for column culture. However, the method of cultivating dwarf fruit vegetables, such as strawberries in the top 2–3 layers of the column and leafy vegetables in the bottom, can be adopted. Because of the special configuration of the column, vegetables cannot grow symmetrically from side to side. Heading vegetables are not suitable for column culture because of their unsightly appearance and poor commerciality.

2. Light

Light is an important environmental factor that affects the yield and quality of stereoscopic culture. In column culture, the lighting intensity decreases with the decrease in the number of culture pot layers, and the light obtained by the plants on the sunny side of the column is better than that on the dark side. According to the measurement, the lighting intensity decreases by an average of 15% for each layer down from the top to the bottom of the column. Except for the highest layer, where the lighting intensity on the dark side is close to that on the sunny side, the lighting intensity on the dark side of other layers is only about 50% of that on the sunny side.

To compensate for the shortage and difference in light, the column must be rotated regularly so that six plants on each layer receive enough sunlight. This is an important method to ensure the orderly growth of crops and increase yield. In addition, the artificial light supplement method can also be adopted. For specific operations, please refer to relevant sections.

[Module Summary]

I. Key and Difficult Points

(1) Mainly master the advantages, disadvantages, and characteristics of deep flow technique;

(2) Master the deep flow culture management technique;

(3) Mainly master the advantages, disadvantages, and characteristics of the nutrient film technique;

(4) Master the management of timer, conductivity meter, pH meter, temperature meter, and other devices used for cultivation with the nutrient film technique;

(5) Master the advantages and disadvantages of aeroponics;

(6) Master the management of aeroponics nutrient solution.

II. Experience and Skills

(1) The deep flow technique has become a practical and efficient soilless culture technique because of its easy management, durable facilities, and less investment in subsequent production materials, and it has been widely promoted and applied in southern provinces and municipalities of China. The construction of deep flow hydroponic device includes the construction of four parts: nutrient solution plantation trough, planting plate (or planting net frame), reservoir, and nutrient solution circulation system and control system. During the deep flow

hydroponic culture, crops that are suitable for hydroponics should be selected for planting in the initial production so as to achieve the success of hydroponics.

(2) The nutrient solution layer of nutrient film hydroponic culture is shallow, which better meets the oxygen requirement of root respiration. In the specific management, we should select stable nutrient solution formulas, adopt targeted liquid supply methods and strengthen the management of liquid temperature.

(3) Aeroponics can solve the problem associated with the root system oxygen supply, and there is almost no phenomenon of poor growth arising from the root system oxygen supply. The utilization rate of nutrients and water is high, and the nutrient supply is fast and effective. Besides, it can make full use of the space in the greenhouse and increase the planting quantity and yield per unit area.

(4) Trough culture is the most common soilless culture form of a solid substrate. The substrate culture facilities are simple in structure and low in cost. Because the substrate has a buffering effect, the changes in nutrients, moisture content, temperature, and other environmental conditions are moderated. The culture technique is very similar to that of traditional soil culture, so it is easy to master. In soilless culture, each place may select the appropriate substrate according to the situation, adopt the culture form suitable for the local economic situation, and choose the appropriate culture trough according to the situation.

(5) Substrate bag culture is widely applied in China. The cultivation bag for holding the substrate can be selected according to local conditions. Shaped plastic bags or discarded cement, flour, or fertilizer packaging bags can be used as containers. The substrate used can also be selected according to local conditions. A single substrate can be selected. For example, organic substrates such as rice husk, mushroom residue, corn husk, or wheat bran; inorganic substrates such as sand, gravel, or slag; or mixed substrates can be selected. The substrate bag culture facilities are simple and easy to manage. The substrate bag culture is suitable for many kinds of plants, so it is recognized by the majority of users.

(6) Stereoscopic substrate culture is relatively rarely applied in actual production, and it is mainly applied in sightseeing agriculture. It can make rational

use of space, improve the utilization rate of facilities, increase the planting area, and enhance the ornamental effect.

[Skill Training]

Skill Training 6-1 Construction of Deep Flow Hydroponic Facilities

I. Purposes and Requirements

Be able to design and construct deep flow hydroponic facilities or key components through on-the-spot investigation of the culture by deep flow hydroponic facilities in combination with the knowledge learned in class.

II. Plan

1. Materials and tools

(1) Materials and chemical agents: red brick, cement, river sand, ϕ 8 steel bars (ϕ 8@20), rigid foam polystyrene board, concrete, waterproof coating, plastic planting cup, rigid plastic pipe, rubber plug, rubber hose, and other facility construction materials; water pump, filter, plastic, dark bucket, foam box, PE or PVC pipe fittings (including ϕ 40 or ϕ 25 plastic pipe, tee, bipass, valve, elbow, plug, and drip irrigation pipe), and other fittings for assembling liquid supply and return systems; nutrient solution.

(2) Instruments and tools: tape measure; steel tape measure; angular instrument (gradiometer), level gauge, and other measuring instruments; pencil, eraser, ruler, and other drawing and recording tools; shovel, rake, hacksaw, hairbrush, scissors, and other construction tools.

2. Implementation plans

The deep flow hydroponic facilities mainly include plantation trough, planting plate, reservoir, and nutrient solution circulation system.

The construction process of deep flow hydroponic facilities is as follows.

(1) Construction of plantation trough facilities: Determine the layout of the plantation trough in the greenhouse, draw a simple construction drawing, measure and mark out according to the construction drawing, and construct a flat plantation trough. The trough is extended north and south, the length of the trough is about 10–20 m, the depth of the trough is controlled at about 12–15 cm, and the deepest point should not exceed 20 cm. Constructing method: First level and firm up the ground, lay a layer of river sand or stone powder of about 3–5 cm thick on the position where the trough is built, and then lay 5-cm-thick concrete on the river sand or stone powder layer as the bottom of the trough, then add a ϕ 8 steel bar every 20 cm in the concrete layer, use cement mortar to lay bricks around the top to form the trough frame, then use high-grade cement mortar to plaster both inside and outside of the plantation trough, and finally add a layer of cement paste to smooth the surface to prevent leakage of nutrient solution. Apply waterproof paint to prevent leakage when the plantation trough frame is built.

(2) Making of planting plate: Make a number of planting holes with a diameter of 5–6 cm on the rigid foam polystyrene board with a thickness of about 2–3 cm according to the planting spacing required for vegetable species to be planted. A plastic planting cup is embedded in each planting hole, with a height of 7.5–8.0 cm. The diameter of the cup mouth is the same as that of the planting hole, and the outer edge of the cup mouth has a lip that is about 5 mm wide so as to be stuck on the planting hole and not fall into the bottom of the trough. Many 3-mm holes are made in the lower half and bottom of the cup. The width of the planting plate is consistent with that of the outer edge of the plantation trough so that both sides of the planting plate can be supported on the wall of the plantation trough, and thus the planting plate can be hung together with the planting cup embedded in the plate hole. The length of the planting plate is generally 150 cm, which can be expanded or contracted for the convenience of work. The planting plate covers the entire plantation trough piece by piece so that the light cannot penetrate into the trough.

(3) Reservoir construction: The underground reservoir is constructed on the general principle of no leakage. When building the reservoir bottom, 10–15-cm-

thick concrete should be used together with ϕ 8 steel bar. The wall of the reservoir should be made of 18–24-cm-thick bricks, and #100 cement mortar should be used for plastering. The cement used for building the reservoir should be of high grade and resistant to plastering. Besides, the surface of the underground reservoir should be 10–20 cm higher than the ground and covered to prevent rain or other debris from falling into the reservoir. The inside of the reservoir should be kept dark to prevent algae from growing.

III. Implementation

(1) Work in groups to formulate the design and construction scheme of deep flow hydroponic facilities, such as plantation trough, planting plate, reservoir, and nutrient solution circulation system.

(2) Conduct an on-the-spot investigation of a deep flow technique, record the types and materials of deep flow hydroponic facilities, the size of plantation trough, the specifications of planting plate, the models of planting holes, planting cups, the volume of the reservoir, as well as the types, specifications, and layout of materials for pipes in the nutrient solution circulation system, and learn about the working principle and application method of key components.

(3) Revise and improve the design and construction scheme of deep flow hydroponic facilities according to the investigation results. Carry out on-site construction according to the construction scheme where conditions permit.

(4) Exchange ideas on the design and construction scheme of deep flow hydroponic facilities.

Skill Training 6-2 Construction of Nutrient Film Hydroponic Facilities

I. Purpose and Requirements

Master the design requirements and construction methods of nutrient film hydroponic facilities.

II. Plan

1. Materials and tools

(1) Materials and chemical agents: red brick, cement, river sand, ϕ8 steel bars (ϕ8@20), hard foam polystyrene board, concrete, waterproof coating, plastic planting cup, hard plastic pipe, rubber plug, rubber hose, polyethylene film, and other facility construction materials; water pump, filter, plastic, dark bucket, foam box, PE or PVC pipe fittings (including ϕ40 or ϕ25 plastic pipe, tee, bipass, valve, elbow, plug, and drip irrigation pipe), and other fittings for assembling liquid supply and return systems; nutrient solution.

(2) Instruments and tools: tape measure; steel tape measure; angular instrument (gradiometer), level gauge, and other measuring instruments; pencil, eraser, ruler, and other drawing and recording tools; shovel, rake, hacksaw, hairbrush, scissors, and other construction tools.

2. Implementation plans

(1) Construction of plantation trough for cultivating large crops: Use 0.1–0.2-mm-thick polyethylene film with a white surface and black bottom to temporarily enclose an isosceles triangular trough. The trough is 20–25 m long, 25–30 cm wide at the bottom, and 20 cm high. That is, take the above film with a width of 75–80 cm and a length of 21–26 m, and lay it on the pre-leveled and compacted ground with a certain slope (about 1∶75), with the long side parallel to the slope direction. When planting, place the seedlings with seedling pots in a row in the center of the film width, and then pull up the two sides of the film so that the width of 20–30 cm in the center of the film is closely attached to the ground, and the pulled up two sides are folded together and clamped with clips to form an isosceles triangular trough with a height of 20 cm. The stems and leaves of the plants extend out of the trough from the crevices at the top of the trough, and the root is placed at the opaque bottom of the trough.

(2) Construction of reservoir: The construction method is the same as that of the reservoir for deep flow hydroponic facilities. The reservoir is generally located

below the ground so that the nutrient solution flowing out of the plantation trough can flow back into the reservoir.

(3) Nutrient solution circulating flow device: Liquid supply pipe: A plastic main pipe is led out from the water pump, and a branch pipe is connected to the main pipe. One of the branch pipes is led back to the reservoir so that a part of the pumped nutrient solution flows back into the reservoir. On the one hand, it plays a role in stirring the nutrient solution to make it more uniform and increase the dissolved oxygen in the solution; on the other hand, the flow can be adjusted through the valve on it. A number of capillary tubes are connected to the branch pipes and delivered to the high end of each plantation trough. The capillary tubes of each trough are provided with a flow-regulating valve, and then small infusion tubes are connected to the capillary tubes and introduced into the plantation trough. Several small infusion tubes of 2–3 mm are set in each large plantation trough, and the number of tubes should be controlled to the extent that the flow rate is controlled within 2–4 L/min in each trough. The purpose of setting up several small infusion tubes is to make sure that at least one or two of them remain unblocked to ensure that there is no water shortage when another one or two of them are blocked. Two small infusion tubes are arranged in each small plantation trough valley to ensure that there is liquid flow in each valley, with the flow rate being 2 L/min per valley.

III. Implementation

(1) Work in groups to formulate the design and construction scheme of the plantation trough, reservoir, nutrient solution circulation system, and other facilities.

(2) Conduct an on-the-spot investigation of a nutrient film technique, record the types and materials of nutrient film hydroponic facilities, the size of the plantation trough, the specifications of the planting plate, the models of planting holes, the volume of the reservoir, as well as the types, specifications, and layout of materials for pipes in the nutrient solution circulation system, and learn about the working principle and application method of key components.

(3) Revise and improve the design and construction scheme of nutrient film hydroponic facilities according to the investigation results. Carry out on-site

construction according to the construction scheme where conditions permit.

(4) Exchange ideas on the design and construction scheme of nutrient film hydroponic facilities.

Skill Training 6-3 Construction of Trough Facilities

I. Purposes and Requirements

Master the design and construction methods of trough substrate culture facilities. Ensure that the design is scientific and reasonable. Make sure that the construction meets the design requirements and that the operation is standardized and skilled.

II. Plan

1. Materials and tools

(1) Materials and chemical agents: red brick, cement, river sand, ϕ 8 steel bars (ϕ 8@20), 2.5-cm-thick densified benzene board, plastic film (0.1–0.2 mm thick), pine board (3–5 cm thick), common bed frame, woven bag, bright oil, and other materials for facility construction; water pump, filter, plastic, dark bucket, foam box, PE or PVC pipe fittings (including ϕ 40 or ϕ 25 plastic pipe, tee, bipass, valve, elbow, plug, and drip irrigation pipe), and other fittings for assembling liquid supply and return systems; peat (coconut chaff), perlite, sterilized and extruded organic fertilizer, and other substrates; substrate heating equipment such as electric heating wire and socket; nitric or phosphoric acid; woven bag.

(2) Instruments and tools: tape measure; steel tape measure; angular instrument (gradiometer), level gauge and other measuring instruments; pencil, eraser, ruler and other drawing and recording tools; shovel, rake, hacksaw, hairbrush, scissors, and other construction tools.

2. Implementation plans

(1) Construction of trough substrate culture facilities: The substrate culture

facilities mainly include reservoir, plantation trough, and drip irrigation system. The reservoir construction method is the same as that of hydroponics, except that no return pipe is buried.

Determine the layout of the plantation trough in the greenhouse, draw a simple construction drawing, then level the ground, compact it, measure and mark out according to the construction drawing, and construct a flat plantation trough:

① The trough is extended north and south, with the length depending on the span of the greenhouse. The inner diameter is 48 cm, the outer diameter is 72 cm, and the trough spacing is 80 cm. The trough frame is made of three layers of standard red bricks of 24 cm × 12 cm × 5 cm stacked on the ground. No mud is needed between the bricks. A U-shaped groove with a width of 20 cm and a depth of 10 cm is made in the middle of the bottom of the trough, and the slope is maintained at 0.5%. A layer of 0.1-mm-thick polyethylene film is laid at the bottom of the trough, and the two sides of the film are clamped between the second and third layers of bricks. To be geared for the needs of soilless production in winter, it is necessary to build an electric hotbed. The construction method is to lay a layer of benzene board at the bottom of the trough, and then lay a layer of 2-cm-thick dry sand on the benzene board. Lay electric heating wire in the sand according to the power of 80–100 W/m^2, and then lay plastic film on it.

② Thoroughly mix the peat (coconut chaff), perlite, and sterilized and extruded organic fertilizer according to the ratio of 6 ∶ 3 ∶ 1, and then disinfect the mixture with formaldehyde.

③ Put about 3–5-cm-thick coarse stone sand that is exposed to the sun for one day in the trough to facilitate drainage, and then lay 1–2 layers of clean woven bags on the coarse stone sand to prevent the crop root system from reaching into the drainage layer. Evenly mix and fill the disinfected substrate of about 12–15 cm thick into the trough, and thoroughly water it. After that, cover the trough with film so that it not only can carry out high-temperature disinfection but also can prevent debris from accumulating in the trough before use. Thus, a common culture trough is constructed. Disinfection of the substrate may also be carried out in the trough. After the plantation trough is constructed, lay the drip irrigation system. Generally, two ϕ 10 mm rigid drip irrigation pipes, liquid supply branch pipes, or capillary pipes are laid in a 48- cm plantation trough, and then a plurality of dropper pipes are led out of them. The liquid supply branch pipe is perforated. One end of the drip irrigation pipe is connected to the liquid supply branch pipe, and the other end is plugged. The drip irrigation pipe is placed flat on the surface of the substrate trough.

(2) Mixing of culture substrates: Thoroughly mix the turf, perlite, and decomposed chicken manure at the ratio of 6 ∶ 3 ∶ 1. Compound fertilizer may be applied during the substrate mixing or after the substrate filling, and the former is generally preferred.

(3) Substrate filling and disinfection: Fill the trough with a 5-cm-thick coarse substrate, then lay a layer of woven bag, and fill the culture substrate of about 12–15 cm into the trough. Water it after the substrate filling. Cover the trough with plastic film for high-temperature solar disinfection, which can also prevent debris from accumulating in the trough before use. It may also be disinfected with a 100-fold dilution of 40% formaldehyde according to the substrate mixing condition.

(4) Planting: Plant in two rows in the plantation trough with a width of 90 cm according to the row spacing of 40–45 cm and plant spacing of 35–40 cm, where about 2,400–3,000 seedlings are to be planted per mu.

(5) Laying of drip irrigation system: After seedlings are planted, lay the drip irrigation tape (or drip irrigation pipe) according to cultivation lines, and integrate it with the liquid supply pipe to form a complete drip irrigation system.

III. Implementation

(1) Discuss in groups and formulate the trough facility construction plan;

(2) Exchange ideas on the plan and revise it;

(3) Construct or make the trough facility model on site as planned;

(4) Summarize and appraise through comparison.

Skill Training 6-4 Construction of Rock Wool Culture Facilities

I. Purpose and Requirements

Master the design requirements and construction methods of rock wool culture facilities.

II. Plan

Construction of open rock wool culture bed and placement of plantation pad.

1. Materials and tools

A 0.2-mm-thick milky white plastic film, rock wool plantation pads, and seedlings with rock wool blocks.

2. Implementation plans

(1) Construction of culture bed: The rock wool culture bed has strict requirements on construction technology. The ground of the culture bed must be flat; otherwise, it will cause uneven liquid supply or even cause salt accumulation and pH value rise, which will affect the cultivation effect. After leveling the ground in the polytunnel, build a turtle-back-shaped culture bed (soil bed) according to specifications, and have it compacted. The specification of the culture bed depends on the crop type. At the points on both sides, 30 cm away from the midpoint of the bed width, it begins to gently incline to form a bed ditch between two beds. The bed length is about 30 m, and the slope of the bed ditch is 1% along the long side. After compacted beds are built on the ground of the polytunnel, a 0.2-mm-thick milky white plastic film will be laid to cover all beds and ditches. The film should be tightly attached to the beds and ditches.

(2) Placement of rock wool pads: Two rows of rock wool plantation pads are placed vertically on the bed back, and the long sides of the pads should be in the same direction as the bed length. Each row is placed on the slope of the bed so that the pad tilts to one side of the bed ditch to facilitate drainage. Considering the

nutrient area occupied by each plant, generally, the rock wool plantation pad is better in a flat rectangle shape, with a thickness of 7–10 cm, a width of 25–30 cm, and a length of about 90 cm. The rock wool block is tightly wrapped with black or black-and-white polyethylene plastic film. Before planting, the planting holes are made on the film for planting seedlings with rock wool blocks.

III. Implementation

(1) Discuss in groups and formulate plans for the construction of an open rock wool culture bed and the placement of plantation pad;

(2) Exchange ideas on the plan and revise it;

(3) Construct and place the rock wool plantation pads on site according to the plan;

(4) Plant the seedlings with rock wool blocks;

(5) Summarize and evaluate through comparison.

[Expansion Task]

I. Review Questions

(1) What is the deep flow technique? What are its characteristics?

(2) What acid is appropriate for the plantation trough or reservoir treatment? Why?

(3) What benefits can be brought by adding non-woven fabric at the bottom of the plantation trough of nutrient film technique?

(4) Why does the nutrient film technique adopt intermittent liquid supply?

(5) What are the similarities and differences between the deep flow technique and the nutrient film technique?

(6) What parts are included in the facility structure of aeroponics? How do we manage them? What is the difference between aeroponics and hydroponics?

(7) Explore the specific application of several soilless culture methods in production.

(8) What are the two common types of bag culture? What problems should we pay attention to during bag culture?

(9) What advantages do you think the substrate culture has over hydroponics?

(10) What are the differences between the facility structures and culture management techniques of various kinds of substrate culture?

(11) What are the characteristics of rock wool culture? What are the differences between open rock wool culture and recycling rock wool culture?

(12) How do we make a rock wool plantation bed? How do we determine the liquid supply amount of rock wool culture?

(13) Do you understand the meaning of stereoscopic culture? Compared with flat culture, what is the application value of stereoscopic culture?

(14) What are the types of stereoscopic culture and their characteristics?

(15) What are the characteristics of the facility structure of column pot soilless culture? How do we construct the facility structure? What problems should we pay attention to during cultivation?

(16) What are the common stereoscopic culture methods in production? What are their characteristics?

II. Case study

Cultivating Vegetables in the Desert with a New Soilless Culture Technique

In Kuwait, the air temperature soars to 60℃ or even higher in summer, and almost no grass grows in such hot and arid conditions, thereby making Kuwait almost completely dependent on imported food. Fortunately, a new technique has made eating locally cultivated vegetables no longer a luxury for Kuwaitis.

Dr. Glenn Carrigan, an agricultural expert from Australia, and Dr. Abdulla from Kuwait applied a new soilless culture technique to ensure that the vegetables can survive the local hot summer. The vegetables are placed in special soilless containers in the greenhouse. These containers are covered with a special light shield, which can reflect about 50% of the sunlight, thereby lowering the temperature by about 10°C and greatly reducing the damage to vegetables caused by the heat of solar radiation.

Dr. Glenn Carrigan said: "We also increased the humidity in the air. This is because the normal humidity in the desert during summer is 7%–11%, which is very dry. Increasing the humidity in the air can lower the temperature. Therefore, by increasing water vapor in the air, it can bring the temperature down from 40–50°C to 20–30°C. This is very good for the growth of vegetables." The water used for humidification is collected in the water tank for recycling, which is necessary in Kuwait where water is an expensive resource. At present, the water used in the farm is collected from 40-m-deep underground and is usually slightly salty. Dr. Abdulla said: "Water is very valuable in Kuwait. In order to improve water quality, we will thoroughly purify the water to get freshwater, and we hope we can use it for fish farming in the near future. The vegetables can be planted outdoors in winter and then be transferred into the greenhouse when the temperature rises in March. These vegetables are grown almost in a sterile environment, so there is no need to spray them with any pesticides."

This is the first farm that adopts the soilless culture technique in Kuwait, and vegetables such as lettuce and celery are planted in a greenhouse of 2,000 m^2. Freshly picked vegetables can be put on the market within three hours for customers to choose from every day. Dr. Abdulla also hopes that the production scale can be expanded as soon as possible so as to increase the yield by 3–4 times.

Module 7　Soilless Culture Protection Facilities and Environmental Control

[Learning Objectives]

I. Knowledge Objectives

(1) Learn the type and main characteristics of soilless culture protection facilities and select the appropriate type of facilities according to production needs;

(2) Learn the characteristics and influencing factors of environmental factors in the facility;

(3) Master the objectives and principles of comprehensive regulation and control of the facility environment;

(4) Master the specific regulation and control measures for environmental factors in the facility environment.

II. Skill Objectives

(1) Be able to correctly observe the changes of various environmental factors in the facility;

(2) Be able to master the environmental change characteristics of the main facility types in this area and carry out effective regulation and control.

[Preparation for Learning]

I. Required Resources

(1) A library and a periodicals reading room;

(2) Environmental regulation and control in the glass greenhouse;

(3) Polytunnel, solar greenhouse, and other cultivation facilities;

(4) Select the appropriate type of protected cultivation.

II. Background Knowledge

(1) Master the basic knowledge related to soilless culture technique;

(2) Master basic knowledge of literature review;

(3) Learn about some basic principles of soilless culture technique;

(4) Understand the knowledge of agricultural meteorology and agricultural regulations.

[Learning Tasks]

Task 1 Selection of Protection Facilities

Soilless culture is an effective technique in protected cultivation. Compared with soil culture, it has high yields, quality, and efficiency, which lies not only in the function of regulating and controlling the crop rhizospheric environment of the soilless culture facility itself, but also in environmental protection facilities such as supporting greenhouses and polytunnels, so that the above-ground growing conditions of crops are in the optimal state as underground. Soilless culture facility itself does not have the function of anti-season and full-year crop production. Anti-season cultivation or full-year supply can only be realized under protection facilities such as greenhouses and polytunnels, thus improving the utilization rate of facilities. Therefore, soilless culture facilities and greenhouse facilities are two different facilities with close association. Soilless culture production is generally carried out in protection facilities under comprehensive environmental regulation and control.

I. Types and Classification of Protection Facilities

Environmental protection facilities refer to cultivation facilities with a transparent covering that allow personnel to enter and regulate and control environmental factors such as temperature, light, moisture, and air in the cultivation space. Protection facilities are generally classified into glass greenhouses and plastic greenhouses by covering material. Plastic greenhouses can be further classified into hard-plastic (such as PC, FRA, FRP, and composite plates) and soft-plastic (such as PVC, PE, and EVA film) greenhouses according to the type of plastic; single-span and multi-span greenhouses by span; double-roof, single-roof, uneven-double-roof, and arched-roof greenhouses by roof.

The currently commonly used environmental protection facilities for soilless culture in China mainly include solar greenhouses, polytunnels, modern greenhouses, canopies, and sunshade nets. Polytunnels have been widely used throughout China, especially in southern China, where polytunnels are developing towards tall multi-span polytunnels on a large scale. Solar greenhouses are primarily promoted and applied in the northern region of 33°N to 46°N and the alpine region of −15°C to −20°C, where production of warmth-loving fruits and vegetables in winter is basically realized without heating. Since the 1990s, large modern greenhouses have been introduced from abroad. Through learning, digestion, and absorption, China has developed its own modern greenhouses which have promoted the scientific and technical progress of greenhouse horticulture and the modernization of agriculture in China, offering great opportunities for the development of soilless culture techniques. Summer protection facilities such as canopies and sunshade nets play an important role in overcoming the threat of catastrophic climate and adverse environments such as frequent high temperatures, rainstorms, typhoons, diseases, and pests in hot summer in southern areas, and provide a simple and effective new way to alleviate the shortage of horticultural products in off seasons (summer and autumn) in southern areas, extending the season of soilless culture from winter and spring to summer and autumn. Due to the different natural, climatic, economic, technical, labor, and other conditions in different places, facilities vary in type, structural material, and scale, and should be selected and sited based on local conditions.

(I) Polytunnel

Generally, only large protection cultivation facilities with plastic film covering and bamboo, wood, cement, or steel-pole frameworks are called plastic film polytunnels, or polytunnels for short. According to roof form, plastic tunnels can be divided into two categories: arched and ridge-shaped, with the former mostly seen. According to the framework material, plastic tunnels can be classified into the following categories: bamboo (wood)-structured, steel-frame (pipe)-structured, steel-bamboo-structured, reinforced-concrete-structured, and inflation-type tunnels; according to the connection method, they can also be divided into single-span, double-span, and multi-span polytunnels. The polytunnel has simple facilities, generally without environmental regulation and control equipment, and relies on natural light for production. Therefore, it develops rapidly in the warm southern region.

At present, polytunnels mainly include bamboo (wood)-structured polytunnels (Fig.7-1), bamboo (wood)-arched polytunnels with suspension of beam and column (Fig.7-2), reinforcement welded columnless polytunnels (Fig.7-3) and fabricated galvanized thin-wall steel pipe polytunnels (Fig.7-4). Among them, the fabricated galvanized thin-wall steel pipe polytunnel has a large inner space, no column and convenient operation, and is a nationally approved product. With uniform specifications, convenient assembly and disassembly, and easy film covering, it is widely used in production.

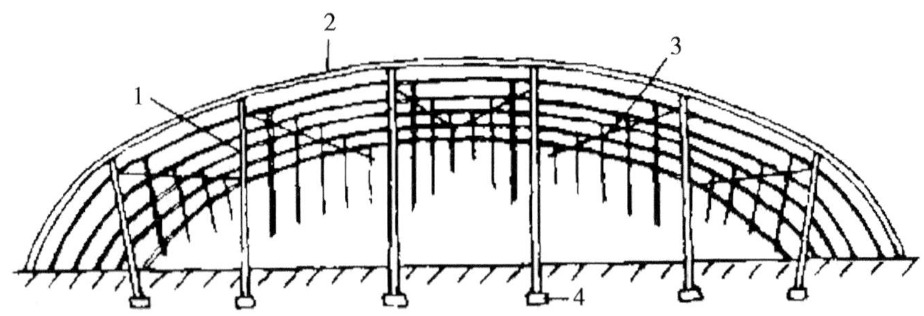

Fig. 7-1 Bamboo (Wood)-structured Polytunnels

1. Column 2. Arch bar 3. Pull rod 4. Column base

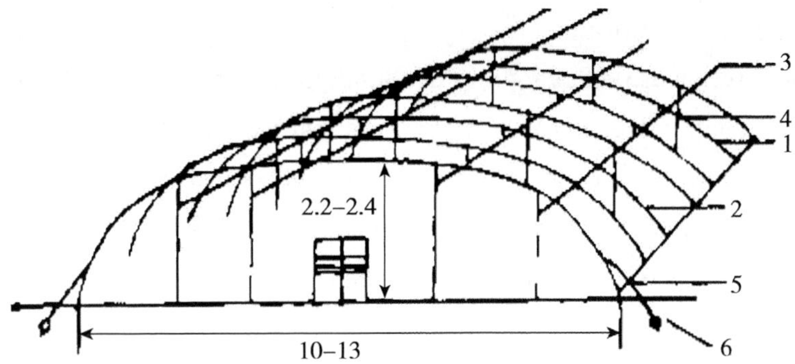

Fig. 7-2 Bamboo (Wood) - arched Polytunnels with Suspension of Beam and Column (Unit: m)

1. Column 2. Arch bar 3. Longitudinal pull rod 4. Suspended column 5. Tension rope 6. Ground anchor

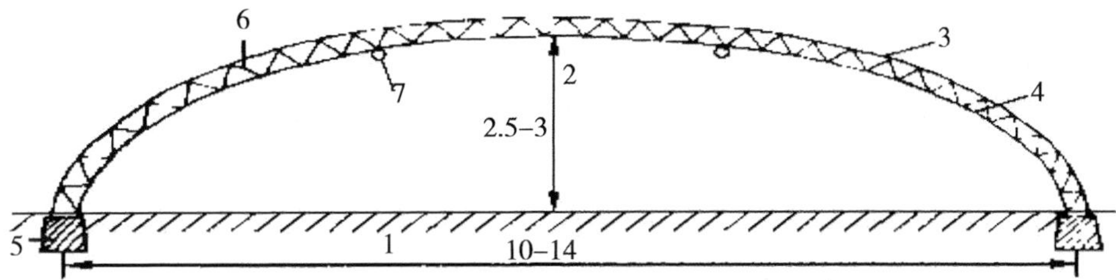

Fig. 7-3 Reinforcement Welded Columnless Polytunnels (Unit: m)

1. Width 2. Middle height 3. Upper chord 4. Lower chord 5. Cement pier 6. λ-shaped steel bar between upper and lower chords 7. Pull rod

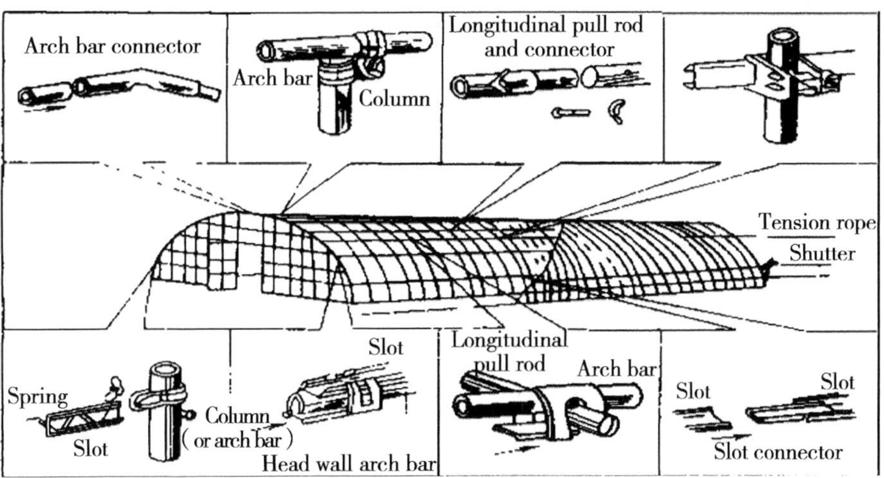

Fig. 7-4 Fabricated Galvanized Thin-wall Steel Pipe Polytunnels and Accessories

(II) Solar greenhouse

Solar greenhouse refers to a horticultural protection facility with three walls, an over 2-m-high ridge, and a span of 6–8 m that mainly relies on solar radiation energy as the heat source (including at nighttime). Most solar greenhouses use plastic film as the lighting covering, and take solar radiation as the heat source. They are provided with a wide range of lighting, thermal insulation, and cold-proof equipment and facilities such as maximum lighting, thickened walls, and back slopes, as well as cold-proof ditches, paper quilts, and straw mats to achieve the effects of heating and heat preservation, thus making full use of photothermal resources and weakening the impact of unfavorable meteorological factors. Solar greenhouses are generally not heated or are only occasionally heated as a supplementary measure. Solar greenhouses mainly include solar greenhouses with low back walls and long back slopes, solar greenhouses with high back wall and short back slope, stringed solar greenhouses, steel-bamboo-structured solar greenhouses, and full-steel-frame columnless solar greenhouses. Energy-saving solar greenhouses covered with plastic film are the main form of vegetable protection facilities in northern China. With low investment and high benefits, it is suitable for the current technical and economic conditions in rural China. It has obvious advantages in lighting, heat retention, low energy consumption and practicability, and it is a protected cultivation method that the northern farmers are willing to apply and whose area continues to grow year by year.

(III) Modern Greenhouse

The modern greenhouse is an advanced type of facility horticulture. The environment in the facility is automatically controlled by computers, basically free from the influence of disastrous weather and adverse environmental conditions, and the facility is a large greenhouse that can produce horticultural crops all-weather and full-year. At present, China's introduction of modern greenhouses mainly includes multi-ridge, multi-span, and small-roof Venlo-type glass greenhouses (Fig. 7-5) that were created in the Netherlands and then became popular worldwide; popular Richel plastic greenhouses created by Richel Group, France; full-open, multi-span, arched-roof plastic greenhouses that allow the film covering the top roof to be rolled up from bottom to top for ventilation; full-open type glass greenhouses (future greenhouses) created by Serre Italia,Italy. Typical modern greenhouses with an automatic control system designed and manufactured in China include double-layer inflatable multi-span plastic greenhouses, double-sloped glass greenhouses, North China large multi-span plastic greenhouses, South China large multi-span plastic greenhouses, arched multi-span plastic greenhouses, LGP-732 multi-span greenhouses, XA and GK greenhouses, and FRP (fiber reinforce plastic) multi-span greenhouses (Fig. 7-6).

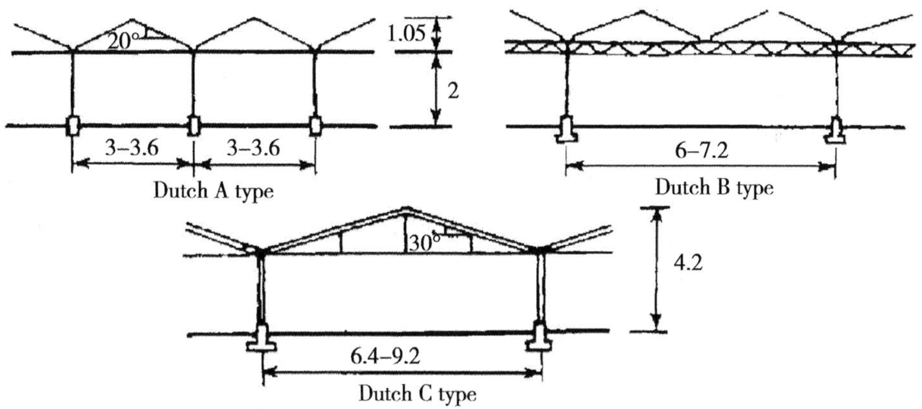

Fig. 7-5 Venlo-type Glass Greenhouses (Unit: m)

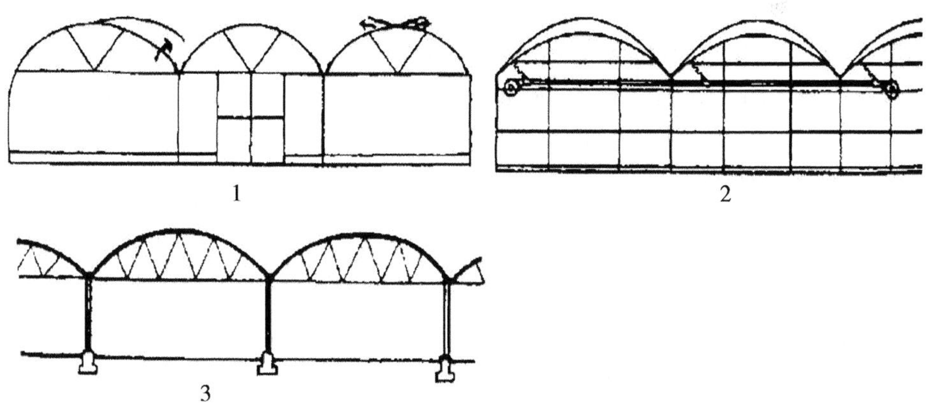

Fig. 7-6 Several Forms of Multi-span Greenhouse

1. Arched multi-span plastic greenhouse 2. Double-layer inflatable multi-span plastic greenhouse 3. FRP multi-span greenhouse

(IV) Plant Factory

A plant factory is a factory-like, fully enclosed facility, in which plants are cultivated with high efficiency, less labor, and great stability using artificial light sources to realize automated control of the environment inside. Plant factories are divided into three categories according to different light sources: artificial light type, natural light type, and combined type. As part of the controlled-environment agriculture, plant factories are at the highest level of horticultural protection facilities, and their management is highly mechanized and automated. Soilless culture and stereo planting of crops are carried out in large facilities. With changes in environmental elements transmitted to the computer through sensors, the operation of various environmental regulating equipment and the production process can be accurately controlled through calculation. The required temperature, moisture, light, water, fertilizer, air, and the like are optimally configured according to the requirements of plant growth. In addition to computer monitoring, fully enclosed production management is carried out by robots and manipulators to realize streamlined production from sowing to harvest, completely removing the natural constraints. Thereby, it makes rational use of space and facilities to achieve economical and efficient production as well as all-weather, seasonless, and pollution-free production. A plant factory is characterized by its high integration, efficient production, high merchantability, and high investment.

Fig. 7-7 shows a medium-sized double-span glass greenhouse from an M-style hydroponics laboratory in Japan. It is 60 m long, 50 m wide, and 4 m high, and each unit area is 3,000 m^2. In terms of environmental control, monitoring is carried out by indoor and outdoor sensors, including sunshine sensors (to measure sunshine time), outdoor temperature sensors, and indoor temperature sensors, as well as wind speed and direction sensors and rainfall sensors, which jointly control the opening and closing of the skylight, side windows, and canopy, as well as the operation of cooling and heating supply units. Humidity sensors control the operation of the mist sprayer and dehumidifier. The nutrient solution is supplied by the nutrient solution tank group (4 pcs) and water pump. Conductivity meter, pH meter, solution temperature sensor, water level meter, etc. control nutrient solution composition, solution temperature, pH value, and balanced timing solution supply. A CO_2 sensor is used to measure the CO_2 concentration in the greenhouse and send a signal to the CO_2 generator to maintain the CO_2 concentration in the greenhouse. All the above data are input into an electronic computer, and various parameters are set according to the cultivated crop types to achieve full automatic control. The similar greenhouse introduced in China is the soilless culture greenhouse in the Nanjing Vegetable Science Research Institute.

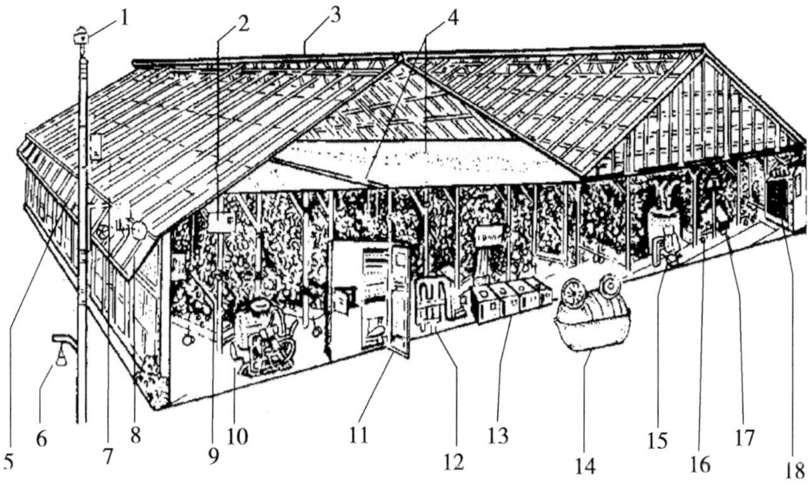

Fig. 7-7 Medium-sized Double-span Glass Greenhouse

1. Sunshine sensor 2. Temperature sensor 3. Skylight and opening device 4. Spray heating devices 5. Side window and opening device 6. Outdoor air temperature sensor 7. Rainfall sensor 8. Wind direction and speed sensor 9. CO_2 sensor 10. Humidifier 11. Computer control cabinet 12. Nutrient solution pump 13. Four liquid fertilizer barrels 14. Nutrient solution tank 15. Cooling and heating air control 16. Room temperature sensor 17. CO_2 generator 18. Nutrient solution heating device

(V) Canopy and Sunshade Net

In summer, abundant rainfall, strong light exposure, high-temperature stress, as well as frequent diseases and pests inhibit the growth of vegetables and other crops, resulting in unstable production. The simplest canopy cultivation is only to cover the top (canopy) and remove the side film (apron) of the polytunnel framework so that it can both prevent rain and realize ventilation in summer. If the top is covered with silver-gray or black sunshade net, the strong light can be weakened, and the substrate temperature in the tunnel can be reduced by 8–12°C at noon in summer, which can effectively reduce the harm caused by high temperatures, and can be used for anti-season cultivation of leaf vegetables and root vegetables in summer. The existence of sunshade nets and canopies, which are auxiliary facilities for other protection facilities in summer, enables soilless culture in polytunnels and solar greenhouses even in summer. For tomatoes, cucumbers, muskmelons, lettuce, spinach, and other vegetables that are difficult to cultivate in summer, their over-summer rainproof cultivation has become a reality.

II. Structure and Performance of Main Environmental Protection Facilities

(I) Polytunnel

1. Structure of polytunnels

A polytunnel has a simple structure that is mainly composed of arch frames, longitudinal beams, columns, gable wall columns, framework connecting clamps, and doors, and varies depending on construction materials. The arch frame is the main component of polytunnels that can withstand the wind, snow, and other loads. It mainly comes into two types: single-rod type and truss type. The longitudinal beam is a component that ensures the longitudinal stability of arch frames by connecting them into one. It also comes into two types: single-rod type and grid type. When the cross-section of the arch frame is small and cannot withstand wind and snow loads, or the span of the arch frame is large and the strength of the supporting structure is insufficient, columns should be set up in the greenhouse to directly support the arch frames and longitudinal beams to improve the overall bearing capacity of the polytunnels. Gable wall columns, i.e. head wall columns, are commonly upright, and arched or diagonal in areas with strong wind. For connections of the assembled polytunnels made of galvanized steel pipes and the RC-structured polytunnels, such as the connections between the arch frame and the gable wall column, between arch frames, and between the longitudinal beam and head wall arch, special prefabricated clamps are used. The polytunnels of a bamboo or wood structure are connected by threads, ropes, or iron wires. The door of the polytunnels is provided for management, transportation, and ventilation. It is generally located in the middle of the head wall of a single-span tunnel. For thermal insulation, the door can be set at the southern head wall. After the temperature rises, another door can be set at the northern end to strengthen ventilation. Such doors can be hinged doors, hanging-track sliding doors, and the like. Polytunnels for soilless culture should have a solid structure with strong resistance to wind and snow, adequate space for growth and a large area, and feature easy ventilation and environmental regulation and control. Preferably, a polytunnel should have a steel structure without or with few columns and should have a span of 8–12 m, a length of 40–60 m, and a ridge height of 2.4–3.0 m. At present, most are prefabricated galvanized thin-wall steel pipe polytunnels.

The polytunnel has approved products since the 1980s, mainly including the GP series developed by the Academy of Agricultural Planning and Engineering and the PGP series developed by the Shijiazhuang Institute of Agricultural Modernization, Chinese Academy of Sciences. The main product specifications are shown in Table 7-1.

Table 7-1 The Main Product Specifications of GP and PGP Series Plastic Tunnel Framework

Model	Structure size (m)					Structure
	Length	Width	Height	Shoulder height	Arch spacing	
GP-Y8-1	42	8.0	3.0	0	0.5	Single arch, 5 longitudinal beams, 2 longitudinal slots
GP-Y825	42	8.0	3.0	–	0.5	Single arch, 5 longitudinal beams, 2 longitudinal slots
GP-Y8.525	39	8.5	3.0	1.0	1.0	Single arch, 5 longitudinal beams, 2 longitudinal slots

continued

Model	Structure size (m)					Structure
	Length	Width	Height	Shoulder height	Arch spacing	
GP-C625-II	30	6.0	2.5	1.2	0.65	Single arch, 3 longitudinal beams, 2 longitudinal slots
GP-C825-II	42	8.0	3.0	1.0	0.5	Single arch, 5 longitudinal beams, 2 longitudinal slots
GP-C1025-S	66	10.0	3.0	1.0	1.0	Double arches with round top and square bottom, 7 longitudinal slots
PGP-5-1	30	5.0	2.1	1.2	0.5	Arch frame pipe diameter: 20 mm × 1.2 mm
PGP-5.5-1	30	5.5	2.6	1.5	0.5	Arch frame pipe diameter: 20 mm × 1.2 mm
PGP-7-1	50	7.0	2.7	1.4	0.5	Arch frame pipe diameter: 25 mm × 1.2 mm
PGP-8-1	42	8.0	2.8	1.3	0.5	Arch frame pipe diameter: 25 mm × 1.2 mm

2. Performance of polytunnels

The polytunnels are characterized by covering plastic film and are well-sealed and receive sufficient light in a uniform manner: short-wave radiation is easy to penetrate, and long-wave radiation is difficult to penetrate.

(1) Light: The vertical lighting intensity in the polytunnels gradually decreases from high to low, becoming the lowest near the ground. For horizontal lighting intensity in the polytunnels, it is relatively uniform in the north-south extended

polytunnels and is higher in the south side of the east-west extended polytunnels than in the middle and north side. Polytunnels of a single-span steel structure or a hard-plastic structure are well-lighted, and those of a single-span bamboo or wood structure or multi-span structure are poorly lighted. In addition, the larger the span of the polytunnel, the higher the frame and the weaker the light in the polytunnel.

(2) Temperature: In sunny weather, the tunnel experiences a rapid temperature rise during the daytime and preserves certain heat at night. Nevertheless, due to the lack of night heating measures and great daily and seasonal variation of temperature, most tunnels with no essential environmental regulation and control equipment have to rely on ventilation windows to adjust the temperature and humidity, resulting in great changes of indoor environmental elements. On sunny days, the temperature in the tunnels rises rapidly 1–2 hours after the sun comes out, and climbs linearly at a rate of 5–8°C per hour after 8–10 hours. During 11:00–13:00, the temperature rise slows down and reaches a peak, which is 20°C higher than the outdoor temperature. Then the indoor temperature begins to drop, and drops rapidly from 15:00 to17:00 at a rate of 5–6°C per hour. With the narrowing of the indoor and outdoor temperature difference, the nighttime drop rate will rapidly decline to about 1°C per hour. Early in the morning, the indoor temperature is only 3–5°C higher than the outdoor temperature. Sometimes, the indoor temperature is lower than the outdoor temperature, which is called temperature inversion, and mostly occurs on a sunny and breezy morning in early spring or late autumn.

(3) Humidity: As the tunnel is well sealed by the plastic film, the humidity indoors is significantly higher than that outdoors, with an average daily increase of 35%–40% (Table 7-2). The northern region with marked seasonal changes has witnessed slow development of polytunnels because of a long low-temperature period in winter and limited utilization (only about one month earlier or later than in the open field). Therefore, polytunnels are widely applied in the southern warm regions or regions with a mild climate. For soilless culture, substrate culture is mainly used for fruit vegetables with long growth

period, and hydroponics is mostly used for producing leaf vegetables with a fast growth rate and a short growth period.

Table 7-2 Daily Variation of Air Humidity Inside and Outside the Polytunnels

Item	Place	Time (hour)												Average
		2	4	6	8	10	12	14	16	18	20	22	24	
Absolute humidity (g/m^2)	Open field	4.5	4.3	4.3	2.7	2.0	1.6	3.7	2.6	5.7	4.7	4.7	4.5	3.8
	Polytunnel	8.2	7.5	6.7	8.8	18.5	22.3	19.8	19	13.7	11.1	10.5	8.8	12.9
Relative humidity (%)	Open field	87	100	100	41	15	10	27	19	55	66	71	77	55.7
	Polytunnel	99	100	94	99	89	71	90	94	95	96	100	96	93.7

(II) Solar Greenhouse

1. Structure of solar greenhouses

The solar greenhouse is a single-roof greenhouse, commonly known as the winter-warm tunnel, which is the largest protection production facility in the north of Huaihe River in China. The framework of the solar greenhouse consists of a back wall, a back slope, a front roof, and two gable walls. The length, width, size, thickness, and material of each part determine its lighting and thermal insulation performance. The overall requirements are good lighting and thermal insulation, with low cost, easy operability, and high efficiency. Table 7-3 shows the structural parameters of common solar greenhouses in China.

Table 7-3 Structural Parameters of Common Solar Greenhouses in China (Unit: m)

Type	Internal span	Ridge height	Height of back wall	Thickness of back wall	Length of rear roof	Height of front window	Location
Beijing improved type	5.15	1.75	1.35	0.7	2.15	1.1	40° N
Tianjin three-fold type	6.50	2.00	1.70	0.7	1.05	0.8	40° N

continued

Type	Internal span	Ridge height	Height of back wall	Thickness of back wall	Length of rear roof	Height of front window	Location
Liaoning Haicheng type	6.00	2.70	0.50	1.5	2.70	0.8	41°N
Shandong Shouguang type	7.20	3.00	2.00	1.8	0.90	0.8	37°N
Anshan type II	6.00	2.80	1.80	0.5	1.60	1.3	41°N
Liaoning shenyang type I	7.50	3.50	2.20	0.5	2.00	0.8	42°N

The basic structure of the solar greenhouse is generally facing south and extending from east to west, with a large light receiving surface and light inlet angle. Solar energy is the main energy source of greenhouses. To improve the thermal insulation, the hollow wall structure can be used for the north, east and west gable walls of the solar greenhouse, and the rear roof can also be built into a thermal insulation structure with multi-layer composite materials. The first-generation solar greenhouses are mostly constructed with bamboo/wood and earth walls at a low cost. The (front, middle, and rear) columns inside are generally arranged every 3 m, and the space between them is called a partition. The second-generation solar greenhouses are mostly made of steel and brick walls, without columns inside and feature enlarged space, better operability, firm structure, improved resistance to wind and snow, longer service life, higher construction cost, and lower depreciation cost compared with the first-generation ones. Some may also be of a steel-bamboo structure, a GRC framework structure or other structures.

2. Performance of solar greenhouses

Solar greenhouses are mainly characterized by thermal insulation. From the perspective of structure, there is less light entering than polytunnels, especially less indoor scattering light. The indoor light is unevenly distributed, with strong light in the front and weak light in the back (Table 7-4). Meanwhile, the coverage of heat preservation straw curtain will affect the lighting time and can reduce it by more than 2–3 h/d in winter. Generally, it is covered with a straw curtain when the temperature is reduced to about 18°C (about 16:00), and the straw curtain is removed until about 8:30 of the next day. The minimum temperature can be maintained at 8–13°C, and the average cooling rate is 0.4–0.6°C/h, which is far lower than that of polytunnel (1°C/h), making the daily temperature range of solar greenhouse significantly lower than that of polytunnel, the minimum temperature increased significantly (5–8°C), and the growth period of crops 35–45 days earlier (in spring) or later (in autumn) than that of polytunnel. The humidity in the solar greenhouse is similar to that in the polytunnel. Due to its small volume, the temperature and humidity are greatly affected by the variation of radiation. The humidity drops rapidly in the quick temperature rise stage in the morning, even to about 15%. In the quick temperature drop stage in the afternoon, the relative humidity increases rapidly, and exceeds 80% around 16:00. Fog often occurs indoors, condensing on plants and inducing diseases. Meanwhile, due to the uneven indoor temperature distribution, there is a certain local humidity difference, and the southern side with fast heat dissipation has higher humidity in the morning and evening.

Table 7-4 Horizontal Distribution of Lighting Intensity in Different Types of Solar Greenhouses

Greenhouse type	Lighting at different indoor parts						Outdoor (10,000 lx)
	Front		Middle		Back		
	Lighting intensity (10,000 lx)	Light transmittance (%)	Lighting intensity (10,000 lx)	Light transmittance (%)	Lighting intensity (10,000 lx)	Light transmittance (%)	
Haicheng type	1.97	85.2	1.48	64.0	1.01	43.7	2.31
Anshan type II	2.93	73.8	2.76	69.5	1.95	49.0	3.97
No back slope type	1.77	85.0	1.44	56.5	0.95	45.8	2.07

(III) Modern Greenhouse

Modern greenhouses can achieve comprehensive control of various environmental factors and meet the requirements for crop growth and development. If polytunnels and solar greenhouses are protection facilities that rely on natural conditions, modern greenhouses are protection facilities that create natural conditions. In addition to the large main structure, there are various environmental regulation devices inside, including heating systems, heat preservation systems, cooling systems, humidification or dehumidification systems, CO_2 application systems, forced ventilation systems and natural ventilation systems, light supplementation or shading systems, fertilizer and water supply systems, drainage and rainwater collection systems, protection systems (insect-proof screens, snow removal equipment), weather stations, power systems, and control systems. The construction cost varies greatly due to the adoption of different systems and equipment models. Expanding the greenhouse scale is the main way to reduce the cost of equipment per unit area, and connecting single-span greenhouses is an effective way to realize that. Using large greenhouses will be the development trend.

Modern greenhouses have a service life of 20–50 years and are mostly truss structures made of hot-dip galvanized steel or aluminum profiles with concrete as foundations. Generally, there is a large lighting area that allows for more lights to enter in winter for those extending in the east-west direction, and a uniform indoor temperature and light distribution for those extending in the south-north direction. Due to different climatic characteristics in different countries and regions, especially the disastrous weather conditions such as wind, snow, and hail, there are great differences in greenhouse design. The greenhouse structures produced by different countries and some domestic manufacturers are shown in Table 7-5.

Table 7-5 Specifications of Greenhouse Structure Produced by Some Manufacturers at Home and Abroad(Unit: m)

Greenhouse type	Length	Span	Ridge height	Shoulder height	Spacing	Roof shape	Producer or designer
North China type	33	8.0	4.5	2.8	3.0	Round arch type	China Agricultural University

continued

Greenhouse type	Length	Span	Ridge height	Shoulder height	Spacing	Roof shape	Producer or designer
Korean type	48	7.0	4.3	2.5	2.0	Round arch type	Korea
Israeli type	–	7.5	5.5	3.75	4.0	Round arch type	AZROM
France	–	8.0	5.4	4.2	5.0	Round arch type	RICHEL
Slope type	48–96	6–8	–	–	6 or 4	Elliptic arch type	Shanxi Agricultural University

Modern greenhouses generally have a large building area and strong environmental regulation and control ability, but their indoor ventilation and cooling in summer are poor. Studies have shown that when the total span of the greenhouse exceeds 40 m, a high-temperature canopy zone will be formed in the middle, affecting CO_2 transmission and plant growth. For this reason, additional forced ventilation equipment or micro-spray system and cooling system (sunshade or spray) are required. In addition, external thermal insulation is difficult for large greenhouses, because the temperature drops rapidly at night, and the heating energy consumption is higher in winter. To save energy, the internal thermal insulation system is often adopted. During the cultivation of flowers and some vegetables, light regulation is required for regulating the anthesis. Therefore, supplementary lighting equipment should be added to the greenhouse. To control the occurrence of diseases, the ground shall be covered with film to reduce the air humidity, and the ground is often hardened in soilless culture, which leads to a lack of CO_2 sources. CO_2 generators are necessary to supply CO_2 and meet the demand for photosynthesis. To prevent insect infestation, insect-proof screens (nylon nets with about 30 mesh) shall be arranged at the air vents and access of the greenhouse. In order to maintain a long-term stable environment, ventilation, humidity regulation, and heating systems with matching power are also required. Comprehensive environmental regulation and control is a complicated and systematic work. Manual control and mechanical control cannot meet the requirements of economic production, and the computer control system is the main form used in modern greenhouses. The modern greenhouse is relatively well-equipped to control the greenhouse environmental factors according to the requirements of crop growth and development so as to achieve full-year production and maintain stable, high-yield, and high-quality crop production in the greenhouse. However, the construction cost of a greenhouse is relatively high. When combined with soilless culture technique, it can better exert the production efficiency of greenhouse facilities, making the greenhouse the main place for the development of soilless culture.

Task 2 Temperature Regulation and Control Technique

I. Characteristics of Temperature Variations in Facilities

The source of temperature in the unheated greenhouse is mainly the greenhouse effect caused by solar radiation. The temperature variation in the greenhouse is characterized by the fact that the temperature in the greenhouse varies with the external solar radiation and temperature, including seasonal and daily variation, and that the temperature difference between day and night is large and the local temperature difference is obvious.

1. Seasonal and daily variations

In the northern region, there are marked seasonal variations in the protection facilities. When compared with open fields, the winter in the solar greenhouse can be shortened by 3–5 months and summer can be extended by 2–3 months, and spring or autumn can also be extended by 20–30 days. Therefore, in the region between 41°N and 33°N, the high-efficiency and energy-saving solar greenhouse (the temperature difference between indoor and outdoor is kept at about 30°C) can produce thermophilous fruit vegetables in four seasons. While the winter in polytunnel is only shortened by about 50 days than the open field, the spring and autumn are only extended by about 20 days shorter than the open field and the summer is barely extended. So, fruit vegetables can only be cultivated in advance in spring and late in autumn, and fruit vegetable production in winter and spring is only possible under multiple covers. From the daily variation, the maximum and minimum temperatures of unheated greenhouses occur in winter and spring in the North slightly later than in the open field, but the daily temperature difference indoors is significantly greater than that in the open field. In northern China, because of its good lighting and heat preservation, the energy-saving solar greenhouse has a daily temperature difference of 15–30°C in winter. It can produce thermophilous fruit vegetables even without heating or basically without heating at about 40°N.

2. Temperature inversion in facilities

Generally, the temperature in greenhouses is higher than that outside, but in plastic tunnels or glass greenhouses without multiple covering, the cooling rate after sunset is often faster than that in the open field. In the case of cold air, especially on the first sunny and breezy night after the strong north wind, the greenhouse and polytunnel radiate heat more intensely outward through covering at night. The indoor temperature is often 1–2°C lower than the outdoor temperature, because the heat cannot be supplemented due to the blocking of the covering. Temperature inversion generally occurs in the early morning and is most harmful in spring.

3. Uneven distribution of air temperature and ground temperature

Generally, the upper part has a higher air temperature than the lower part, the middle part has a higher air temperature than the surrounding parts, and the north side of the solar greenhouse in the north region has a higher air temperature than the south side at night. The smaller the area of protection facilities, the larger the proportion of low-temperature areas and the more uneven the distribution. However, the variation in ground temperature, regardless of seasonal and daily variation, is smaller than the variation in air temperature. The ground temperature around the greenhouse is lower than that in the middle; the temperature variation on the ground surface is greater than that in the ground, but with the increase of soil depth, the variation of ground temperature becomes smaller and smaller.

II. Regulation and Control of Temperature Conditions in Facilities

Temperature is the primary environmental condition for the protected cultivation of crops. Any crop requires a certain temperature range, i.e. the minimum and maximum temperatures for its growth and development and the maintenance of life; the temperature and day-night temperature variation will affect the growth and development, plant morphology, as well as the yield and quality of crops. Therefore, temperature is the primary environmental condition for the protected cultivation of crops, and it is used by producers as the primary means of controlling the growth of greenhouse crops. However, considering all factors, the management temperature of economic production is different from the optimal temperature of crop growth, and the management temperature shall be determined to make the crop production suitable for the market demand, so as to obtain the maximum benefit.

Manually creating a stable temperature environment is an important guarantee for stable growth and long-season production of crops. The size, orientation, interception of light energy, wind speed, and temperature of the site will affect the stability of greenhouse temperature. The temperature environment in the facility is generally regulated and controlled through thermal insulation, heating, cooling, and other means.

(I) Thermal Insulation Measures

1. Thermal insulation of solar greenhouse

(1) Use multiple covering: It is relatively simple and easy to provide tunnels and a second curtain in the greenhouse, or cover the transparent covering with straw curtains, paper quilts, thermal insulation quilts, cotton quilts, etc. for external thermal insulation.

(2) Improve the light transmittance of greenhouse: A reasonable design of greenhouse orientation and roof slope can keep the roof clean and minimize the shadow of construction materials. Covering materials with high light transmittance shall be used to improve the light transmittance of greenhouses and increase the total heat storage capacity of facilities.

(3) Increase the thermal insulation ratio: The thermal insulation ratio is the ratio of the land area to the surface area of the protection facilities covering and the surrounding materials. That is, the higher the protection facilities, the smaller

the thermal insulation ratio and the worse the thermal insulation effect. However, because of the thick back wall and back slope of the solar greenhouse (similar to land), increasing the height of the solar greenhouse has less influence on the thermal insulation ratio, but it is conducive to adjusting the roof angle, improving the light transmission and increasing the indoor solar radiation, thus raising the temperature.

(4) Provide thermal insulating walls: Reinforce back slopes and insulate with polystyrene foam board.

(5) Provide cold-proof ditch: The cold-proof ditch is provided around the solar greenhouse. Generally, the cold-proof ditch is 30–50 cm wide and 50–70 cm deep. The ditch is filled with rice husk, wormwood, and other materials that have low thermal conductivity.

(6) Keep the greenhouse relatively closed: Minimize ventilation.

2. Thermal insulation of large greenhouse

(1) Use double-layer gas-filled film or double-layer polyethylene board: Thermal insulation of transparent roof takes advantage of the low thermal conductivity of still air.

(2) Provide two-layer insulation curtain: The development and application of two-layer insulation curtains play an important role in the thermal insulation of large greenhouses. Insulation curtain materials include film, fiber, textile materials, non-textile materials (non-woven fabrics), and their combination. At present, the insulation curtain materials used mainly include three raw materials: polyethylene, polyester fiber, and acrylic fiber. There are two main ways to hang the insulation curtain, namely permanent curtain and semi-fixed curtain system (partially movable) or movable curtain system (equally movable).

(3) Provide vertical insulation curtain: In recent years, the thermal insulation (vertical curtain) on the sides of the greenhouse has also attracted attention. Aluminum foil reflective materials are used abroad to make a folding curtain with folds or a rolling and sliding curtain. In China, double-layer film or glass is used around the greenhouse, and a wall structure is used on the north side for thermal insulation.

(II) Heating Measures

When the temperature of production facilities is low and the growth of crops is slow in winter, appropriate heating measures can be taken, such as air heating, substrate heating, and nutrient solution heating.

1. Air heating

The heating methods include hot water heating, steam heating, flue heating, hot air heating, etc. Hot water heating is characterized by good heat stability, stable and uniform room temperature, small fluctuation, safe and reliable production, and large heating loads, making it a commonly used heating method in large and medium-sized greenhouses; steam heating and hot air heating work rapidly but show poor temperature stability; flue heating feature low construction and operation costs, but also low thermal efficiency.

2. Substrate heating

Substrate heating methods include heating with fermented materials, electric heating, and water heating, and electric heating is most commonly used. Special electric heating wires are used for electric heating, which are convenient to bury and remove with high utilization efficiency of thermal energy. It is easy to realize high-precision control with a temperature controller, but the power consumption is high. Electric heating wires have a short service life and are generally used in seedbeds.

3. Nutrient solution heating

Too high or too low solution temperature will inhibit the growth of crops. Improving the growth conditions of crops by adjusting the solution temperature is much more economical than heating or cooling the whole polytunnel or greenhouse. For NFT hydroponics in winter, a stainless steel spiral pipe can be installed in the reservoir to heat the nutrient solution with hot air, or an electric heating tube can be used to heat the nutrient solution, with a temperature controller to control the temperature of the nutrient solution. In winter and spring, heating the nutrient solution can generally increase the yield by more than 5%. In order to ensure the stable temperature of the nutrient solution and save energy, the solution supply pipeline shall be subject to heat insulation treatment, i.e. coating the pipeline with tinfoil, rock wool and others.

In addition to the above heating methods, geothermal, factory waste heat, underground latent heat, municipal waste fermentation heat, solar energy, and other methods can also be used to heat the facilities. Sometimes temporary heating can be used, such as burning charcoal, sawdust, or smoke.

(III) Cooling Measures

In summer, facilities can be cooled by reducing the heat input and increasing the heat output, such as shading with sunshade nets, spraying coating on transparent roofs, ventilation, spraying, installing wet curtain fan systems,etc.

1. Shade cooling method

In summer, strong light and high temperature are the restrictive factors for crop growth. Shade cooling can be achieved by shading with sunshade nets or shading curtains, including internal shading and external shading. External shading is to hang a shading curtain at a distance of about 40 cm outside the greenhouse and polytunnel roof, which is very effective for cooling the greenhouse. When the shading rate is 20%–30%, the room temperature can be reduced by 4–6°C accordingly. Internal shading is to install a sunshade net in the greenhouse for cooling.

2. Roof running water cooling method

The running water layer formed on the roof can absorb about 8% of the solar radiation projected on the roof, and can absorb heat to cool the roof. The room temperature can generally be reduced by 3–4°C. In areas with hard water, soften the water before use.

3. Evaporative cooling method

The air is cooled by water evaporation before being sent to the greenhouse for cooling. There are roughly three types.

(1) Wet-curtain fan cooling system: A 10-cm-thick paper pad window or coir pad window is provided in the air inlet of the greenhouse, which is continuously wetted by water. The other end of the greenhouse is provided with an exhaust fan so that the air is cooled through the wet pad window before entering the greenhouse(Fig. 7-8). Experiments have shown that the wet-curtain fan cooling system can reduce the room temperature by 5–6°C. The disadvantage of the

wet curtain cooling system lies in the fact that it will generate dirt and algae on the wet curtain and cause certain temperature and humidity differences in the greenhouse. In regions with high humidity, its cooling effect will be significantly reduced.

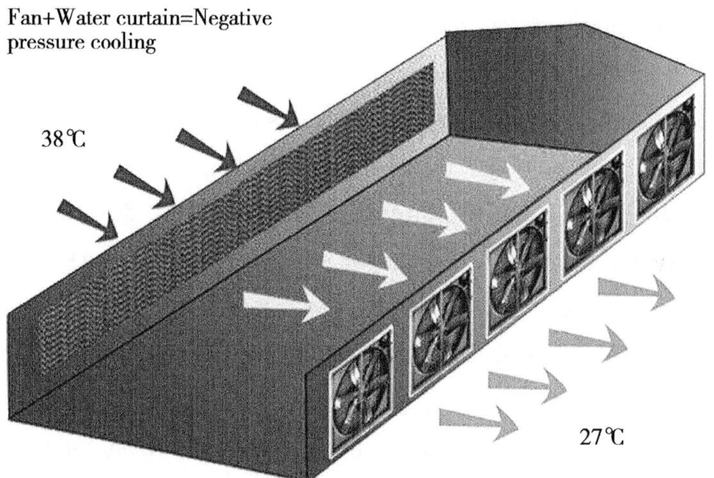

Fig. 7-8 Greenhouse Wet-curtain Fan Cooling System

(2) Fine mist cooling method: The floating fine mist with a diameter of less than 0.5 mm is sprayed indoors, and then evaporated to cool the full greenhouse by forced ventilation airflow. The indoor temperature can be reduced evenly when the mist is appropriate. This cooling method has a better effect than the wet-curtain fan cooling system, especially for some foliage plants, many of which will be "burnt" in the environment of high-temperature airflow generated by fans. Note that fine mist cooling is only suitable for high-temperature-resistant vegetables or flower crops.

(3) Roof spraying method: Continuously spray and wet the external surface of the whole roof so that the temperature of the roof is reduced close to the outdoor wet-bulb temperature, and the cooled air is convected downward under the roof.

When the water quality in the above method is not good, the scale left after evaporation will block the nozzle and wet pad, and the water quality shall be subject to treatment.

4. Ventilation

Ventilation is an important means of cooling. The principle of natural ventilation is ventilating from small vents to large ones, from middle vents to top ones, and finally to bottom ones (these vents are closed in reverse order). The principle of forced ventilation is that the airflow should be kept away from the plants to reduce their impact on them and that many small vents are better than a few large ones. In winter, an exhaust fan is used to realize outward heat dissipation, thus preventing cold air from blowing the plants and frostbiting them. In summer, the cold air can be evenly sent to the vicinity of the plant through the perforated pipeline. During ventilation, the crop can also be sprayed directly to reduce the temperature of the crop surface through the evaporation of leaf water.

5. Nutrient solution cooling

Stainless steel spiral pipes can be installed in the reservoir to cool down by circulating groundwater.

Temperature is the primary environmental condition for the protected cultivation of crops. Any crop requires a certain temperature range, i.e. the minimum and maximum temperatures for its growth and development and the maintenance of life. Temperature is directly related to the growth, flower bud differentiation, and flowering of crops, and day-night temperature variation will affect plant morphology, product yield, and quality. Therefore, producers should first take temperature as the main means of controlling the growth of greenhouse crops. Considering all factors, it is clear that the optimal temperature of crop growth is different from that of economic production, and the management temperature shall be determined to make the crop production suitable for the market demand so as to obtain the maximum benefit.

A stable temperature environment is an important guarantee for stable growth and long-season production of crops. The size, orientation, interception of light energy, wind speed, and temperature of the site will affect the stability of greenhouse temperature. The temperature environment in the facility is generally regulated and controlled through thermal insulation, heating, cooling, and other means.

Task 3 Light Regulation and Control Technique

I. Light Conditions and Influencing Factors of Facilities

(I) Light Conditions of Protection Facilities

The light conditions in the protection facilities include light intensity, light quality, light duration, and light distribution, which have different influences on the growth and development of greenhouse crops. When compared with light conditions in the open field, the light conditions in the facility have the following characteristics.

1. Low total radiation and weak lighting intensity

The total radiation in the facility mainly depends on the lighting intensity and light duration. The lighting intensity in the facility varies correspondingly with the external light in the daytime. It is relatively weakly affected by the type, aging degree, and cleanliness of transparent covering materials. During the protected production in winter and spring, for cold-proof and thermal insulation, opaque covering materials such as cattail mats and straw mats are required. Affected by the uncovering time, the light-receiving hours in facilities such as greenhouses are obviously less than those in the open field. Therefore, due to the weak lighting intensity and short light duration, the total radiation in the protection facilities is low, only 50%–80% of the outdoor radiation. This phenomenon often becomes the main limiting factor for producing heliophilous crops in winter.

2. Great variation in light quality

Light quality refers to spectral components. In open cultivation, the sunlight directly irradiates the crops, and the light components are consistent without differences in light quality. However, the composition of the radiation wavelength in the protection facility is very different from that outdoors due to the different transmittance of transparent covering materials to light radiation of different wavelengths. Generally, the transmittance of ultraviolet light is low. But when short-wave solar radiation enters the facility and is absorbed by crops, soil, or substrate and then radiates outwards in the form of long waves, it is blocked by covering materials such as glass or film. Therefore, the infrared long-wave radiation in the whole facility increases, which is also an important reason for the thermal insulation of the facility. The characteristics of covering materials such as plastic film, glass, and rigid plastic sheets directly affect the light quality components in the facility.

3. Extremely uneven light distribution

The amount of solar radiation in the greenhouse, especially the total daily amount of direct light, is extremely unevenly distributed in different parts of the greenhouse, at different orientations and at different times and seasons. The uneven distribution of light in the facilities in time and space causes inconsistent growth and development of crops. Especially for facilities in high latitudes, where the light is weak and the light duration is short in winter, the growth and development of crops are more seriously affected.

(II) Factors Affecting Light Conditions in Facilities

1. Roof angle and roof shape

When other factors are consistent, the roof angle (the angle between the line connecting the front window and the ridge and the ground plane) has a great influence on lighting. For a particular area, the solar altitude changes regularly, so the roof angle becomes the most closely related factor of the incident amount. Especially for the solar greenhouse with a small capacity, more light will enter the greenhouse at a larger roof angle, but at the same time, the ridge will rise to increase the construction cost and the corresponding heat dissipation area, which is not conducive to thermal insulation. The roof angle of solar greenhouse in China is generally 18°–28°, and the average roof angle in North China should reach more than 25°.

2. Greenhouse orientation

A greenhouse orientation is the orientation of a greenhouse. It has a great impact on the lighting of greenhouses. From the perspective of the medium- and high-temperature latitudes where greenhouses are distributed in China, the light transmittance in east-west orientation greenhouses is the highest in winter, followed by the east-west orientation multi-span greenhouses, and the light transmittance in north-south orientation greenhouses is less than that in east-west orientation in greenhouses winter. However, in summer, this relationship is reversed. Therefore, from the perspective of light transmittance, the east-west orientation is better than the north-south orientation; while from the perspective of indoor light distribution, the north-south orientation greenhouse has more uniform light distribution than the east-west orientation.

3. Transparent covering materials

The light transmittance of different covering materials for greenhouse varies greatly. As a transparent covering material, plastic film is widely used in Asian, African, and Mediterranean countries due to its good light transmittance, light weight, low price, flexibility and bendability, loose requirements for framework materials, as well as low design and construction costs. According to the masterbatch, plastic films can be divided into three categories: polyethylene (PE), polyvinyl chloride (PVC) film, and ethylene-vinyl acetate (EVA) film. The comprehensive performance of the last one is better than the former two (Fig. 7-9). The light transmittance of new glass can reach more than 90%, which is better than that of plastic film. In addition, the glass is easy to clean and can maintain high light transmittance, good thermal insulation, and fast growth of crops. However, the construction cost is high. Therefore, it is generally used in high-end greenhouses and widely used in European countries. As one of the hard material plates, the polycarbonate (PC) resin plate has the advantages of good light transmittance, long service life, and easy installation. It is a new generation transparent covering material and has been developed rapidly, with the problems of high price and easy aging.

In addition, the aging degree, cleanliness, and condensation on the inner surface of covering materials also impact the greenhouse lighting. For example, the light transmittance of the new plastic film can reach more than 80%; after two days of use, the light transmittance can be reduced by 14%, and even by 25% after ten days of use due to dust pollution. After aging, the light transmittance of the plastic film will be reduced by 20%–40%.

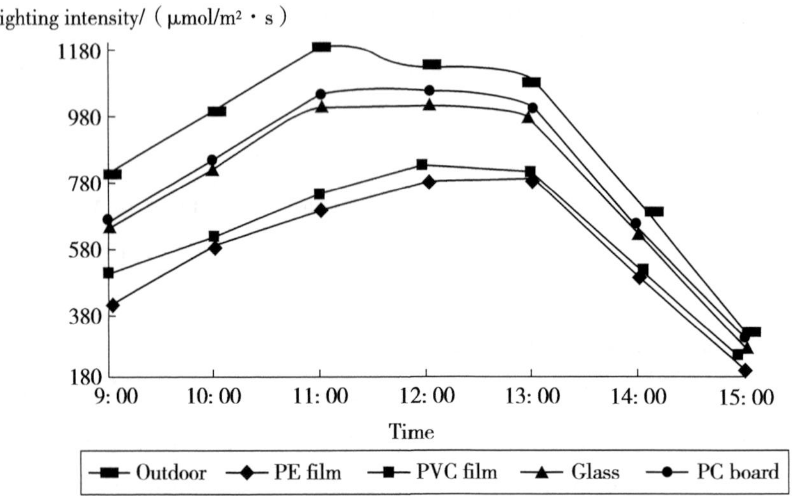

Fig. 7-9 Comparison of Light Intensity Inside and Outside Greenhouses with Different Covering Materials

4. Framework material

The size, quantity, shape, and use direction of the facility framework material affect the light transmittance and light distribution in the facility. The more, larger, and thicker framework materials are used, the larger the shading area will be; the smaller the angle of incidence is, the smaller the shading area of the framework is; the smaller the solar altitude is, the larger the shading area of the framework is. Therefore, if structural safety allows, thin and solid building materials shall be used as the framework materials where possible.

5. Number of greenhouse spans

The more spans in the east-west orientation greenhouses, the lower the light transmittance. However, when exceeding five greenhouses, the light transmittance changes little. The light transmittance in the north-south orientation greenhouse is not related to the number of spans.

6. Spacing between adjacent greenhouses or polytunnels

The spacing between adjacent greenhouses extended from east to west shall not be less than 2–2.5 times the sum of the ridge height of the greenhouse and the height of the straw curtain, while the spacing between adjacent greenhouses extended from north to south shall be about one time the ridge height, so as to ensure sufficient light in the greenhouse before and after winter solstice when the sun is the lowest.

7. Impact of crop population structure

Crop population structure has a great impact on the light distribution inside the facility. For example, the light at 20 cm below the top of eggplant plants (60 cm in height) population decreased by 50%–60% compared with the top. In the case of small row spacing, the light distribution in the crop population of the north-south beds is uniform compared with that of the east-west beds, and the crops have good growth and high yield.

II. Regulation and Control of Light Conditions in Facilities

Light is the basic condition of crop growth, and it has a light effect, heat effect, and morphological effect on the growth of greenhouse crops. Therefore, it is necessary and essential to strengthen the reasonable regulation and control of lighting conditions in the facilities and to meet the requirements of light environment for the growth and development of crops where possible. Specific regulation and control measures include the following aspects.

1. Reasonable facility structure

The greenhouse should face south and extend from east to west. From the perspective of strengthening daylighting, in addition to the modern multi-span greenhouses, single-span greenhouses with a proper span, height, and roof angle, and reasonable spacing from adjacent greenhouses should be selected where possible. Transparent covering materials that are dust-proof, drip-proof, age-proof, and with strong light transmittance should be selected. Currently, EVA film is preferred, followed by PE film and PVC film. Fine and solid framework materials should be selected where possible to improve the indoor lighting quantity and reduce the shading area of greenhouse structural materials.

2. Strengthening facility management

Frequently cleaning and washing are required to maintain the high light transmittance of transparent covering materials on the roof. Under the premise of maintaining room temperature, opaque internal and external coverings (insulation curtains, straw mats, etc.) should be removed early and added late where possible so as to prolong the light duration and increase the light transmittance; VMPET reflection curtains should be hung in the greenhouse or the glass greenhouse roof should be whitewashed to increase the light intensity and the uniformity of light distribution.

3. Strengthening cultivation management

Select a proper planting density, pay attention to the row direction (generally south-north direction is better), expand the row spacing and reduce the plant spacing, and remove the lateral branches and old leaves at the base of seedlings to increase the light transmittance of the crops.

4. Timely supplementing light

Artificial light supplementing is to adjust the light period with low-intensity light, known as cultivation under electric light; or to promote photosynthesis and supplement the deficiency of natural light, which requires light intensity above the light compensation point. There are three requirements for the electric light source used for this purpose: ① It must have a certain level of lighting intensity (the lighting intensity on the bed surface should be above the light compensation point and below the saturation point). ②The lighting intensity should be adjustable to a certain extent. ③ It must have a certain spectral energy distribution that can simulate the natural light with a continuous spectrum. Incandescent lamps are mostly used for cultivation under electric light, high-pressure gas discharge lamps for cultivation by supplementing light, and fluorescent lamps for both types of cultivation. The supplemental lighting is set below the inner insulation layer, and the reflective film is often used around the greenhouse to improve the light supplementing effect. The supplemental light intensity varies from crop to crop. Owing to the high cost of equipment, power consumption, and operation cost, light supplementing is only used for flowers with high economic value or seasonal production of seedlings. Fluorescent lamps and iodine-tungsten lamps are commonly used as supplementary light sources in greenhouses.

5. Shading and darkening as required

Horticultural plants should be shaded or kept in darkness if they are short-day plants or need to oversummer or be blanched. In production, 25%–85% sunshade nets or aluminum foil composite materials are generally selected in accordance with the lighting condition. They are required to have certain light transmittance, high reflectivity, and low absorption rate, and should better be movable. A balance should be kept between temperature and light by opening and closing such coverings as appropriate. Glass greenhouse can also be sprayed with lime and other special reflective materials on the top to weaken the lighting intensity, but such materials must be washed off after summer. Materials such as black PE film and black braided fabrics can be used to keep the facility in darkness.

Task 4 Humidity Regulation and Control Technique

The horticultural facility is a closed or semi-closed system with relatively small space and relatively stable air flow, so the air humidity inside the facility has different characteristics from that of the open field.

I. Characteristics and Influencing Factors of Air Humidity in Facilities

(I) Characteristics of Air Humidity in Facilities

Due to the relatively small space and relatively stable air flow in the closed or semi-closed facility, the air humidity inside the facility has different characteristics from that of the open field. The variation of air humidity inside the facility has the following characteristics.

1. High humidity

The relative humidity and absolute humidity in the greenhouse and polytunnel are higher than those in the open field. The average relative humidity is usually about 90%, and 100% saturation especially at night. In particular, in solar greenhouses and medium-sized and small-sized tunnels, the space in the facilities is relatively small. In addition, ventilation is rarely made in winter and spring for thermal insulation, so the air humidity often reaches 100%.

2. Marked seasonal and daily variations

Generally, relative humidity is high in the low-temperature season and low in the high-temperature season; it is high at night and low during the day, and the humidity is the lowest around noon during the day. The smaller the facility space, the more marked the variation.

3. Uniform humidity distribution

Relative humidity distribution also varies due to the variations in temperature distribution inside the facility. Under normal circumstances, the relative humidity is relatively high in the parts with low temperatures, and it often leads to condensation in local low-temperature parts, adversely affecting the facility environment, plant growth, and plant development.

(II) Influencing Factors of Air Humidity in Facilities

1. Tightness of the facility

Under relative conditions, the better the tightness of the facility environment is, the harder the moisture in the air is discharged, and the higher the internal air humidity is.

2. Temperature in the facility

The effect of temperature on humidity in the facility: On the one hand, the temperature rises to increase the soil moisture evaporation and plant transpiration, thus increasing the moisture content in the air, and then increasing the relative humidity. On the other hand, the leaf surface temperature affects the saturated moisture content in the air, the higher the temperature is, the lower the air humidity will be, and vice versa. The movement of moisture in the greenhouse is shown in Fig. 7-10.

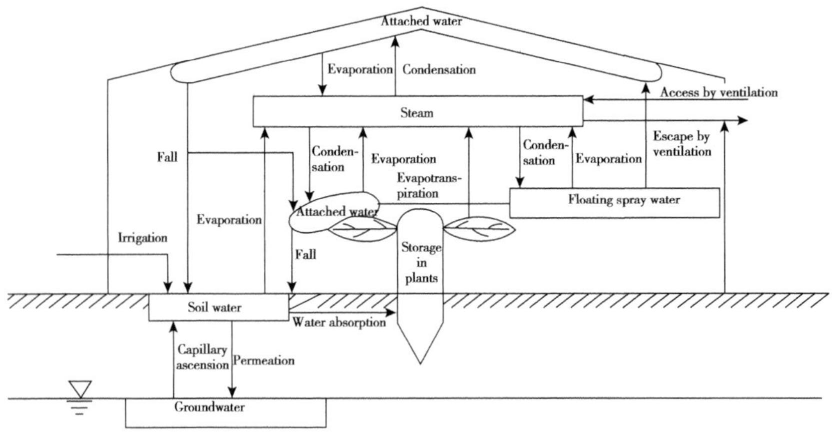

Fig. 7-10 The Movement of Moisture in the Greenhouse

II. Regulation and Control of Air Humidity

Air humidity mainly affects the stomatal opening and closing and leaf transpiration of horticultural crops, and directly affects the growth and development of crops. If the air humidity is too low, the plant leaves will be too small and thick, the mechanical tissue will increase, the flowering and fruit setting will be poor, and the fruit growth will be slow; if the humidity is too high, the crops will be prone to overgrowth of stems and leaves, poor flowering and fruit setting, weakening of physiological function, poor resistance, and nutritional deficiency symptom, which will affect the yield and quality. The relative air humidity suitable for the growth and development of most vegetable crops is between 50% and 85% (Table 7-6). In addition, many diseases are closely related to air humidity. Most diseases occur in high-humidity conditions. Under high humidity and low temperatures, condensation on the surface of plants and the dripping of condensation on covering materials onto plants will aggravate the occurrence and spread of diseases. Under low humidity, especially under high temperatures and drought conditions, diseases are easy to occur. Therefore, from the perspective of creating suitable conditions for plant growth and development, controlling disease occurrence, saving energy, improving yield and quality, and increasing economic benefits, the air humidity should be controlled between 70% and 90%.

Humidity regulation mainly includes controlling moisture source, temperature, ventilation, and use of moisture absorbent. Under the condition of protected cultivation, excessive air humidity often occurs in the facility, so dehumidification is the main content of humidity regulation and control.

Table 7-6　Basic Requirements for Air Humidity of Vegetable Crops

Type	Vegetable species	Suitable relative humidity (%)
High-humidity type	Cucumber, Chinese cabbage, green leaf vegetables, aquatic vegetables	85–90
Medium-humidity type	Potato, pea, broad bean, root vegetables (excluding carrots)	70–80

continued

Type	Vegetable species	Suitable relative humidity (%)
Low-humidity type	Eggplant, legume (excluding pea and broad bean)	55–65
Dry-type	Watermelon, muskmelon, carrot, scallion and garlic, pumpkin	45–55

(I) Dehumidification

Dehumidification is mainly intended to prevent crops from getting wet and reduce air humidity so as to adjust plant physiological status and prevent diseases. There are active dehumidification and passive dehumidification based on whether power is required.

1. Active dehumidification

Heating and ventilation (especially forced ventilation) are mainly used to reduce indoor humidity. The heat-exchange dehumidifier uses forced ventilation to reduce the air temperature. The working principle is to take in low-temperature and low-humidity air from the outside through two ventilation fans (one for suction and one for exhaust) in the heat exchange. After entering the greenhouse, the air first becomes high-temperature and low-humidity air, then absorbs moisture to become high-temperature and high-humidity air. Then it is discharged outside the greenhouse as low-temperature and high-humidity air, thus eliminating condensation on plants at night after sunrise in the morning.

2. Passive dehumidification

The common methods currently in use are as follows.

(1) Natural ventilation: The humidity in the facility is reduced by opening ventilation windows, uncovering films, scratching seams, etc.

(2) Ground hardening and mulching: The ground of the greenhouse is hardened or covered with mulching film, which can reduce the evaporation of surface moisture and reduce the air humidity from 95%–100% to 75%–80%, thus reducing the air moisture content inside the facility.

(3) Scientific liquid supply: The use of drip irrigation, infiltration irrigation,

and underground irrigation, especially drip irrigation under film mulch, can effectively reduce air humidity. The relative humidity can also be reduced by reducing the frequency and the amount of liquid supplies.

(4) Use moisture-absorbing materials: For example, the transparent covering material of the facility is a drop-free durable film, and the two-layer curtain are made of non-woven fabric. The ground is paved with straw, quicklime, silica gel, lithium chloride, etc., which are used to absorb moisture in the air or collect water drops from the film, thereby effectively preventing excessive air humidity and wetting of the crops.

(5) Spray anti-transpirant: To reduce absolute humidity.

(6) Plant adjustment: Plant adjustment is conducive to ventilation and light transmission between rows, reducing transpiration and humidity.

(II) Humidification

Under high temperatures and strong light in summer, excessively low air humidity is detrimental to crop growth and may cause plant withering or death in severe cases. For the cultivation of some flowers and vegetables requiring high humidity, it is necessary to increase the relative humidity when it is less than 40%. Common methods include atomizing and sprinkling with equipment such as electric atomizing humidifiers, air washers, centrifugal atomizers, and ultrasonic atomizers. The wet curtain cooling system can also increase relative humidity. In addition, lowering the room temperature or the light intensity can raise relative humidity or reduce transpiration intensity. Air humidity can also be increased by increasing the frequency and amount of watering and reducing ventilation.

Task 5 CO_2 Regulation and Control Technique

I. Technical Principle and Main Function of Applying Supplemental CO_2 Fertilizer

CO_2 is an important factor in crop photosynthesis. Insufficient supply of CO_2

will directly affect the normal photosynthesis of crops, resulting in the reduction of production and revenue. Therefore, it plays an equal role in crop growth and development with water and fertilizer. Studies have shown that the photosynthetic efficiency of plants can more than double if CO_2 concentration is increased from atmospheric concentration (about 300 ppm) to 1,000 ppm, while photosynthesis stops due to a lack of raw materials if CO_2 concentration is reduced to 50 PPM. When CO_2 concentration is within 100–2,000 ppm, crop yield increases with the CO_2 concentration. Applying supplemental CO_2 fertilizer to strawberries, watermelons, eggplants, cucumbers, tomatoes, and pumpkins mainly has five functions: First, improve the photosynthetic efficiency of plants and make them grow strong; Second, improve the internal and external quality of agricultural products and increase the benefits; Third, increase the yield, especially the early yield of melons and fruits; Fourth, advance the marketing date; Fifth, enhance the resistance of plants and improve the storability of products.

II. Characteristics and Influencing Factors of CO_2 in Facilities

(I) Characteristics of CO_2 Concentration in Facilities

1. Daily variation of CO_2 in facilities

The protection facilities covered by plastic film and glass are in a relatively closed state, and the daily variation of internal CO_2 concentration is much higher than that of external CO_2. Fig. 7-11 shows the daily variation curve of CO_2 concentration

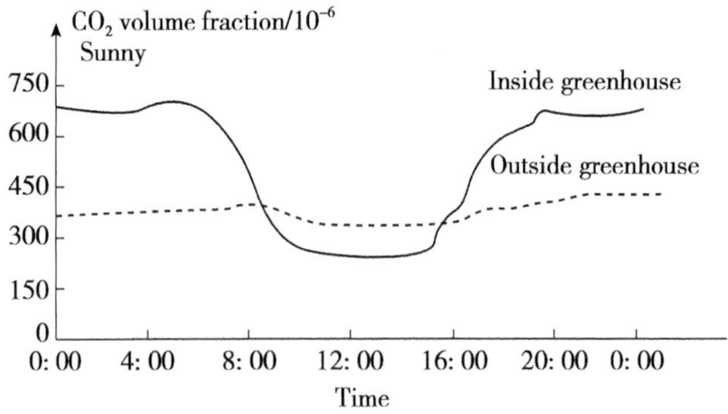

Fig. 7-11 Daily Variation Curve of CO_2 Volume Fraction Inside and Outside the Greenhouse

in the greenhouse. It can be seen that the CO_2 concentration is relatively high due to the accumulation of CO_2 released by crop respiration at night before sunrise in the morning; 1–2 hours after sunrise, a large amount of CO_2 is absorbed by crop photosynthesis, and the CO_2 concentration in the greenhouse rapidly drops to a very low level; and the concentration begins to rise slowly in the evening. On sunny days, the photosynthesis of crops is strong, and the CO_2 concentration is low, while on cloudy days, the CO_2 concentration is high.

2. Distribution of CO_2 concentration at different parts of the facility

The distribution of CO_2 concentration at different parts of the facility is not uniform. From the distribution of CO_2 concentration at noon, the CO_2 concentration in the air at the upper part of the population fertile layer and near the channel and surface is higher, and the CO_2 concentration in the fertile layer is lower. Therefore, the CO_2 supply mainly comes from soil and external air. At night, CO_2 concentration near the surface is often quite high, with higher CO_2 concentration in the fertile layer and lower concentration in the upper layer.

(II) Influencing Factors of CO_2 Concentration in the Facility

The CO_2 concentration inside the facility varies greatly depending on the type of facility, the space, the ventilation, the type of crop cultivated, the growth stage, and the condition of the culture beds. The soil conditions in the facility have a significant impact on the CO_2 environment. If the application of manure or other organic matter is increased, the CO_2 concentration inside the facility will increase. In soilless culture facilities, CO_2 emissions from the substrate are extremely low, especially in winter with low ventilation, and the CO_2 deficit is more serious. Additional CO_2 fertilizer shall be applied to ensure crop production.

III. Regulation and Control of CO_2 Concentration in Facilities

CO_2 is an important raw material for crop photosynthesis. However, the limited CO_2 content in the facility is far from meeting the needs of crop photosynthesis, which limits the growth rate of crops. The limited space and tightness of the greenhouse make it possible for CO_2 application (gas fertilization). Regardless of light conditions, the application of CO_2 during the day promotes crop growth.

In North China, the greenhouse is tightly closed in winter with less ventilation, so the CO_2 deficit is serious indoors. Currently, the promoted CO_2 fertilization technique has achieved remarkable results. In general, CO_2 fertilization increases the yield of fruits and vegetables such as cucumbers, tomatoes, and peppers by 20%–30% on average, and can improve the quality; CO_2 fertilization to fresh-cut flowers can increase the number of flowers, increase the number of flowers, increase the number and thickness of side branches, and improve the quality of flowers. CO_2 fertilization can not only improve the yield per unit area, but also improve the utilization rate of facilities, energy, and light energy.

(I) Sources and Application Methods of CO_2

1. Liquid CO_2

It does not contain harmful substances.It is safe and reliable to use, but has a high cost. It is usually packed in a high-pressure cylinder. During fertilization, open the cylinder plug to directly release the fertilizer, and transport it through a pipeline, which makes it easy to control the dosage and time of fertilization. The main sources include the brewing industry, by-products of the chemical industry, air separation, and underground CO_2 storage.

2. Dry ice burial

A pit is dug per square meter in the polytunnel, and a small amount of dry ice is buried in each pit, so that CO_2 can be slowly released into the polytunnel. This method is costly, labor intensive, and cannot be timed and quantified.

3. Fuel combustion

In Europe, America, Japan, and other countries, low-sulfur fuels such as natural gas, white kerosene, paraffin, and propane are combusted to release CO_2. The application is convenient and easy to control. White kerosene is a liquid under normal temperature and pressure, which is convenient for storage and transportation. Complete combustion of 1 kg white kerosene can produce 3 kg CO_2. The foreign device mainly includes a CO_2 generator and central boiler system. In China, some people further modified the coal-fired stoves, added purification devices for flue gas, and output pure CO_2 into the facility. The device uses coke, charcoal, coal pellets, and coal blocks as fuel with low raw material costs, and the application time and concentration are easy to regulate and control. In addition, in some areas, biogas stoves or alcohol burners fueled by biogas or alcohol are also used for CO_2 fertilization. When using CO_2 generated after combustion, attention shall be paid to the hazard to crops caused by incomplete combustion or impurities in fuel, such as ethylene (C_2H_4), propylene (C_3H_6), hydrogen sulfide (H_2S), carbon monoxide (CO), and sulfur dioxide (SO_2).

4. CO_2 granular fertilizer

The solid granular fertilizer developed by the Shandong Academy of Agricultural Sciences is made of calcium carbonate as the base material, organic acid as the conditioning agent and inorganic acid as the carrier. It is extruded under high temperature and pressure, and can slowly release CO_2 under the comprehensive physicochemical and biochemical action after being applied to the soil. This type of fertilizer source is convenient and safe to use, but the storage conditions are extremely strict, and the release rate is difficult to be controlled manually.

5. Chemical reaction

Among reactions of strong acid with carbonate to produce CO_2, the reaction method between sulfuric acid and ammonium carbonate is the most widely used. In recent years, several package CO_2 fertilization devices have been successively developed, mainly consisting of acid storage tanks, reaction barrels, CO_2 purification and absorption barrels and air ducts. The CO_2 generation is controlled by sulfuric acid supply. The method is simple, safe to operate, and has a good application effect.

6. Other methods

In addition to the above-mentioned methods to increase CO_2 concentration in the facility, forced or natural ventilation, applying supplemental organic fertilizers, bioecological methods, and other methods can also be used to increase and supplement CO_2 in the facility. Additional organic fertilizers in the production of greenhouse substrate culture in the facility have a significant effect on mitigating CO_2 insufficiency and improving production. Production of edible fungi below the culture bed can keep the indoor CO_2 between 800 and 980 μmol/mol.

(II) Application Concentration of CO_2

For general horticultural crops, the economical and effective CO_2 concentration is about five times the atmospheric concentration. Japanese scholars proposed that the suitable concentration of CO_2 in the greenhouse should be 0.01%, but the dosage is mostly maintained between 0.0045% and 0.005% in Dutch greenhouse production, so as to avoid excessive internal and external concentration and escape of CO_2 during ventilation, which is not economical. The optimal concentration and dosage of CO_2 fertilizer are related to crop characteristics and environmental conditions. Generally, the CO_2 concentration shall be increased accordingly with the increase of lighting intensity. CO_2 application on cloudy days can improve the utilization of scattered light by plants; CO_2 application during light supplementation has an obvious collaborative effect.

The saturation point of CO_2 in crop photosynthesis is very high and varies with environmental factors, so the application concentration of CO_2 should be based on the purpose and principle of economic production. Excessive CO_2 concentration not only increases costs, but also causes premature senescence or morphological changes of crops.

(III) Application Time of CO_2

Theoretically, CO_2 fertilization should be carried out during the period of strongest photosynthesis in a crop's lifetime and during the hours with the best lighting conditions in a day. CO_2 fertilization in the seedling stage should be carried out as early as possible. The CO_2 fertilization time after field planting depends on the crop type, cultivation season, facility condition, and type of fertilizer source. Generally, fertilization is not conducted for fruit vegetables after field planting and before flowering, but starts after flowering and fruit setting. The most important reason is to prevent excessive vegetative growth. Leaf vegetables are fertilized immediately after field planting. In the Netherlands, boiler gas is used and CO_2 fertilization is often throughout the whole developing period of crops.

The CO_2 fertilization time of each day shall be determined according to the variation rule of CO_2 in the facility and the photosynthetic characteristics of the plant. In Japan and China, CO_2 fertilization mostly starts from sunrise or 0.5–1 h after sunrise and ends before ventilation. In severe cold seasons or cloudy days without ventilation, fertilization may be stopped at noon. In Nordic, Netherlands, and other countries, CO_2 fertilization is carried out throughout the day, and automatically stops at a certain time after opening the window for ventilation at noon.

(IV) Environmental Regulation During CO_2 Fertilization

(1) Light: CO_2 fertilization can improve the utilization of light energy and make up for the loss of weak light. Studies have shown that increasing the concentration of CO_2 under strong light is more beneficial to increasing the photosynthetic rate of crops. Therefore, attention should be paid to improving the light-receiving conditions of the population during CO_2 fertilization.

(2) Temperature: From the perspective of photosynthesis, when the lighting intensity is a non-limiting factor, the increase of photosynthesis by increasing CO_2 concentration is related to temperature, and the optical temperature for photosynthesis rises under high CO_2 concentration. Therefore, it is necessary to increase the level of temperature management while CO_2 fertilization.

(3) Fertilizer-water: CO_2 fertilization promotes the growth and development of crops and increases the demand for water and mineral nutrients. Therefore, water and nutrient supply must be increased when applying CO_2 fertilization to meet the physiological and metabolic needs of crops.

Task 6 Integrated Intelligent Regulation and Control Technique of Facility Environment

The environmental elements of the greenhouse have a comprehensive impact on crops, and the environmental elements are interrelated and have linkage effects. Therefore, although we can control the variation of a certain factor in one day through sensors and equipment (e.g. the linkage between hygrometer and spraying equipment to maintain the minimum air humidity or the linkage between temperature controller and time controller for temperature-varying management), they appear to be mechanical or uneconomical to some extent.

With the continuous development and application of computer techniques, complex calculations and analysis can be made quickly, creating conditions for the comprehensive regulation and control of greenhouse environment elements, which is changed from static management to dynamic management and from single-factor control to multi-factor integrated control. The computer is connected to indoor and outdoor meteorological stations and control equipment for indoor environmental factors, forming a computer control system to realize automatic control of environmental factors such as temperature, light, and gas in the greenhouse, which is characterized by powerful function, strong reliability, high universality, flexibility and convenience, high efficiency, energy conservation, and income increase, etc. In general, the reasonable parameters of temperature and humidity in the greenhouse are determined first according to the daily sunlight exposure and cultivation type, and then the intelligent control equipment is started to continuously and automatically observe and record the variations of indoor and outdoor environmental meteorological elements and equipment operation. Through comparison of yield and quality, scientific analysis is carried out to adjust the original design procedures and change the regulation and control mode, thus achieving economical production. In recent years, the progress of integrated control techniques in the Netherlands has increased the tomato yield from 40 kg/m^2 to 75 kg/m^2, and the production costs such as energy consumption and labor have been significantly reduced, thus substantially improving the economic benefits of greenhouse production. The computer system can also be equipped with an early warning device, which can handle major changes in environmental factors in time. In the case of power failure, water failure, insufficient pump power, and motor failure, it can give an alarm in time and record it to provide the basis for future adjustment and improvement. The development and application of the computer control system for greenhouse environments make the complicated greenhouse management become simplified, normalized, and scientific.

The facility environment intelligent greenhouse monitoring system collects environmental parameters such as air temperature, humidity, lighting, soil temperature, and soil moisture in the facility in real time, makes real-time intelligent decisions according to the needs of crop growth, and automatically turns on or turns off the designated environmental regulation equipment. The deployment and implementation of this system can provide a scientific basis and effective means for automatic monitoring of agro-ecological information, automatic control, and intelligent management of facilities.

The polytunnel monitoring and intelligent control solution collects environmental parameters closely related to crop growth in the greenhouse, such as temperature, humidity, lighting, soil temperature, soil moisture, and CO_2 concentration, through various wireless sensors and network transmission equipment that can be flexibly deployed in the greenhouse, store the real-time monitoring data on the data server for intelligent analysis and decision, and automatically turns on or off the designated equipment (such as remote control of irrigation, drawing of roller shutter,etc).

I. Equipment Requirements

Wireless air temperature and humidity sensors, wireless soil temperature sensors, wireless soil moisture sensors, wireless light sensors, wireless CO_2 sensors, etc. are deployed in each intelligent facility to monitor environmental parameters such as air temperature and humidity, soil temperature, soil moisture, light and CO_2 concentration in the greenhouse. To facilitate deployment and position adjustment, all sensors shall be battery-powered and use wireless data transmission. AC 220V mains supply is only necessary for a few fixed locations in the greenhouse (e.g. fans, water pumps, heaters, electric roller shutters,etc.).

One set of collection and transmission equipment (including routing nodes, long-distance wireless gateway nodes, Wi-Fi wireless gateway, etc.) is deployed in each facility area to cover all facilities in the area, and transmit sensor data and equipment control instruction data of facilities in the area to the Internet for interacting with platform servers.

Intelligent control equipment (including integrated controller, extended control distribution box, solenoid valve, power conversion adaptation equipment, etc.) is installed in each polytunnel in need of an intelligent control function to receive control instructions and respond to control execution equipment, thus realizing the electric roller shutter, intelligent water spraying, intelligent ventilation, and other behaviors in the greenhouse.

II. Architecture Design of Intelligent Greenhouse Monitoring System

1. Overall architecture

The overall architecture of the intelligent greenhouse monitoring system is divided into four parts: onsite data collection, network transmission, intelligent data processing platform, and remote control.

2. Two typical configuration structures of intelligent greenhouse monitoring system

(1) Two-tier network: The two-tier network system consists of two types of nodes: wireless sensor nodes, including the wireless air temperature and humidity sensor, wireless soil temperature sensor, wireless soil moisture sensor, wireless light sensor, wireless CO_2 sensor, etc; wireless gateway nodes, including Wi-Fi wireless gateway or GPRS wireless gateway.

This structure is applicable to scenarios where the area already has Wi-Fi LAN coverage or can directly upload data with GPRS. In this structure, the wireless gateways are only necessary in the appropriate area to realize the collection and uploading of sensor data.

(2) Three-tier network: The three-tier network system consists of three types of nodes: wireless sensor nodes, including the wireless air temperature and humidity sensor, wireless soil temperature sensor, wireless soil moisture sensor, wireless light sensor, wireless CO_2 sensor, etc.; wireless gateway nodes; data routers.

This structure is applicable to scenarios where the area has no Wi-Fi LAN coverage and is not intended to directly upload data with GPRS. In this structure, it is necessary to deploy data routing nodes and wireless gateways, between which long-distance wireless communication is used for data exchange. When the area is large and the communication distance between nodes is insufficient, the wireless gateways can also automatically relay data to each other to expand the coverage of the monitoring network.

3. Collection of sensor information

In the monitoring network, the wireless air temperature and humidity sensor, wireless soil temperature sensor, wireless soil moisture sensor, wireless light sensor, wireless CO_2 sensor, and other sensors all support low-power run, and can work for a long time with cheap dry battery. In addition, all wireless sensor nodes run the low-power multi-hop ad hoc network protocol, which can provide automatic relay forwarding of data for other nodes to expand the coverage of the monitoring network and increase deployment flexibility.

The sensor data is transmitted to the wireless gateway nodes through the protocol, and then the wireless gateway nodes send the sensor data to the server of the data platform directly or through the data routing node. Users can access the data platform through a wired/wireless network to monitor the sensor parameters on the polytunnel site in real time and control the relevant equipment on the polytunnel site.

III. Layout on the Polytunnel Site

The polytunnel site is mainly for the collection of environmental parameters inside the polytunnel and the execution of control equipment. The data collected mainly includes the parameters required for agricultural production, such as light, air temperature, air humidity, soil temperature, soil moisture, and CO_2 concentration.

Low-power wireless transmission mode is adopted for uploading sensor data. The sensor data is transmitted wirelessly to the wireless gateway nodes through the wireless transmission module by protocol. The control instructions transmitted between the user terminal and the integrated controller are also transmitted to the central node through the wireless transmission module, which saves the deployment of communication cables. The central node then packs the sensor data and control instructions through the edge gateway and sends them to the system business platform located on the Internet. Users can access the system business platform through a wired/wireless network to monitor the sensor parameters on the polytunnel site in real time and control the relevant equipment on the polytunnel site. The low-power wireless transmission mode makes the deployment of sensors on the polytunnel site flexible and convenient for expansion.

The control system mainly consists of an integrated controller, execution equipment, and relevant lines. Through the integrated controller, various agricultural production execution equipment can be freely controlled, including the water spraying system and air conditioning system. The water spraying system can support various spraying and drip irrigation equipment, and the air conditioning system can support equipment such as roller shuttes and fans.

The business platform is used to showcase all functions of the intelligent polytunnel to users, including environmental data monitoring, data spatial-temporal distribution, historical data, ultra-threshold alarm, and remote control. Users can also add video equipment as required to realize remote video surveillance functions. The data spatial/temporal distribution shows the temporal distribution (line chart) and spatial distribution (field chart) of the collected values to the user in an intuitive form, and the historical data can provide the user with the numerical display of a historical period; the ultra-threshold alarm allows the user to define a customized data range and reflect out-of-range circumstance to the user.

IV. Platform Software

The system platform software consists of the following four parts.

(1) Data collection and storage software

Collect, analyze, and classify the sensor data, and finally store it in the database in the preset format.

(2) Display and decision-making software

Visualize the interface to read corresponding data from the database, display sensor data in the form of tables and curves, and support several query and display manners. The control object and decision-making algorithm of the decision-making system can be customized, which are interconnected with the object control software to realize automatic control.

(3) Remote control software

Complete the operation of onsite control objects, visualize the operation interface, support the redefining of remote switch names and other information, and interface with decision-making software to realize automatic control.

(4) Secondary development kit

With the development kit, users can completely use their familiar development platform to develop the data display and decision-making platform of independent intellectual property rights, so that users can complete the development of an environmental monitoring application system without knowing the hardware and other underlying information of the system.

V. Common Sensors

1. Air temperature and humidity sensors

They are used to detect the ambient temperature and humidity of air in protected agriculture. Generally, the effective temperature range is 0–50°C and the effective humidity range is 30%–90%. Most air temperature and humidity sensors are installed in greenhouses, polytunnels, or livestock and poultry houses with good air ventilation and shading. Generally, 1–4 sensors are installed according to the length of greenhouses, polytunnels, or livestock and poultry houses to avoid local microclimate effects caused by poor air ventilation.

2. Soil temperature sensors

They are used to detect soil temperature. Generally, the effective temperature range is 10–40°C (the soil heat capacity is large and the temperature change is not obvious). They are installed in the soil at the root of the crop to measure the soil temperature of crop growth and development and the soil temperature variation after watering. 2–4 sensors are installed according to the length of greenhouses or polytunnels, and the buried depth in the soil is determined during installation according to the root system depth of different crops.

3. Soil moisture sensors

They are used to detect the moisture content in the soil for timely and appropriate watering. At present, there are two expressions: volumetric moisture content (V/V), and mass moisture content (M/M). Most products are expressed by volumetric moisture content, with a general effective range of 10%–70%. Different soil textures can hold different volumes of water, so different soil textures will display different volumetric moisture contents after watering equal amounts of water. 2–4 sensors are installed according to the length of greenhouses or polytunnels, and the buried depth in the soil is determined during installation according to the root system depth of different crops.

4. CO_2 content sensors

They are used to detect the CO_2 content in the environment so as to decide whether to supplement fertilizer or ventilate. Generally, the unit is ppm, and the effective range is 100–1,000 ppm. They can be used in greenhouses, polytunnels, or closed/semi-closed livestock and poultry houses. In the greenhouses and polytunnels, they are mainly used to detect whether the CO_2 content under light is lower than the optimal concentration for crop photosynthesis, and in the livestock and poultry houses, they are mainly used to detect whether the CO_2 concentration in the closed environment exceeds the maximum concentration affecting the growth and development of livestock and poultry so as to facilitate timely ventilation. One sensor is enough for a single-span greenhouse and polytunnel.

[Module Summary]

I. Key and Difficult Points

(1) Facility horticulture can regulate and control internal environmental factors and improve the internal crop growing environment, which is a carrier to break regional, climatic, and environmental differences and create an environment for normal crop growth. In the cultivation of facility environment, environmental factors which are composed of temperature, light, moisture, gas, etc., may also be influenced by certain factors. Mastering the change law of each facility factor is of great significance in regulating and controlling the environmental change of the facility and meeting the needs of plants for each environmental factor.

(2) By comprehensively applying various knowledge and mastering environmental regulation and control techniques for facility temperature, light, moisture, gas, and soil, the parameters and indicators of each environmental factor in different growth stages and periods of crops can be scientifically regulated and

controlled to achieve the goal of high yields, high efficiency, and high income in protected cultivation.

(3) The facility environment intelligent regulation and control system collects environmental parameters such as air temperature, humidity, lighting, soil temperature, and soil moisture in the facility in real time, makes real-time intelligent decisions according to the needs of crop growth, and automatically turns on or off the designated environmental regulation equipment, thus improving the production volume and production level of crops and promoting the development of information agriculture construction.

II. Summary of Experience and Skills

(1) For integrated environmental regulation and control of solar greenhouse, taking any vegetable as an example, the integrated environmental control scheme shall be formulated and implemented by using instruments such as the radiation illuminometer, temperature and humidity sensor, and CO_2 sensor, as well as regulation and control equipment and methods for light, temperature, humidity, and CO_2 in solar greenhouses according to the requirements of vegetables on the environment.

(2) In integrated environmental regulation and control of modern greenhouse, first know about the integrated environmental management system for modern greenhouse management, set the parameters of the integrated environmental management system according to the requirements of vegetable crops on environmental conditions, and observe and record the actual implementation.

[Skill Training]

Skill Training 7-1 Fertilization with CO_2 Fertilizer

I. Purposes and Requirements

Since polytunnels, solar greenhouses, multi-span greenhouses, and other facilities are relatively closed, CO_2 deficiency often occurs during plant growth. Generally, the CO_2 concentration is relatively high in the morning (up to 500–600 ppm in the early morning) and evening. With the progress of photosynthesis, CO_2 in the facilities is absorbed by the leaves, resulting in a continuous drop of concentration. By around noon, the CO_2 concentration is very low (generally less than 100 ppm), which cannot meet the concentration requirement for the photosynthesis of vegetables in the facilities. The CO_2 in the facilities is often in short supply, seriously affecting photosynthesis as well as the growth and development of crops. Therefore, applying supplemental CO_2 in protected cultivation has become an important means to reach high yields, high quality, and high efficiency in protected cultivation.

II. Plan

1. Materials and tools

CO_2 gas testers, CO_2 cylinders, rulers, etc.

2. Implementation plans

(1) Application time of CO_2: CO_2 can be applied in the early stage of crop growth. Applying supplemental CO_2 in the seedling stage of fruit vegetables has a significant effect on shortening seedling age, promoting flower bud differentiation, and cultivating strong seedlings. The fruit setting and fruit swelling stages of fruit vegetables are the optimal periods for the additional application of CO_2. The time of CO_2 application varies in different seasons, generally 2 h after sunrise from November to next February, 1 h after sunrise from March to mid-April, and half an hour after sunrise from late April to May. The application time in the seedling stage is generally 1.5 h after sunrise. CO_2 is generally applied once per day, and it is necessary to close the polytunnel 1.5–2 h upon completion of the application before ventilation.

Seedling stage: The concentration of CO_2 to be applied should be 0.5 mL/L at a time when the plants have 3–5 true leaves and can be gradually increased to 0.8 mL/L with the growth of plants. The application should be stopped 5–7 d before planting, and significant results can be achieved by continuous application for 18–20 d.

After planting: CO_2 should be applied 7–10 d after planting in polytunnels (recovery) or 15–20 d after planting in greenhouses for 30–35 d continuously, and the yield can be increased to the maximum. The application concentration of CO_2 should be 1 mL/L in polytunnels and 0.8–1 mL/L in greenhouses. CO_2 should be applied according to the required concentration on sunny days, reduced concentration on cloudy days and stopped on rainy and snowy days.

The application time on each day should be subject to the daily variation of CO_2 concentration in the facility. CO_2 should be applied half an hour after straw mats are removed in the morning, continuously applied for more than 2 h on sunny days, and maintained at a high concentration and stopped 1 h before ventilation. The application shall be stopped on cloudy or rainy days.

(2) Application concentration of CO_2: In closed greenhouses with sufficient light and vigorous crop growth, CO_2 deficiency often occurs; when the concentration is less than 80–100 mg/L, the normal growth of vegetables will be seriously restricted. The optimal application concentration of CO_2 is closely related to the crop type, growth stage, and weather conditions. Under suitable temperature, light, water, and fertilizer conditions, the photosynthetic rate of vegetable crops is generally fastest at 600–1,500 ppm CO_2 concentration. To be specific, the optimal CO_2 concentration is 1,000–1,500 ppm for fruit vegetables and 1,000 ppm for leaf vegetables.

III. Implementation

(1) Discuss in groups, consult data, and prepare CO_2 fertilization plans.

(2) Discuss and revise the plan.

(3) Apply CO_2 as planned.

(4) Compare parameters.

(5) Summarize and evaluate through comparison.

[Expansion Task]

I. Review Questions

(1) What are the measures for warming and insulation in horticultural facilities?

(2) What are the factors affecting the lighting environment in facilities?

(3) Please briefly describe the measures of humidity control in greenhouses from the perspectives of humidification and dehumidification.

(4) Please briefly describe the application method of CO_2.

(5) Combine production practice and understand the objectives and principles of integrated regulation and control of protected cultivation environment.

II. Case Study

"Internet+" Intelligent Management System in Protected Agriculture

With the development of the Internet of Things (IoT), the technique has been preliminarily applied in the field of agricultural informatization, including sensor techniques in precision agriculture, intelligent expert management systems, remote monitoring systems, and agricultural product safety tracing systems. This contributes greatly to the promotion of agricultural techniques, the monitoring, prediction, prevention, control, and treatment of diseases and pests for agricultural plants, as well as to the scientific decision-making regarding agricultural production and operation.

Based on the actual needs and with the use of cutting-edge information techniques such as cloud computing, IoT, knowledge engineering, and 3S, an open, low-cost, intelligent, and wide-coverage area intelligent application solution, i.e. protected agriculture intelligent application management system, has been developed. With the use of IoT technologies, this system helps to create a relatively independent crop growth environment and collects agricultural and ecological information about the area to realize intelligent management in combination with intelligent control and intelligent video surveillance techniques. Through the application of modern agricultural technologies and management methods, a sustainable agricultural development model aiming at realizing high yields, high efficiency, high quality, and ecological safety has been promoted to help realize intelligent, scientific, and intensive agricultural production and promote transformation and upgrading of modern agriculture, taking the development of protected agriculture to a new level.

Based on integrated communication technology, wireless network, and Zigbee technique, this system remotely collects real-time data, including air temperature and humidity, soil moisture and temperature, CO_2 concentration, lighting intensity, and videos and images in the facilities through the sensor network, data acquisition module, and video surveillance module for model analysis through intelligent control, and automatically controls equipment such as wet curtain fan, spray and drip irrigation equipment, internal and external shades, roof windows and side windows, warming and light supplementing equipment, and integrated water-fertilizer equipment, to ensure optimal greenhouse environment for crop growth, and create conditions for achieving high quality, high yields, high efficiency, and safety of crops. This system can also release real-time monitoring information and warning messages through information terminals such as mobile phones, PDAs, and computers.

It can be widely applied in protected agriculture, horticulture, and flower cultivation, or in places requiring a special environment, to timely provide a scientific basis for realizing the healthy growth of ecological crops and timely adjusting cultivation, management, and other measures, and to realize automatic monitoring and management. The following systems have been established and improved: a greenhouse intelligent control and acquisition system, an audio/video diagnosis and monitoring system, a peripheral video surveillance system, a regional environmental monitoring and display system, an integrated water-fertilizer control system, and a comprehensive display service center. Based on integrated communication technique, wireless network, and Zigbee technique, these systems can realize real-time display of data such as environmental factors including air temperature and humidity, soil temperature and humidity, CO_2 concentration as well as video surveillance data; control the operation of polytunnel roller shutter, ventilation, and irrigation equipment; send video information and intercom signals to the management center for remote video diagnosis, thus realizing remote intelligent management, training, and display; realize real-time two-way audio/video communication, integrate videos and intercom signals, initiate real-time interaction between the management service center and an operator in any polytunnel, and provide on-site technical guidance and training. The greenhouse data, video diagnosis data, and video surveillance data can be exchanged between the display service center and the comprehensive display center through high-speed network links to realize ICT-based management, centralized display, real-time preview, and view of video surveillance data of the facility area. Therefore, the interconnection and linkage of intelligent greenhouses can be truly realized, which means one person can manage fifty or even more greenhouses by controlling the greenhouses separately or together to meet the growth management requirements of different crops, which greatly reduces manpower and material inputs, creates conditions for achieving high yields, high quality, high efficiency, eco-friendliness and safety of crops, and realizes the intensive, networked and large-scale production management of the whole demonstration base. In addition, with the use of a computer-based automatic control system, the information can be collected, processed, analyzed, and released more quickly and accurately, remarkably reducing the workload. As the crop growing environment can be accurately controlled, the crop yield and quality can be significantly improved. The agricultural information provided for government departments serves as the basis for policy-making, guides agricultural production and management, and reduces agricultural risks. Through the application of modern agricultural techniques and management methods, a sustainable agricultural development model aiming at achieving high yields, high efficiency, high quality, and ecological safety has been promoted to help realize intelligent, scientific, and intensive agricultural production and promote transformation and upgrading of modern agriculture, taking the development of protected agriculture to a new process. In addition, considerable economic and social benefits have been created, saving costs and increasing efficiency by over 10%, reducing labor by over 10% on average, and improving yield and benefits by over 10% on average.

Module 8 Examples of Soilless Culture Production

[Learning Objectives]

I. Knowledge Objectives

(1) Understand the relationship of common fruit vegetables, leaf vegetables, flowers, strawberries, and grapes with nutrient solutions and environmental conditions;

(2) Master the soilless culture techniques of common fruit vegetables, leaf vegetables, flowers, strawberries, and grapes;

(3) Master the production conditions, procedures, techniques, and cause analysis of common problems and solutions of sprout vegetables.

II. Skill Objectives

(1) Be able to comprehensively apply the theoretical knowledge and skills learned, and independently engage in the soilless culture production and management of vegetables, flowers, and fruit trees;

(2) Be able to independently work on the production and management of sprout vegetables, master the techniques of seed pre-sowing treatment, tray-stacking pregermination, environmental regulation and control, and harvest.

[Preparation for Learning]

I. Required Resources

(1) Relevant learning materials about the cultivation of vegetables, flowers and fruit trees;

(2) Production of sprout vegetables;

(3) Formulation laboratory of nutrient solution;

(4) Electrical conductivity meter and pH meter required for nutrient solution management;

(5) Standardized soilless culture production base.

II. Background Knowledge

(1) Master relevant knowledge of soilless seedling culture;

(2) Understand the biological characteristics of common vegetables, flowers, and fruit trees;

(3) Master the key points in the cultivation and management of common vegetables, flowers, and fruit trees;

(4) Master the basic knowledge of literature review.

[Learning Tasks]

Task 1 Fruit Vegetables

I. Eco-organic Soilless Culture of Tomatoes

(I) Facility Structure

1. Culture trough

Culture trough framework may be constructed of brick, cement board, plastic foam board, wood board, etc. But in general, the brick has low cost, is convenient for operation and management, and is easy to observe the growth of the root system. The culture trough runs from north to south and is built with an inner diameter of 48 cm and a border height of 24 cm (four layers of bricks are placed horizontally). Alternatively, a semi-underground culture trough with a depth of 12 cm can be directly excavated on the ground and then built with two layers of bricks on both sides. The trough spacing is 70–80 cm, reserving an 80-cm-wide walkway in the north and a 30-cm-wide walkway in the south of the greenhouse. The walkway between troughs can be isolated from the soil by cement brick, red brick, woven fabric, plastic film, sand, etc. to keep the cultivation system clean. The substrate is separated from the soil by a plastic film at the bottom of the culture trough, and the edge of the film is compacted by the topmost brick. The plastic film is paved with a layer of disinfected coarse substrate with a thickness of about 5 cm for water and gas storage. The coarse substrate can be coarse sand, gravel, coarse slag, etc. A layer of water-permeable plastic woven fabric is paved on the coarse substrate, and the cultivation substrate is paved on the plastic woven fabric. The plastic woven fabric may be replaced by ordinary woven bags cut open, which can effectively reduce the cost.

2. Cultivation substrate

The organic substrates include cornstalk, mushroom dregs, cottonseed hull, sunflower stalk, corn cob, bagasse, sawdust, bark, shavings, etc. The selected substrates are fermented and disinfected before use. A certain proportion of inorganic substrates, such as slag, sand, and perlite, is added for mixing. For example, use a mixture of turf, cornstalk, and slag in the ratio of 2 : 6 : 2 for cultivation. In addition, basal fertilizer is required for the cultivation substrate before field planting, and the solid fertilizer special for eco-organic soilless culture can be used. The fertilizer applied per cubic meter of substrates shall contain 1.5–2.0 kg of total nitrogen (N), 0.5–0.8 kg of total phosphorus (P_2O_5), and 0.8–2.4 kg of total potassium (K_2O). For example, 10 kg of decomposed disinfected chicken manure, 1 kg of diammonium phosphate, 1.5 kg of thiamin, and 1.5 kg of potassium sulfate are added to each m^3 of substrates as basal fertilizer, which is sufficient for the nutrient requirement of 8,000–10,000 kg per mu of tomatoes. The trough can be filled after the substrate is evenly mixed with fertilizer. The substrate can be disinfected after harvesting each crop, and the substrate renewal period is generally 3–5 years.

3. Irrigation facilities

A reservoir is built in the greenhouse, and the external pipeline, the main pipeline in the polytunnel and the drip irrigation tape in the culture trough can be plastic pipelines. 1–2 drip irrigation tapes shall be laid in the trough, and a layer of plastic film shall be paved on the drip irrigation tapes to prevent the drip irrigation water from spraying outside for irrigation under the film.

(II) Production Techniques

1. Variety selection

Select varieties with strong resistance, early maturity, high yields, good quality, long fruit-bearing stage, and storability, such as Zhong Za series, Golden Crown series, Jia Fen series,etc.

2. Seedling culture

In order to kill the pathogens and insect eggs that may be carried on the seeds, the seeds must be disinfected before pregermination, and the commonly used disinfection methods include soaking in hot water and chemical treatment. The water temperature for soaking seeds in hot water shall be 50–55°C. The seed shall be stirred continuously during soaking, and warm water shall be supplemented at any time. After the water temperature is kept at 50–55°C for 10–15 minutes, stop the stirring when the water temperature naturally drops to 30°C. Continue soaking for 8–10 h as required. The chemical treatment includes mixing the seed with chemical powder and soaking the seed in a chemical solution. Common fungicides such as carbendazim and captan and pesticides such as trichlorfon are used for seed mixing. The dosage is 0.2%–0.3% of the seed mass. Soak the seeds with 1% potassium permanganate solution for 10–15 min, wash them with clean water, and then soak them.

Scrub the seeds after soaking, rinse them with clean water 2–3 times, take them out of the water, dry them slightly, wrap them with wet gauze, towel, or burlap, and put them at 25–30°C to promote germination for 2–4 d. During the pregermination, the seeds shall be checked and turned over frequently to ensure that the seeds are heated evenly. Wash the seeds with clean water, stop the pregermination, and sow the seeds after most of the seeds (60%–70%) show the white buds.

Prepare the substrate with a ratio of turf to vermiculite of 3 : 1 before sowing, and add 5 kg of disinfected chicken manure and 0.5 kg of special fertilizer for eco-organic soilless culture into each cubic meter of substrate. After mixing evenly, fill them into the plug tray, which shall be separated from the ground. Sowing can be started after thoroughly watering. Each hole has one seed, covered with 1 cm vermiculite, and the surface is slightly watered. Cover with plastic film in winter and spring for heat preservation, and with newspaper for moisturizing and cooling in summer and autumn. Remove the coverings in time after seedling emergence. Before emergence, the temperature shall be kept at 25–30°C, and the temperature shall be 22–25°C in the daytime and 10–15°C at night after emergence. The seedling tray shall be kept wet for about 30 d. When the seedling has 2–3 true leaves, the temperature shall be kept at 20°C in the daytime and 13°C at night for 10–12 d, which is conducive to flower bud differentiation. When the seedling has 3–4 true leaves, they can be taken out of the tray for field planting.

3. Planting

Before field planting, blocking and hardening seedlings shall be carried out properly first, which is conducive to the cultivation of strong seedlings. In addition, the substrate shall be turned and leveled. The substrate in each culture trough shall be flooded with water to make the substrate fully absorb water. After water infiltration, tomatoes shall be planted in two rows in each trough, with a plant spacing of 30 cm. Each row of plants shall be about 10 cm away from the inner edge of the culture trough. After field planting, each plant shall be immediately irrigated with 200 mL water for field planting to facilitate close contact between the substrate and the root system.

4. Management after planting

(1) Fertilizer and water management

When the tomato is cultivated by soilless culture, watering should be carried out according to the morphology of the plant and the external climate. Attention should be paid to controlling water in the early stage after field planting. The substrate humidity should be maintained at 60%–65% before flowering and fruit setting, and at 70%–80% after flowering and fruit setting. Generally, watering should be carried out 5 d after field planting to keep the rhizospheric substrate moist. The plant should not be overgrown or controlled to be a small old seedling. Watering should be carried out more frequently after fruit setting, generally once in the morning and once in the afternoon on sunny days, and the duration is 15–20 min. On cloudy days, watering may be less or canceled depending on the specific situation. The top dressing generally starts 20 d after field planting, and then once every 10–15 d, with 10–15 g of topdressing for each plant. The topdressing shall be carried out once every 7 d after fruit setting, with 25 g for each plant. The fertilizer should be evenly spread 5 cm away from the root and can infiltrate into the substrate with drip irrigation. According to the fact of CO_2 deficiency in the greenhouse, CO_2 topdressing can be carried out in the polytunnel to enhance the stress resistance and increase the yield of tomatoes.

(2) Temperature and light management

After the field planting of the tomato, the temperature should be maintained at 22–25°C during the day and 10–15°C at night. The temperature is increased after fruit setting, maintaining 25–28°C during the day and about 12°C at night. In late winter, the temperature in the polytunnel can reach 30°C in a short time, and large ventilation is not allowed for cooling to prevent too low humidity. After the severe winter, normal temperature management should be restored. Tomato is very heliophilous. During the whole cultivation period, as long as the normal room temperature is ensured and the temperature in the polytunnel is not excessively low, the straw mat should be pulled down early and pulled back late to make the plant exposed to more light if possible.

(3) Plant adjustment

When the plant height reaches about 30 cm, a trellis should be built, or the vine shall be hung in time. For tomatoes cultivated on the trellis, it is necessary to bind the vine. Attention should be paid to the growth vigor of plants when binding the vine, which helps to give full play to their apical dominance and enhance the growth vigor of plants.

Most tomatoes cultivated by soilless culture in the greenhouse are growth-indeterminate varieties, and the pruning method is generally single-trunk pruning; that is, only the main vine is reserved for growth and fruit setting, and the lateral branches in all leaf axils are removed. To ensure that the plants grow well, the pruning shall be carried out when the lateral branches are 10–15 cm long. Generally, the spring crops are topped, leaving 6–8 fruit clusters, and the autumn crops are topped, leaving 4–5 fruit clusters, so as to facilitate the early maturity of the fruit and the timely uprooting without affecting the substrate preparation of the subsequent crop.

The diseased leaves, aged leaves, and yellow leaves are removed during the growth period, which is conducive to ventilation and light transmission at the lower part of the plant, reducing the occurrence and spread of diseases, reducing nutrient consumption, and promoting the development of the plant. The suitable period for leaf removal is in the middle and late stages of growth. The leaves with dark green bases and then yellowed, and those that have severe disease and lose assimilation function are removed. Leaf removal shall be conducted on a sunny morning by cutting off a small part of the petiole with scissors. The infection of pathogenic bacteria should also be considered during operation, and the scissors should be disinfected after the diseased leaves are cut off. The removed aged and diseased leaves should be collected in a special collection bag for residual leaves and branches and then transported out of the greenhouse for treatment so as to prevent the spread of diseases and pests.

(4) Flower and fruit management

There are many ways of tomato pollination, such as hormone treatment, mechanical pollination and insect-assisted pollination. Commonly hormones include PCPA, 2,4-D, which have good effects in treating the flowers about to bloom. To save labor, one inflorescence may be sprayed just once when the first flower is in bloom and some of the other flowers are still in buds. The concentration of auxin solution formulated should be strictly in accordance with the instructions for use. Generally, the concentration of PCPA is 20–30 mg/L, and the concentration of 2,4-D is 10–20 mg/L. The best timing for auxin treatment is on sunny days. Mechanical pollination includes manual oscillation pollination and oscillator oscillation pollination. After flowering, oscillation pollination is carried out every morning from 10:00 to 11:00, and the pollination effect is better than that of hormone treatment. Insect-assisted pollination is also a good policy for large-scale vegetable production.

For large fruit varieties, 3–4 fruits are reserved per cluster, and for medium fruit cultivars, 4–5 fruits are reserved per cluster. The thinning of flowers and fruits is carried out in two steps. When most of the clusters blossom, remove the deformed flowers and the small flowers; secondly, after the fruit setting, remove the fruits that are not well developed and of non-standard shape.

5. Harvest

After harvesting, if the fruits need long-distance transportation for 1–2 d, they can be harvested during the veraison stage. At this time, most part of the fruit is whitish-green, the top is red, and the fruits are hard and resistant to storage and transportation. If the fruits are sold locally after harvesting, they can be harvested in the maturity stage. At this time, 1/3 of the fruit turns red, the fruit is not softened, and the taste is the best.

II. Rock Wool Culture of Tomatoes

(I) Seedling Culture

In terms of tomato variety selection, most are cultivars from the Netherlands, including large, medium, small and micro fruits in red, pink, yellow, orange, and other colors.

The seedlings were raised with a 10×24 hole polystyrene plug tray and a rock wool plug with the same diameter was placed in each hole. Before sowing, soak the plug tray filled with rock wool plug in a nutrient solution with an EC value of 1.5–2.0 to absorb enough moisture, and then sow the seed into the rock wool plug. At 25°C, promote germination in closed and shaded conditions. After germination of the seed, cover with 1-mm-thick vermiculite to promote the root system to extend downward. Then the cotyledon removes its hull and comes up. The temperature in the seedling stage should be kept at about 25°C so that more roots can be grown after 10 days, and the flower bud differentiation of tomatoes can start after the ninth leaf comes out. If the temperature is lower (about 18°C), the root is less, the biological seedling age is delayed by one week, and the tomato bud differentiates early in the seventh leaf, thus affecting the later yield of the tomato.

(II) Transplanting

When the tomato seedlings have two true leaves, the rock wool plug with the seedlings needs to be transplanted into a double-hole rock wool block. It takes about 20 days from seedling culture to transplanting. For Dutch Venlo glass greenhouse, the standard span is 6.4 m, allowing four rows of rock wool strips per span for tomato planting, and the row width is 1.6 m. A heating pipeline is between each row of rock wool strips, which also serves as the track for picking vehicles. Before transplanting the seedlings, place the rock wool strips, place the double-hole rock wool blocks on the rock wool strips in the same direction, insert the drip plug into the rock wool blocks, and soak the rock wool blocks using a nutrient solution with an EC value of 2.5. Then, transplant the seedlings into rock wool stocks to promote the formation of more roots, thus ensuring high tomato yields in the future.

After the tomato is transplanted into the rock wool block and before it is planted into the rock wool strip, it is not necessary to supply too much water. Keeping 80% water and 20% air in the rock wool block can promote the growth of the root system, and supply water when no water can be squeezed out of the rock wool block. One week after transplanting, check the growth of tomato seedlings, remove the seedlings with poor growth and shorter leaf internodes, and resow the robust seedlings.

(III) Planting

When the first spike of tomato begins to bear fruit, the rock wool blocks are planted into the planting holes of the rock wool strips to extend the root system into the rock wool strips and expand the nutrient area of the root system (Fig. 8-1). Tomato rock wool culture requires a control period of vegetative growth before flowering. If the rock wool block is directly placed into the planting hole of the rock wool strip during transplanting, the tomato vegetative growth will be vigorous and affect the reproductive growth.

Before planting, soak the rock wool strip in a nutrient solution with an EC value of 3.0, and reserve drainage outlets at both ends of the rock wool strip for drainage. When planting, the rock wool block should be placed into the planting hole of the rock wool strip in the same direction. The operation should be minimized within one week after planting to avoid shaking the plant and damaging the root system.

Fig. 8-1 Rock Wool Culture of Tomatoes

(IV) Cultivation Management

1. Nutrient solution management

The formula for drip irrigation rock wool culture of Dutch Greenhouse Crop Research Institute can be used. The main part of tomato rock wool culture is the nutrient solution supply. The formula of the nutrient solution should be adjusted in time in different stages of tomato growth. Increase the content of N in the seedling stage to promote vegetative growth, increase the content of K in the fruit setting stage to promote root growth, increase the content of P and K. During irrigation, the computer control system can be used to adjust the drip irrigation time, control the nutrient solution content in the rock wool, maintain 80% content during the day and 60% at night, ensure that there is a certain air in the rock wool to promote the development of the root system.

2. Temperature control

The suitable temperature for tomato growth is controlled by the computer automatic control system. In winter, keep the temperature at about 20°C to promote vegetative growth, and keep it at 18°C on cloudy days to prevent excessive growth. In the reproductive growth stage, adopt the alternating temperature treatment: keep it at 20°C in winter during the daytime, keep it at 16°C from sunset to 00:00, and keep it at 18°C from 00:00 to sunrise. Alternating temperature treatment in summer: 28°C during the daytime and 20°C at night.

3. Humidity regulation and control

Tomatoes require a relative air humidity of 60%–80%. If the air humidity in the greenhouse is relatively high, ventilation can reduce the relative air humidity in summer. In winter, moisture in hot air after heating will condense on the cold surface of the glass and flow into the condensation collecting tank, thus reducing the air humidity.

4. Gas regulation and control

In the process of culture, CO_2 content in the greenhouse should be focused, and the appropriate CO_2 concentration is 600–1,000 mg/L. In summer, CO_2 in the greenhouse is supplemented by frequent ventilation. In winter, if there is no CO_2 generation device, natural ventilation should be strengthened to supplement the indoor CO_2 content at noon with a higher temperature, which can increase the yield by 10%–15%. In production, the reaction between sulfuric acid and ammonium bicarbonate is used to produce CO_2. Approximately 2.2 kg of concentrated sulphuric acid (diluted with 3 times water before use) and 3.36 kg of ammonium bicarbonate are required per mu of greenhouse per day. It is applied 0.5 h after sunrise every day for about 2 h. The CO_2 may be liquefied CO_2 of about 2 kg, or be generated by burning coal. It should be noted that CO_2 gas should be piped evenly to the upper greenhouse space.

5. Pruning and pollination

Most tomato varieties are growth-indeterminate types with strong branching ability. When the tomato grows to a height of about 30 cm, it is necessary to start hanging the vine and prune the branches in time. In winter, since the light is weak, fewer branches shall be reserved. In spring, with the increase of light, more branches may be reserved appropriately, so that more leaves can transpire in summer to reduce the indoor temperature. However, do not leave too many branches to avoid affecting the size of fruits.

Tomato rock wool culture is a long-season culture, generally one crop a year, and each tomato can bear 20 or more fruit clusters on average. To keep the uniform fruit size, the measure of thinning flowers and fruits can be taken to maintain 4–6 fruits per cluster (some tomato varieties do not need fruit thinning because their fruit size is uniform under balanced nutrition conditions). To improve the toughness of the tomato stalk, squeeze the stalk of newly blooming inflorescence with hands, which will be thickened after the wound is healed. After harvesting each fruit cluster, the aged leaves should be cut off in time, and the vine should be released.

The pollination method can refer to the eco-organic soilless culture method of tomato.

III. Deep Flow Hydroponics of Eggplant

(I) Variety Selection

Eggplant is a perennial plant of the eggplant genus in the Solanaceae family. There are three main types of eggplants: round eggplant, oval eggplant, and long eggplant. The appearance color of the fruit is purple, light green, or milky white. At present, the varieties for soilless culture mainly have a bright appearance, uniform coloration, smaller flower stalk and pedicel, vigorous growth, and a higher proportion of normal flowers.

(II) Seedling Culture and Seedling Stage Management

The seeds can be soaked in hot water. Before soaking, the seeds can be pre-soaked in warm water at 20–30°C. Then add warm water at 55–60°C to the container, pour eggplant seeds into the container, and keep stirring. Hot water can be added appropriately to maintain 55–60°C for 15 minutes, and then add cold water to reduce the water temperature to about 30°C. The chemical treatment is soak to seeds with 1,000-fold dilution of 50% carbendazim for 20 minutes, or 10% trisodium phosphate solution for 20 minutes, or 0.2% potassium permanganate solution for 30 minutes. After the chemical treatment, rinse away the chemicals with clean water, and soak the seeds in 20–30°C clean water for 20–24 h. Then alternately germinated at 28–30°C for 12–18 h and 16–18°C for 12–6 h. Generally, after 5–6 d, the seeds can be sown immediately after they show the white buds.

The eggplant can be sown in the plug tray, and the substrate preparation and sowing methods are the same as those of sweet pepper seedlings. Pay attention to heat insulation after sowing. The temperature suitable for the substrate during seedling emergence is above 20°C, and at least 18–20°C should be ensured. After 5–6 d, seedlings emerge successively. When about 70% of seedlings emerge, the covering on the plug tray should be removed. Generally, eggplants do not overgrow in the stage from cotyledon emergence to true leaf breaking. The substrate temperature can be maintained at 18–20°C, the air temperature in daytime is 25–28°C, and the air temperature at night is 16–17°C, not lower than 15°C. The light shall be increased to ensure more than six hours of direct light every day by removing the straw mat in advance and adding the straw mat late. In terms of moisture management, the substrate shall be watered once the surface becomes dry. No nutrient solution is required before growing two true leaves. After growing two true leaves, 1/2 dosage of the fruit vegetable formula of South China Agricultural University can be used for spraying, and seedlings can be transplanted for planting after growing more than 4 true leaves. In high-temperature seasons like summer, it is better to choose young eggplant seedlings for late cultivation in autumn or overwintering cultivation, and choose aged eggplant seedlings with 7–8 true leaves for spring cultivation.

(III) Planting and Post Management

1. Planting

When transferring seedlings, take the seedlings together with the seedling culture substrate out of the plug tray, put them into the planting cup, and fix the seedlings with gravel. Before placing the seedlings in the planting cup, pave a layer of non-woven fabric at the bottom of the cup to prevent the substrate in the cup from falling into the nutrient solution in the plantation trough from the bottom. Note that the temperature difference between the substrate temperature of the plug tray and the temperature of the nutrient solution in the plantation trough shall not exceed 5°C during planting. Otherwise, the root will be damaged due to the large temperature difference when transplanting the seedlings from the plug tray to the plantation trough. The plant spacing is 40–50 cm, and the row spacing is 50–60 cm, which means that 6 plants can be planted on the foam board for deep flow hydroponics with a length of 150 cm and a width of 100 cm, and about 1,300–1,500 plants can be planted per mu of polytunnel.

2. Management after planting

(1) Nutrient solution management. The formula of the nutrient solution can be 1/2 dosage of the Japanese garden test formula or fruit vegetable formula of the South China Agricultural University. After eggplants are planted, the concentration of nutrient solution is kept within 1.0–1.2 mS/cm, and after bearing fruits, the concentration of nutrient solution could be increased to 1.2–1.5 mS/cm. After the fourth branching, the plant grows vigorously and the need for nutrients increases. At this time, the concentration of nutrient solution should be increased to 1.5–1.8 mS/cm, and the nutrient supplementation should be stopped only one week before the end of harvest.

The root system of eggplant is very developed, and a thick layer of root mat will be formed in the hydroponics plantation trough, so it is required to ensure the circulation supply of nutrient solution and prevent the root system from rotting. Before the flowering of plants, the nutrient solution shall be supplied for 15 min and stopped for 45 min every hour during the day; it shall be supplied for 15 min and stopped for 75 min every half hour during the night. After flowering, the root system is more developed. During the day, the nutrient solution is supplied for 15 min every 45 min, and at night, it is supplied for 15 min every hour (same as the supply during the day before flowering). Of course, these parameters should be adjusted according to plant growth vigor and climate conditions.

(2) Temperature and light management. The eggplants are thermophilous. In the morning, the temperature shall be within 25–30°C, and proper ventilation shall be conducted when the temperature exceeds 30°C; in the afternoon, the temperature shall be within 28–20°C, and the ventilation shall be stopped to maintain above 20°C when the temperature is below 25°C; at night, the temperature shall be kept above 15°C. The temperature of the nutrient solution shall be kept at 15–20°C in

winter, not lower than 13°C, and not higher than 28–30°C in summer.

Eggplant is heliophilous, and various measures should be taken to supplement light. If a reflective curtain is hung on the rear wall of the greenhouse, pay attention to cleaning the film. The straw mat should be removed early and added late under the premise of ensuring the temperature.

(3) Plant adjustment. The double-branch pruning method is adopted, reserving only two primary lateral branches, and the lateral branches of nodes below are removed as early as possible. After flowering and setting of opposite fruits, on the lateral branch bearing fruits, conduct topping and reserve two leaves before fruiting, and release the non-bearing branches. Repeatedly treat the subsequent lateral branches, and reserve only two trunks growing upwards. The shoots can be fixed using nylon rope to hang the seedlings. With the growth of the plant, the diseased leaves in the lower part shall be properly removed to facilitate ventilation and light transmission.

(4) Flower and fruit retention. For spring eggplant production in the greenhouse, the low indoor temperature and weak light make it not easy to set fruit. The measure of increasing fruit setting rate is to strengthen management and create environmental conditions suitable for plant growth. In addition, the growth regulator can be used for treatment, and 30–40 mg/L PCPA spraying can be used for treatment during anthesis.

(5) Harvest. The criterion for harvesting eggplants that reach commercial maturity is the disappearance of the “eye” (a light zone under the sepals), which indicates that the fruit growth is slowing down and can be harvested.

Task 2 Soilless Culture of Leaf Vegetables

I. Lettuce

Lettuce, also known as leaf lettuce, is a variety of the *Lactuca* genus in the Compositae family. It is a biennial herb native to China, India, the Near East, and the Mediterranean coast. The lettuce includes three varieties, namely, long-leaf lettuce, wrinkled-leaf lettuce, and head lettuce. The leaves of long-leaf lettuce are long, narrow, and erect, and generally do not form a head or curls in a cylindrical shape. The leaf surface of the wrinkled-leaf lettuce is wrinkled, the leaf edge is deeply split, and there is no head. It can be further divided into green wrinkled-leaf lettuce and purple wrinkled-leaf lettuce according to the leaf color. The top leaves of the head lettuce form a leaf head, which is round or oblate. It can be further divided into soft-leaf head lettuce and crisp-leaf head lettuce according to the leaf texture. The soft-leaf head lettuce has thin, yellowish-green, and soft leaves, and the leaf head is small and resistant to extrusion in transportation. The crispy-leaf lettuce has crisp, tender, and green leaves. The midrib of the leave is large, and the head is not tight enough, easy to break, and not resistant to extrusion during transportation.

(I) Structure of Culture Facility

The protected nutrient film culture facilities mainly include reservoir, culture bed, infusion pipe, water pump, timer, etc.

1. Reservoir construction

This content refers to the construction of deep flow hydroponics facility.

2. Culture bed

The culture bed is made of brick, cement, or hard plastic. The slope of the culture bed is reduced by 1 m (80 : 1–100 : 1) for every 80–100 m. The bed is covered with plastic film to prevent leakage. A thin layer of nutrient solution (2–3 mm) is often kept on the culture bed. The culture bed is covered with a polystyrene board with planting holes. When the seedlings are inserted into the planting holes, the root system will hang or stand upright in the culture bed, and a thin layer of nutrient solution at the bottom of the bed will flow continuously and slowly, so that the root system growing in the planting bed will be in a dark environment with water, gas, and nutrients.

3. Solution supply system

It consists of the solution inlet pipe, solution return trough, water pump, timer, and some pipe fittings. The timer controls the working time of the water pump, and the nutrient solution is regularly pumped out from the reservoir, and enters the culture bed through the solution inlet pipe for absorption and utilization by crops, and then flows back to the reservoir through the solution return trough. Intermittent solution supply is adopted to meet the needs of crops for oxygen, water, and nutrients.

(II) Crop Number Arrangement

The growth period of loose-leaf lettuce and wrinkled-leaf lettuce is relatively short, and there are no strict harvesting criteria, so as many as ten crops can be produced in one year. The following specific arrangements can be used for reference and application.

One crop is sown every month from February to May, with a seedling stage of 25–35 days. They are planted from late March to early June, harvested 30–40 days after planting and supplied from early April to early July. Three crops are sown from late June to late August, with a seedling stage of 15–25 days. They are planted from late July to mid-September, harvested 25–35 days after planting, and supplied from late August to mid-October. Two crops are sown from late September to November, with a seedling stage of 30 days. They are planted from mid-December to mid-February, harvested 55–60 days after planting, and supplied from mid-December to mid-February. The growth period of head lettuce is relatively long, so the number of crops shall be appropriately reduced.

(III) Culture Technique

1. Variety selection

Lettuce is cryophilic, and a temperature up to 25°C will cause difficulty in head formation for head lettuce. Therefore, early-maturing, heat-tolerant, late bolting, and adaptive varieties should be selected for soilless culture of lettuce, such as Kaiser, Great Lakes 366, Crsipy, Great Lakes 659, etc., which are relatively ideal varieties for soilless culture of lettuce.

2. Sowing and seedling culture

(1) Preparation before sowing: Prepare 3-cm-thick loose sponge blocks, cut them into 3 cm^2 pieces, and leave a little connection between each other to facilitate stacking. After the sponge blocks are cleaned, stack them flatly in the water-tight seedling tray.

(2) Sowing: Directly apply the disinfected seeds on the surface of the sponge blocks by hand, apply 2–3 seeds on each sponge block, add sufficient water to the

seedling tray until the surface of the sponge blocks is soaked, and cover a layer of non-woven fabric after sowing.

(3) Seedling stage management: It is very important to keep the seeds wet after sowing. The seeds shall be sprayed with a watering can 1–2 times every day to keep the surface wet. Under normal conditions, the seedlings can emerge evenly within about 3 days. In winter, the seedlings can emerge evenly within 5–7 days, which is slightly late. When the first true leaf is growing, a small amount of nutrient solution can be poured on the sponge block, and the concentration can be 1/3–1/2 of the standard solution concentration. Seedling thinning is conducted when there are both true leaves and growing points, leaving only one seedling on each sponge block. The seedling age of lettuce is generally about 15–30 days.

3. Planting management

(1) Preparation for planting: After the culture bed is prepared, install the solution supply system and check the sealing performance of the nutrient solution circulation system with clean water. After that, the facilities shall be disinfected, and formaldehyde-potassium permanganate can be used for disinfection and sterilization in the empty polytunnel. After the prepared nutrient solution is injected into the reservoir, the seedlings are planted in the culture hole of the PVC plate at a density of 30–40 plants/m^2 (Fig. 8-2).

Fig. 8-2 Hydroponics of Lettuce

(2) Nutrient solution management: Nutrient solution can be selected from the formula of Japanese Yamazaki Lettuce. The concentration of the nutrient solution used for planting seedlings within one week can still be 1/2 concentration of the standard solution. One week after planting, the concentration of the nutrient solution can be adjusted to 2/3 concentration of the standard solution. In the later growth period, the standard solution is used. EC value is controlled at 1.4–1.8 mS/cm. For head lettuce in the heading stage, EC value can be increased to 2.0–2.5 mS/cm, pH value is controlled at 6.0–7, and nutrient solution is circulated 4–5 times every day.

(3) Temperature management: The suitable growing temperature for lettuce is 15–20°C, and it is most suitable for growing in an environment with large day-night temperature differences and low nighttime temperatures. The daytime temperature is controlled at 18–20°C, and the nighttime temperature is maintained at 10–12°C. The temperature of the nutrient solution should be 15–18°C.

(4) Applying supplemental CO_2 fertilizer: CO_2 cylinder or generator can be used to apply supplemental CO_2 fertilizer to supplement CO_2 in the greenhouse, promote the photosynthesis of plants and improve yield.

4. Harvesting period management

The harvesting time of lettuce varies with season. Generally, lettuce grows slowly in winter and spring, requiring a longer time from sowing to harvesting, while lettuce grows quicker in summer. Be sure to note the harvesting time of lettuce, so as to achieve timely harvesting. Otherwise, if harvested too early, the yield will be affected; if harvested too late, it will bolt. The lettuce should be harvested in stages, and when harvested, it should be uprooted and sold with roots to indicate that the product is a green vegetable by soilless culture.

After this crop is harvested, the prepared seedlings shall be placed in the planting holes for planting, i.e. sow upon harvest. After 3–4 crops, the culture bed shall be flushed to thoroughly remove the residual roots, leaves, and dusts, and the solution supply system shall be cleaned. The culture facility shall be disinfected and re-injected with nutrient solution for the production of the next crop.

5. Pest and disease control

Soilless culture of lettuce rarely has pests or diseases. However, sometimes pests and pathogenic bacteria will be introduced due to human factors or greenhouse ventilation, thus causing diseases and pests in the lettuce. The diseased and insect seedlings shall be found and removed in time, and treated with chemicals.

II. Crown Daisy

It is also known as *Glebionis coronaria* or *Chrysanthemum coronarium*, and is an annual or biennial herb of the *Glebionis* genus in the Compositae family. Its roots, stems, leaves, and flowers can be used as medicine, with the effects of clearing blood, nourishing heart, moistening lungs, and clearing phlegm. The tender stems and leaves can be stir-fried, tossed, or made into soup, with a special fragrance. Both substrate culture and hydroponics can be adopted. Here is an introduction to the three-layer hydroponic method.

(I) Facility Construction

1. Facility structure

For convenient operation and better lighting of plants, the culture shelf is 1.6–1.7 m high and 60 cm wide, and its length depends on the cultivation space of the greenhouse.The three-layer culture bed is designed (Fig. 8-3), with 55–60 cm between two layers and 50 cm between the bottom layer and the ground. The culture

Fig. 8-3 The Three-layer Culture Bed

bed is made of polystyrene foam board, with a depth of 7–10 cm. A layer of black PVC film is paved at the bottom of the bed to prevent leakage of nutrient solution. The planting plate is composed of small foam boards, which are convenient for cleaning and disinfection. The nutrient solution first reaches the top layer through the solution supply pipeline, then flows through the second layer and the bottom layer, and finally flows back to the reservoir through the return pipeline. Valves are installed on each solution supply branch pipe to control the supply of nutrient solution.

2. Preparation of culture bed

Lay a thick layer of non-woven fabric on the planting plate, with both ends of the non-woven fabric penetrating deep into the nutrient solution of the culture bed from the bottom of the planting plate, so that a certain amount of water can be absorbed and the non-woven fabric can be wetted (Fig. 8-4). Before sowing, the non-woven fabric shall be wetted with water thoroughly, and then disinfected small gravels shall be evenly spread on the non-woven fabric to make the non-woven fabric better attached to the planting plate. Gravel shall not be too much; otherwise the foam culture bed tends to be damaged under heavy loads. In this culture form, in addition to the crown daisy, other leaf vegetables such as water spinach, pakchoi, lettuce, and endive can also be planted. It should be noted that crown daisy needs to be planted into the culture hole in a single plant (Fig. 8-5).

Fig. 8-4 Preparation of Soilless Culture Bed of Crown Daisy

Fig. 8-5 Sowing in Soilless Culture of Crown Daisy

(II) Production Techniques

1. Types of varieties

According to the leaf size, it can be divided into two types, i.e. big leaf crown daisy and small leaf crown daisy. The former is also known as round leaf crown daisy or broad leaf crown daisy, with broad and thick leaves, fewer and shallow incisions, thick and short stems, few fibers, dense internodes, high yields and good quality. It is poorly cold-resistant, relatively heat-resistant, and slightly late-maturing, and its cultivation is relatively common. The small leaf crown daisy is also known as fine leaf crown daisy or floral leaf crown daisy, with narrow and thin leaves, more and deep incisions, and thin and tall stems. It grows fast, is cold-resistant but not heat-resistant, and is suitable for cultivation in northern regions.

2. Seeding and seedling management

The seeds may be sown directly on wet non-woven fabrics without pregermination, then covered with a layer of perlite, and finally covered with a layer of film for moisturizing. If the temperature is high and the sunlight is strong, the sunshade net is required for shading. The root system of plant seedlings grows directly in non-woven fabrics, and the perlite has poor fixation ability, so the plant is prone to lodging, and the seeding density shall be higher than that in soil. When 70% of the seeds come out from the covering perlite, the film shall be removed in time to prevent baking the seedlings. At this time, the root system of the seedlings is still relatively weak. In the case of strong light at noon, the non-woven fabric will soon lose water and dry up, and the seedlings will wither. Therefore, in the seedling stage, the culture bed shall be sprayed with water in the morning of sunny days until the root system is developed and can absorb water from the nutrient solution. At that time, water spraying is no longer needed.

3. Cultivation management

(1) Temperature and light management.

Crown daisy is cryophilic and not resistant to high temperatures, with a suitable growth temperature of 17–20°C. It grows slowly below 12°C and grows poorly above 29°C. Its requirement for lighting is not strict, and it can grow normally under weaker light, so it is suitable for shelf culture.

(2) Nutrient solution management.

① Formula of nutrient solution: The standard formula of Japanese Yamazaki Crown Daisy can be selected.

② Management of nutrient solution concentration: Generally, the management concentration should be an EC value of 1.5–2.0 mS/cm, with 0.7–1.0 mS/cm in the initial stage of planting and gradually increased in the later stage.

③ pH control of nutrient solution: The pH value of the nutrient solution shall be measured regularly once a week. The appropriate pH value is 6.0–6.5. If the pH value of the nutrient solution is higher or lower than this range, it should be adjusted in time.

④ Solution supply method: The water pump shall be started regularly 2–3 times in the morning and afternoon every day for solution supply. The solution supply shall last for 15 min each time, and the water pump shall not be started at night.

4. Harvest

Generally, the plants can be harvested 40–50 days after sowing when they grow to 18–20 cm. If harvested too late, the stem skin will age and the quality will decrease. Harvesting can be divided into one-time harvesting and staged harvesting. For staged harvesting, 4–5 leaves or 1–2 lateral branches can be reserved at the base of the main stem, and the upper young stems and leaves can be cut off with a knife. After 20–30 days, when the lateral branches at the base sprout, they can be harvested again.

III. *Gynura bicolor*

It is also known as blood dishes or Okinawan spinach, and it is a perennial herb of the *Gynura* genus in the Compositae family. *Gynura bicolor* is rich in flavonoids and trace elements such as iron, manganese, and zinc, which are beneficial to the human body and have high healthcare value. Its tender stems and leaves are eatable as salad, soup, or fried dishes, with a soft and smooth taste and unique flavor. The soilless culture methods of *Gynura bicolor* include substrate culture, DFT hydroponics, static hydroponics, and stereoscopic culture.

(I) Types of Varieties

There are two types of *Gynura bicolor,* namely red-leaf *Gynura bicolor* and purple-stem green-leaf *Gynura bicolor*. The leaf back and stem of the red-leaf *Gynura bicolor* are purplish-red, including the new leaves. As the stems mature, they gradually turn green. According to the leaf size, it is further divided into large leaf and small leaf types. The large leaf type has large and slender leaves, pointed tips, more mucus, and long stem node, and its leaf back and stem are purplish-red; the small leaf type has fewer leaves, less mucus, and longer node, and its stem is purplish-red, and it is resistant to low temperature and suitable for cold regions in winter. Purple-stem green-leaf *Gynura bicolor* has a pale purple stem base, short node, poor branching performance, small elliptic dense green leaves, taper tips, short down, less mucus, poor texture, but strong heat and moisture resistance.

(II) Propagation Methods

There are three propagation methods: cuttage propagation, division propagation, and sowing propagation.

1. Cuttage propagation

Adventitious roots are easy to grow on the stem nodes of *Gynura bicolor*, and the cuttings are easy to survive, making them suitable for cuttage propagation, which is also a common propagation method in production. It is generally carried out in February–March and September–October. Select healthy and disease-free plants, cut 6–8 cm tender shoots, each with 3–5 leaves, remove 1–2 leaves from the base of each, and insert them into the pre-prepared seedbed or nutrition bowl. The substrate in the seedbed can be prepared by mixing sand or turf with vermiculite in a ratio of 2 : 1. Before cutting, the substrate shall be watered thoroughly. After cutting, a small tunnel shall be erected, and covered with used film or non-woven fabric to shade and moisturize it. The temperature in the seedling stage shall be controlled at 20–25°C, and the substrate in the seedbed shall be kept wet. Generally, the seedling can survive about 2 weeks.

2. Division propagation

Generally, division propagation is carried out after the plant enters dormancy or before recovery of growth (mostly before spring germination in the southern region) by cutting the underground perennial root, which shall be planted immediately after cutting. However, division propagation has a low propagation coefficient, and the growth vigor of plants after division propagation is weak, so it is generally not used in production.

3. Sowing propagation

When the temperature is stable above 12°C, the seeds shall be sown, and the seedlings can emerge 8–10 days after sowing. The seedlings shall be planted after reaching 10–15 cm. The advantage of seed propagation is that the seedlings are almost virus-free.

(III) Culture Technique of Column Pot Substrate of *Gynura bicolor*

1. Facility construction

(1) Level the ground and erect columns: First, level and compact the ground in the tunnel, and then build a culture trough with bricks and cement, with a length of 5 m, a width of 1.2 m, and a depth of 0.1 m. Two rows of columns are erected in the trough, with an axial distance of 0.7 m. Nutrient solution drips from the upper end of the culture column, seeps into the culture trough, and finally flows back to the reservoir. The culture column consists of the culture pot, planting cup, and PVC tube. There are five culture holes around the culture pot, and the culture pot has an internal hole diameter of 5 cm, so that the culture pot can be erected in the culture trough through a PVC tube. The space outside the hole diameter is filled with perlite wrapped in non-woven fabric, which is used to supply plant roots with water and

nutrient solution, promoting plant growth. The planting cup is filled with a complex substrate of turf, vermiculite, and perlite (in a ratio of 2 ∶ 1 ∶ 1). The seedlings of the plant should be planted in the planting cup in advance and then placed in the culture hole. The culture pot can rotate freely so that all seedlings on the column can receive light normally (Fig. 8-6).

Fig. 8-6 Culture Column of *Gynura bicolor*

(2) Composition of circulation system: The circulation system consists of the solution supply pipeline, solution return pipeline, water pump, timer, and other devices (Fig. 8-7). The reservoir is made of cement and brick to store clean water and nutrient solution.

Fig. 8-7 Circulation System for Soilless Culture of *Gynura bicolor*

One end of the main solution supply pipe is connected to the water pump of the reservoir through valves, and the other end is respectively connected to the solution supply branch pipes of each row of columns through valves. A hole is drilled at the corresponding solution supply branch pipe above each column to install the solution supply drip irrigation pipe (capillary), which is coiled in the culture pot substrate at the top of the column. During solution supply, the nutrient solution flows through the solution supply main pipe → solution supply branch pipe → capillary → culture pot, and the flow is controlled by the valves of the solution supply branch pipe. The excess water in the culture pot seeps into the culture trough from top to bottom through the small holes at the bottom of each culture pot, and finally the excess water flows back to the reservoir.

2. Substrate selection and disinfection

There are many types of substrates for soilless culture. The bulk density of substrates suitable for column culture cannot be too high; otherwise, the structure of the column is easily destroyed. We use a complex substrate mixed with turf, vermiculite and perlite (in a ratio of 2 ∶ 1 ∶ 1). During mixing, 80–100 g/m^3 carbendazim can be added and watered to mix evenly. The water content of the substrate is controlled at 60%.

3. Planting

For cuttage propagation of *Gynura bicolor*, when the root system of seedlings is developed, they are transferred to the planting cup for pre-culture for 10–15 days. During planting, pay attention to not damaging the root system of the plant, and the substrate in the planting cup shall not be too tight, as long as the lower end does not fall off. During the pre-culture period, pay attention to shading and moisturizing, and water according to the season and plant growth to keep the substrate in the planting cup wet. When the root system at the bottom of the planting cup is exposed, the cup can be placed on columns. Before that, start the water pump and water each column through the drip irrigation system to make the internal substrate wet.

4. Cultivation management

(1) Temperature and humidity regulation and control: The *Gynura bicolor* is resistant to heat but fears of frost, so pay attention to heat insulation during the low-temperature season in winter and spring (November to March of the following year) for the protected cultivation. In summer and autumn, the temperature is relatively high, so sunshade nets should be used, and the film should be removed for ventilation to reduce the temperature and humidity in the tunnel, reduce the occurrence of pests and diseases, and improve the quality and yield.

(2) Pruning adjustment: After planting, when the plant is 15 cm high, the growing point shall be removed to promote the germination of its axillary buds into the main vegetative branches. Generally, 4–6 shoots shall be reserved on each plant for column culture.

(3) Nutrient solution management:

① Formula of nutrient solution: The formula of general nutrient solution for leaf vegetables of South China Agricultural University can be selected.

② Management of nutrient solution concentration: The nutrient solution for the substrate culture of *Gynura bicolor* is widely applied, and it can grow within the EC value range of 1.5–3.0 mS/cm, without physiological abnormality. Generally, the management concentration should be an EC value of 2.0–2.5 mS/cm, which is 0.7–1.0 mS/cm in the initial stage of planting and gradually increased in the later stage. The EC value of the nutrient solution in the reservoir shall be measured once a week. When the EC value is low, the nutrient solution fertilizer shall be added in time.

③ pH control of nutrient solution: During the growth period of the plants on the columns, the pH value shall be measured once a week, and the appropriate pH value shall be 6.5–6.8.

④ Solution supply method: The amount and frequency of solution supply are related to the culture season. In high-temperature season, the solution is supplied 2–3 times a day, 10–15 min each time. In winter, the solution is supplied once every 3–5 days, 10–15 min each time.

(4) Regularly rotate the culture pot: Generally, the culture pot is rotated every 2–3 days so that all seedlings on each culture pot can receive normal light, and the growth vigor of seedlings is as consistent as possible.

(5) Regular salt washing treatment: During the culture process of *Gynura bicolor*, clean water should be regularly used through the drip irrigation system to clean the nutrient elements adsorbed on the surface of the substrate in the planting cup, so as to avoid salt accumulation poisoning the root system. Generally, it is cleaned once a month and watered with clean water by drip irrigation every half a month in high-temperature seasons like summer.

5. Harvest

20–25 days after planting, the seedling is about 20 cm high, and can be harvested before the top leaf is unfolded. When harvesting, cut the tender shoots of 10–15 cm, with 5–6 tender leaves at the tip and 2–4 leaves at the base, so as to germinate new lateral branches. After that, harvesting shall be conducted once every 10–15 days all year round.

IV. Celery

Celery is a biennial vegetable of the *Apium* genus in the Apiaceae family, and originates from the Mediterranean coast and the swamp in Sweden, Egypt, and other regions. Celery is rich in vitamins and minerals as well as volatile celery oil, with a strong flavor. The leaves taste bitter, so the edible part is the petiole in general. It can be fried, tossed, or used as stuffing, and some people use celery petiole to make juice. Recent medical studies have shown that celery has certain auxiliary therapeutic effects on cardiovascular diseases such as coronary heart disease, atherosclerosis, and hypertension.

(I) Variety Selection

In cultivation, according to petiole morphology, celery is usually divided into two types: local celery and western celery. Western celery with thick petiole is mostly chosen for soilless culture. Western celery has low crude fiber content, crisp taste, and higher yields, but its fragrance is not as strong as local celery. Representative varieties include Cornell 619, Florida 683, Tall Utah, Dutch Celery, California King, etc.

(II) Culture Season

Although celery can be cultivated in several crops for one year, from the perspective of protected soilless culture, the autumn and winter crops with the highest yields and maximum benefit should be the main ones. Generally, seeds are sown in early August, planted from early September to early October, and harvested around December.

(III) Soilless Culture Methods

NFT, DFT, rock wool culture, or other substrate culture methods can be used for celery. Here we introduce the DFT culture of celery (Fig. 8-8).

Fig. 8-8 DFT Culture of Celery

1. Culture facilities

See the facility construction of deep flow hydroponics.

2. Seedling culture and planting

Celery likes cold and humid environments, with a suitable growth temperature of 15–20°C, and it will grow poorly above 26°C, with high fiber content, bitter taste and poor quality. The seeds begin to germinate at 4°C, the optimal germination temperature is 15–20°C, and the germination is slow and uneven at high temperatures.

Before sowing celery seeds, soak the seeds in hot water at 48°C for 30 min, which has the function of disinfection and sterilization. Then soak the seeds in cold water for 24 h, wrap the seeds with wet cloth, and put them at 15–22°C to promote germination. Turn over the seeds 1–2 times every day to receive light, and rinse them with cold water. After 7–12 days, when over 70% of seedlings emerge, they can be sown. The seedlings can be raised in plug trays or substrate seedbeds. The turf, vermiculite and perlite are mixed in a ratio of 2∶1∶1, plated, watered, sown, spread with a thin layer of substrate, and covered with a film for moisturizing. The sowing density shall not be too high; otherwise the root damage will be serious when cleaning the root system of seedlings in the future.

When the seedlings grow to 5–6 cm, they can be transferred into the planting cup. Take the seedlings out of the substrate, rinse off the root substrate with clean water and disinfect them with 500-fold dilution of carbendazim. Then, put the seedlings into the planting cup. Gently take out the root system from the hole of the planting cup to contact the nutrient solution.

3. Cultivation Management

(1) Selection of nutrient solution formula: Leaf vegetable formula A of South China Agricultural University can be selected.

(2) Temperature management: Celery is a cryophilic vegetable. Generally speaking, it has strong adaptability to temperature in the seedling stage, while in the product formation stage, it has relatively strict temperature requirements. In general, the suitable temperature is maintained at 18–25°C during the day and 12–15°C at night. The temperature of the nutrient solution is 18–20°C. If the root temperature is

lower than 15°C for a long time, it is not conducive to growth.

(3) Nutrient solution management: Nutrient solution concentration varies in different growth stages of celery. Immediately after planting, EC value is controlled at 1.0–1.5 mS/cm, and increased to 1.5–2.0 mS/cm after one month of growth. After 2–3 weeks of further growth, the concentration of nutrient solution could be increased to 1.8–2.5 mS/cm, which lasted until one week before harvest, and no additional nutrient is required for harvesting. The growth pH value during this period is controlled within 6.4–7.2.

The circulation time of the water pump is controlled at 10–15 min/h in the daytime before the plant covers the row and 15–20 min/h after that, and it shall be controlled at 10–15 min/2h at night.

4. Timely harvest

About 50–60 days after planting of celery seedlings, the plant height is about 70–80 cm, and they can be harvested in due time according to the market demand.

Task 3 Sprout Vegetables

Edible sprouts, protocorms, buds, tender buds, young stems, or young shoots that are grown in dark or light conditions by using crop seeds or other nutritional storage organs (such as rhizomes and shoots) can be called sprout vegetables. Sprout vegetables are fast-growing, clean, nutrient-rich, high-quality, healthy, and suitable for various cultivation methods, with low investment, low cost, quick effect, and high economic benefit, and they are convenient for industrialized and intensive production.

According to the different nutrition sources used in the formation of sprout vegetable products, they can be divided into seed sprout vegetables and body sprout vegetables. Seed sprout vegetables refer to the buds or sprouts directly cultured using the nutrients stored in the seed, such as the soya bean sprout, green bean sprout, red bean sprout, broad bean sprout, Chinese toon sprout, pea sprout, radish sprout, buckwheat sprout, and alfalfa sprout. Body sprout vegetables refer to the protocorms, tender buds, young stems, or young shoots cultured using nutrients accumulated in the biennial or perennial crop's roots, fleshy tap roots, rhizomes, or shoots, for example, endive protocorms cultured from fleshy tap roots, dandelion buds and bitter *Ixeris* buds cultured from persistent roots, ginger buds cultured from rhizomes, as well as Chinese toon tree buds, pea shoots, and pepper shoots cultured from shoots.

I. Soilless Culture of Pea Seedlings

Pea seedlings, also known as Long Xu, have thick mesophyll, less fiber, tender and smooth taste and pleasant fragrance, and they are praised as a delicacy. Pea seedlings are produced indoors using a seedling tray, which has a high yield, simple operation, and good benefits.

1. Production equipment

The production of pea seedlings has low requirements on the site, and polytunnel or idle houses can be used. In order to improve the utilization rate of the site and make full use of the space, a multi-layer stereoscopic culture shelf can be placed in the pregermination room, with a height of 1.5–1.8 m, a length of 1.5 m, a width of 50 cm and a layer-to-layer spacing of 30–40 cm. The seedling trays are placed on the shelf, which is a black plastic hard tray with an outer diameter of 60 cm, a width of 24 cm, a height of 4–6 cm, and a perforated flat bottom.

2. Variety selection

The pea seeds used for producing sprout vegetables shall have high purity, good cleanliness, large grain size and high germination rate, so as to ensure good quality, disease resistance, and high yields of the pea seedlings produced. Generally, peas for vegetables easily rot in the production process, so green peas, colored peas, gray peas, brown peas, spotted peas, and other grain peas can be selected. In summer, it is best to choose peas that are resistant to high temperatures and viruses. Besides, when purchasing seeds, in addition to the quality of seeds, it is also necessary to consider whether the supply is sufficient and stable, and whether the seeds are clean and pollution-free.

3. Seed cleaning and soaking

The growth uniformity, commodity rate, and yield of pea seedlings are closely related to the seed quality. Therefore, in addition to using high-quality seeds, the seeds used for pea seedling production must be disinfected before sowing, and the insect-eaten, residual, deformed, mildewed, empty, extra small, and germinated seeds shall be removed.

In order to promote seed germination, the cleaned seeds need to be soaked. Generally, the seeds are first washed with clean water at 20–30°C for 2–3 times and then soaked for 20–24 hours, with a water volume of 2–3 times the seed volume. After soaking, the seeds shall be washed 2–3 times again, and then the seeds shall be taken out, and the surface water shall be drained before sowing.

4. Sowing and tray-stacking pregermination

Before sowing, the seedling tray shall be washed clean and disinfected with limestone solution or bleach solution, and then washed clean with clean water. After that, a layer of paper shall be paved on the bottom of the tray to sow seeds. 350–450 g of peas shall be sown evenly on each tray so that the sprouts can grow evenly.

After sowing, the seedling trays shall be stacked together and placed on a flat ground to promote germination. Note that the stacking height of seedling trays shall not exceed 100 cm, and the spacing between stacks shall be 2–3 cm, so as to avoid uneven emergence of seedlings caused by excessive closure and poor ventilation. In order to maintain proper air humidity, the stacked trays shall be covered with black film or double-layer sunshade net. Pregermination shall be carried out in a pregermination room with relatively stable humidity conditions, and the temperature shall be kept at 20–22°C to improve the germination rate. Water shall be sprayed once a day during the stacking-tray pregermination period, and the water volume shall not be too large to avoid bud rot. During water spraying, tray reversing shall be carried out once to change the positions of the surrounding seedling trays so as to make the culture environment of the seedling trays as even as possible and promote the even growth of seedlings. Under normal conditions, stacking-tray pregermination can be completed for about 4 days, and the trays are scattered on the culture shelf for greening. The pea sprouts are about 1–2 cm high when put on the shelf. The temperature in summer is relatively high, and the accumulated temperature during the germination of seeds is easy to cause rotten seeds. They can be put on shelves earlier to reduce rotten seeds. The temperature in winter is low, and the accumulated temperature during the germination of seeds is conducive to growth. They can be put on shelves later. As long as the long seedlings do not top the tray, stacking-tray pregermination is available.

5. Greening period management

(1) Light management. In order to ensure the safe transition of pea seedlings from a dark and high humidity environment for stacking-tray pregermination to a new culture environment, when the seedlings are transferred to the culture shelf, there should be a transition period of 1 day with relatively stable air humidity and weak light to avoid wilting of sprouts. To produce "green products", the seedling tray should be placed in an area with strong light for 2–3 days before the pea seedlings are put on the market so that the sprouts can be greened better. In summary, pea seedlings can grow into green large-leaf seedlings under strong light, yellow thin seedlings under no light, or light green seedlings under weak light.

(2) Temperature and ventilation management. The suitable growth temperature of pea seedlings is about 20°C. The growth is hindered when the temperature exceeds 30°C, and is slow when the temperature is lower than 14°C.

Ventilation is one of the important measures to regulate the temperature of the culture room and reduce the rot of seed buds. On the premise of ensuring the indoor temperature, ventilation shall be carried out at least once or twice each day. Even when the indoor temperature is low, ventilation shall be carried out for a while.

(3) Water management. Since sprout vegetables are cultivated in a different way from the seedling tray paper bed culture method in general soilless culture, and the sprouts themselves are tender and juicy, so they should be frequently supplemented with water. The principle of water supplement is watering frequently in a small amount, i.e. spray water twice a day in winter and 4–5 times a day in summer. The watering should be even and from the upper layer to the lower layer. The amount of watering should be controlled to the extent that the bottom of the seedling tray has no large amount of dripping water after spraying. In addition, the water amount shall be small in the early growth stage and slightly larger in the late

growth stage; that is, less watering is required before the seedlings reach 5 cm to wet the seeds and paper at the bottom of the seedling tray, and more watering is required after the seedlings reach 5 cm. The water amount should be small in rainy and low-temperature weather and slightly larger in high-temperature weather. When the indoor air humidity is large and the evaporation capacity is small, the water amount should be smaller.

6. Harvest

Under normal culture and management conditions, pea seedlings can be harvested 8–9 days after sowing when the height of seedlings is about 10–15 cm. At the time of harvesting, the top leaflets have been unfolded. The tips are cut 7–9 cm for consumption, and each tray can yield 350–500 g. After the first harvesting is completed, quickly place the seedling trays under a strong light for culture. After the new sprout is germinated, place the trays under 2.0–3.0 klx light for culture. When the seedling height reaches 10–15 cm, they can be harvested again.

7. Precautions

(1) After two harvests, clean the seedling trays in time, remove the adherent impurities, then dry and collect them for use in the next crops.

(2) After producing 1–2 lots of sprout vegetables, the production site and supplies shall be disinfected with bleaching powder or calcium carbonate spray.

(3) Keep the water source and site clean throughout the production.

II. Soilless Culture of Broad Bean Sprouts

Broad bean seedlings contain calcium, zinc, manganese, phospholipids, etc., which regulate the brain and nerve tissue. It is a new healthy sprout vegetable with abundant choline, which can enhance memory and invigorate the brain. Because of their good taste, high nutritional value, green pollution-free, and low fiber content, broad bean sprouts are favored by consumers.

1. Production equipment

The production site and equipment of broad bean sprouts are prepared in the same way as that of pea seedlings.

2. Variety selection

There are many varieties of broad beans everywhere, including flesh red, grayish-white, brown, purple, and milky white types classified by color; or large-grain, medium-grain, and small-grain types classified by the thousand-seed weight. It is better to use the small-grain seeds to produce broad bean sprouts.

3. Seed cleaning and soaking

The seeds shall be fresh seeds with uniform size, high maturity, good germination rate, plump and disease-free grain, and the preservation period shall not exceed 12–20 months. Before soaking, the insect-eaten seeds, residual seeds, and other poor-quality seeds shall be removed.

After cleaning of seeds, they shall be disinfected to avoid mildew during pregermination, and then soaked in warm water for 20–24 hours.

4. Pregermination and sowing

After the seeds are soaked, clean and place them at 20–25°C to promote germination, and sow them after 80% of the seeds show the white buds. Broad bean sprouts can be produced by sand culture.

First, clean and disinfect the seedling trays, lay a layer of white paper at the bottom of the trays, and spread a layer of 2.5-cm-thick disinfected fine sand on the white paper. Second, evenly spread the germinated broad bean seeds, cover a layer of fine sand, and spray water immediately. Third, place the seedling trays on the culture shelf.

5. On-shelf management

On-shelf management is similar to that of pea seedlings, but due to the certain water retention of fine sand, the watering frequency per day depends on the specific condition. Generally, 1–2 times per day in winter and 3–4 times per day in summer are enough. The other general principle of watering is the same as that of pea seedling production management.

6. Harvest

In general, the first crop of bean seedlings can be harvested 10–15 days after sowing. The large-leaf seedlings can be harvested when they are 10 cm high. The thin seedlings are generally harvested when they are about 15 cm high.

Task 4 Soilless Culture of Fruit Trees

I. Elevated Substrate Culture of Strawberries

Strawberries are the main species in Chinese facility horticulture. Compared with other species, strawberry protected cultivation has the characteristics of short period, quick effect, high benefit, long harvest period, and good quality. However, the current main production is protected ground ridge cultivation covered by film. Long-term continuous cropping easily causes the accumulation of various pathogens and secondary salinization of the soil, and the soil-borne diseases produced seriously restrict the yield and quality of strawberries and affect economic benefits. Moreover, strawberry planting requires a large amount of labor. In addition to ground preparation before planting, seedling culture, planting of small seedlings, removal of aged leaves, thinning of flowers and fruits, and fruit harvest all require personnel to bend over, with very high labor intensity. Production of 1 hm^2 strawberry requires about 20,000–20,500 hours of labor time. With the aging of the agricultural population, increasing production costs, and shortage of labor force in China, it is urgent to develop and popularize the labor-saving cultivation mode of strawberry. Stereoscopic facility culture is a culture method to alleviate the obstacles of continuous cropping and realize the labor-saving culture and clean production of strawberries. It aims to improve the natural environment and production conditions through human efforts, utilize and exert the integration effect, expand the effective utilization space and time of resources, improve the utilization rate of resources per unit area and time, realize the productive potential of limited land, and obtain remarkable results. Stereoscopic facility culture emerged in Japan in the late 1990s. At present, the application area accounts for about 1/10 of the strawberry cultivation area in Japan. In recent years, it has also been applied in Taiwan, Beijing, Shanghai, Zhejiang, Jiangsu, and other regions of China. With the rapid development of China's economy, people's consumption ability is enhanced, and the market prospect of strawberry stereoscopic facility culture characterized by high quality, safer for eating, leisure, and sightseeing is broad. It is also of great significance to introduce, apply, and innovate this technique.

(I) Significance of Strawberry Elevated Substrate Culture

1. Space-saving and wide application regions

From the perspective of cultivation volume per unit area, the cultivation density of strawberries in the traditional cultivation mode is 120,000–150,000 plants/hm^2; for the stereoscopic soilless culture mode, the total culture volume can reach 300,000–450,000 plants/hm^2. From the perspective of unit area yield and value, the average yield of conventional greenhouse-cultivated strawberries is about 45,000 kg/hm^2, the average value is CNY 270,000–300,000/hm^2, the maximum yield is 90,000 kg/hm^2, and the maximum value is about CNY 600,000/hm^2; for stereoscopic soilless culture, the average yield is about 180,000 kg/hm^2, the value is nearly CNY 900,000/hm^2, and the benefit has increased significantly. In addition, in terms of soil conditions, strawberries have high requirements for fertilizer and water. When using traditional methods of cultivation, it is advisable to select fertile soil plots. When using stereoscopic soilless culture, the nutrient substrate can be selected freely according to the requirements and is basically not limited by soil conditions. Regardless of whether the soil is fertile or not, it can be developed even on barren slopes, barren waste, or cement land that is not suitable for planting.

2. Effectively improve quality, suitable for developing leisure and sightseeing agriculture

The traditional cultivation mode of strawberries is usually continuous cultivation on the same land, which is easy to cause continuous cropping obstacles. For many farmers, continuous cultivation of strawberries requires changing plots or rotation. The substrates used in stereoscopic soilless culture of strawberries are mainly mixtures of peat soil, turf, and perlite, which can not only produce high yields and high efficiency, but also supplement nutrition in continuous cropping, thus effectively alleviating the difficulties in continuous cropping of strawberries. Meanwhile, according to the physiological characteristics of strawberries and the demand for water and fertilizer in different growth stages, the concentration of nutrient solution can be adjusted, and the growing environment such as ventilation, heat energy, and humidity can be controlled to keep the growth and development of strawberries always in the best state. The occurrence of powdery mildew, gray mold, anthracnose, mites, thrips, and other diseases and pests should be suppressed, so as to reduce the application of pesticides and increase the proportion of high-quality fruits. Besides, strawberries can be picked without bending, and the green leaves and red fruits are also of great ornamental value, which is more suitable for the development of leisure and sightseeing agriculture.

(II) Facility Construction

1. Erection of culture shelf

(1) Principles of erection. Shelves can be erected in paddy fields after the ground is dried, compacted, and hardened, and can be erected in dry land after the land is shallowly plowed, leveled, and hardened. Shelves should be erected in the north-south direction under uniform light, and shall be consistent with the direction of the polytunnel. The spacing between shelves shall be 70–80 cm, which is convenient for walking operation. For single-span polytunnel with a width of 7 m or a multi-span polytunnel with a width of 6 m, each span can hold five rows of shelves respectively. The width of a shelf is controlled within 35 cm, and the length is controlled within 50 m. If the length exceeds 50 m, two sections of shelves can be erected separately.

(2) Type of culture shelf. The elevated strawberry culture emerged in Japan in the late 1990s. It is a production model where strawberry seedlings are planted and managed in beds at a certain distance from the ground using special equipment. The planting equipment can be divided into bracket culture and hanging trough culture, single-stack horizontal culture, double-stack stereoscopic culture, etc. The following mainly introduces the bracket-type elevated substrate culture shelf (Fig. 8-9) and the prefabricated-type elevated cultivation device (Fig. 8-10). The former, which is cost-effective and easy to install, can effectively solve the problems of weak water retention capacity of the substrate surface in the early stage of substrate culture, low planting survival rate, and stunted growth. The latter features an adjustable culture shelf, and combines the advantages of substrate culture and deep flow culture technique, enabling effective control of the substrate water retention capacity, which

in turn achieves accurate control of growth vigor, water and fertilizer saving, easy operation and management, and effective improvements in yields and quality.

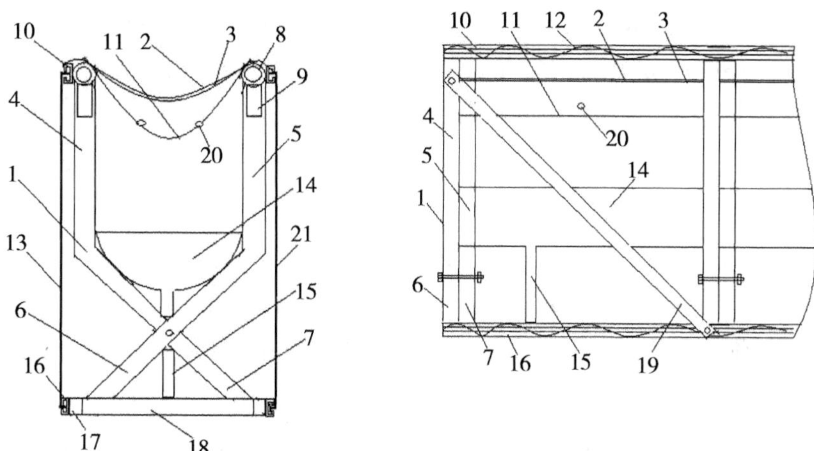

Fig. 8-9 Bracket-type Elevated Substrate Culture Shelf

1.Fixing bracket 2. Non-woven fabric 3. Insect-proof screen 4 & 5. Vertical pipes 6 & 7. Supporting legs 8. Horizontal pipe 9. Connecting rod 10 & 16. C-shaped pipes 11. Film 12. Clamping strip 13 & 21. Humidity control films 14. Recovery trough 15. Water return pipe 17. Side supporting rod 18. End supporting rod 19. Diagonal bracing rod 20. Leakage hole

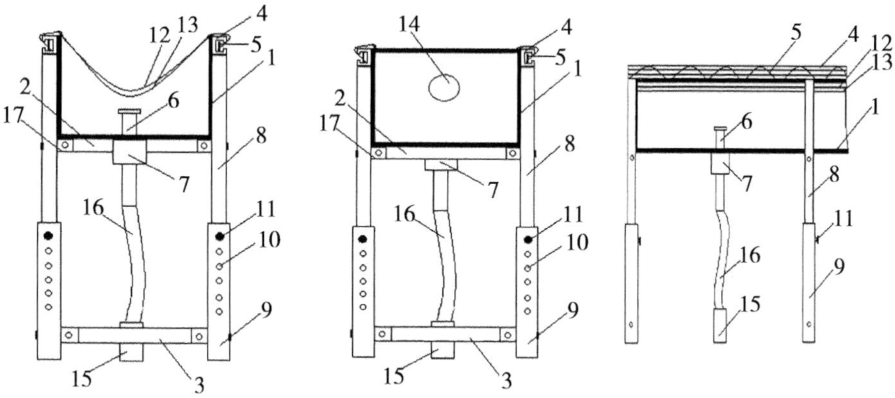

Fig. 8-10 Prefabricated-type Elevated Cultivation Device

1. Culture trough 2. First cross bar 3. Second cross bar 4. C-shaped pipe 5. Clamping strip 6. Guide pipe 7. Pipe sleeve 8. Upper pipe 9. Lower pipe 10. Second through hole 11. Lock pin 12. Non-woven fabric layer 13. Insect-proof screen layer 14. Vent hole 15. Drain outlet 16. Hose 17. Clamping piece

2. Culture substrate

(1) Principles of substrate formulation. A good substrate should have the following characteristics: stable physicochemical properties make it less prone to decomposition, rotting, and degradation; strong water conservation capacity and good permeability; cost-effective, easily accessible; small specific gravity and light-weight; safe and environmental-friendly. To meet the foregoing requirements, the substrate is usually a mix of organic materials such as peat, special organic fertilizers as the main ingredients, inorganic materials such as perlite and vermiculite and some fused calcium magnesium phosphate fertilizer. The formulation technique meets the needs of strawberry growth.

(2) Method of substrate formulation. The formula of strawberry culture substrate is peat：vermiculite：perlite=2：1：1 and appropriate commercial organic fertilizer and calcium-magnesium-phosphate fertilizer are added. This formula has good physicochemical properties, and the strawberry has a high survival rate after transplantation and strong growth vigor.

(3) Substrate loading. Stir the substrate evenly and transport it to the elevated culture shelf for loading through the loading vehicle.

3. Irrigation facilities

The technique of drip irrigation under film is used for elevated substrate culture. The drip irrigation pipe can be hard rubber pipe or soft plastic drip irrigation tape. The spacing between drip irrigation holes is 10–15 cm. Due to the large substrate cavity, strong vertical permeability of moisture, and poor horizontal diffusivity, a layer of non-woven fabric can be paved on the substrate surface, and then a drip irrigation pipe can be laid on the non-woven fabric to uniformly permeate water into the substrate through the diffusion and permeation of non-woven fabric.

(III) Production Techniques

1. Variety selection

Varieties suitable for shelf culture are mostly selected, such as Red Cheek, Akihime, Jiuxiang, Ningyu, and other high-quality early-maturing varieties.

2. Planting

(1) Time for planting. It depends on the flower bud differentiation of strawberry

seedlings. Generally, it is appropriate to plant from late August to early- and mid-September. The seedlings are plug seedlings of about 1.0-cm-thick rhizomes, short but strong, and free of diseases and pests.

(2) Density of planting. Single-type elevated culture trough adopts double-row culture, with a plant spacing of 15–20 cm, a small row spacing of 25–30 cm, and about 5 cm from the trough edge, generally 6,000 plants/mu.

(3) Methods of planting. To facilitate management and harvesting, the strawberry seedlings must be planted in a manner that the stem bow back is oriented towards the culture trough side. In this way, the well-ventilated and well-lighted culture trough sides provide a favorable growing environment for the strawberry fruits with brighter color and higher quality, and flowers and fruits grow on both sides of the culture trough to reduce humidity and the occurrence of pests and diseases.

(4) Depth of planting. Depth of planting is one of the keys to the survival of strawberry substrate culture. During planting, the seedling base shall be flush with the substrate surface and the substrate around the root system shall be compacted. Shallow planting will make the root system exposed and easy to dry because of the difficulty of absorbing water; deep planting with the growing point buried in the soil can hinder the development of new leaves and eventually kill the plant.

3. Management after planting

(1) Water management. The most critical period of water management is 1–7 days after the planting of strawberry elevated substrate culture. Because of the limited water retention capacity of the substrate surface, the substrate surface is extremely prone to being dry, and at this time the new roots of strawberry seedlings have not been generated. Therefore, micro-sprinkler irrigation shall be used for the whole trough, with a small amount and high frequency. In addition, the liquid phase device in the culture trough shall be adjusted to the highest water level, keeping the culture substrate in a saturated humidity environment to ensure the wet substrate surface. When new strawberry roots and new leaves emerge in 7–10 days, water control is necessary to ensure the substrate is wet, and both moisture and oxygen

are sufficient in the substrate to balance the water and air in the substrate. When the strawberry roots extend out of the insect-proof screen and towards the liquid, the liquid phase control valve shall be gradually turned down to ensure that there is enough space for oxygen supply in the lower part of the culture trough.

(2) Environmental humidity management. Around 1–3 days after planting, the strawberry enters the rooting stage, and the new roots have not yet grown. At this time, the strawberry mainly depends on the moisture in the air to maintain its life. Therefore, the humidity in the polytunnel shall be above 80%, which is conducive to strawberry rooting. Around 4–10 days, the humidity of the culture environment can be gradually reduced and controlled at about 70%. About 10 days after strawberry rooting, the environmental humidity shall be controlled, and the daytime relative humidity shall be below 60%, especially in the anthesis.

(3) Light management. Around 1–3 days after planting, strawberries can be shaded all day to reduce evaporation of water from leaves brought by sunlight, which affects the seedling recovery stage and survival rate of strawberries. In the first 4–6 days after planting, light exposure is allowed in the morning and evening to promote new root formation and leaf photosynthesis of strawberries. After survival, shading is no longer required.

(4) Temperature management. After planting and before survival, increase the environmental temperature (25–28°C during the day and 20–25°C at night) to promote the formation of new roots; in the growth stage after survival, 26–28°C during the day and 10–15°C at night; in the squaring stage, 25–28°C during the day and 8–12°C at night; in the anthesis, 22–25°C during the day and 8–10°C at night; in the fruit expansion stage and maturity stage, 20–25°C during the day and 5–10°C at night, forming a large temperature difference to improve the fruit quality.

(5) Fertilization management. Detection of nutrient solution concentration in the substrate: Readings (EC values) of the leakage solution leaked under the shelf are detected by using the ion electrical conductivity meter. When the EC value is higher than the target concentration of each growth stage, it indicates that the nutrient concentration of the culture substrate is too high and can be diluted just by dripping water; when the EC value is lower than the target

concentration, it indicates that the nutrient of the culture substrate is insufficient and the nutrient solution with appropriate concentration must be prepared for supplementation. The most suitable pH value for strawberry growth is 5.0–6.5. When the pH value < 4 or pH value > 8, the growth and development of the strawberry are impeded. See Table 8-1 for details.

Table 8-1　EC Value and pH Value of Nutrient Solution Concentration at Different Growth Stages

Growth stage	After planting and before flowering	Anthesis	Fruit-bearing stage
EC value (mS/cm)	0.5–0.6	0.6–0.7	0.8
pH value	Maintained at 5.0–6.5 throughout the growth period		

(6) Plant management.

① Remove lateral buds: Red Cheek, Akihime, and other strawberries have both one-bud and multi-bud plants. Generally, two buds are reserved in the axil, and three buds at most are reserved for plant pruning. Other redundant lateral buds should be removed from the parent plant in time.

② Remove aged leaves: Two weeks after planting, five unfolded leaves are reserved, and the aged leaves are removed to promote the development of primary roots in the rhizomes. Five leaves are still reserved for pruning before and after film mulching. Five unfolded leaves shall be ensured when the terminal inflorescence is budding. Functional leaves should be reserved as much as possible during harvesting, except that aged leaves should be removed. However, in order to inhibit the occurrence of red spider and calyx blight obstruction during the second budding of the second axillary inflorescence, the drooping leaves shall be removed as much as possible at the end of the terminal infructescence harvest.

③ Remove flowers: Small flowers are removed carefully according to their conditions, which can increase the proportion of large fruits and is conducive to marketing. Ten low-order small flowers are reserved for the terminal inflorescence; 7–10 small flowers are reserved for the first batch of axillary inflorescences with only one bud and 5–7 small flowers are reserved for each with two buds; five small flowers are reserved for the second batch of axillary inflorescences, and the redundant ones are removed. Besides, from the perspective of fruit quality, the content of soluble solids in fruit is also different when the flowers are removed. Comparing the contents of soluble solids in fruits of the same order from the areas with or without flower removal, it can be found that the content of soluble solids in fruits is the highest in those treated with flower removal and the lowest in those bearing more fruits with the development of inflorescence. Therefore, flower removal can promote fruit expansion and is very important to improve fruit quality.

④ Petal and pedicel removal: Petals scattered on fruits and sepals after flowering are a good medium for diseases and pests and are easy to cause virus occurrence. After flowering and fruit setting, they shall be regularly blown off with a mist sprayer or other air supply machinery. After harvesting a batch of infructescences, the pedicels will consume a large amount of nutrients in the plant, so they shall be removed as soon as possible.

(7) Assisted pollination. In winter, strawberry greenhouse has low temperature, high humidity and less ventilation, resulting in poor pollination of strawberries and a large number of deformed fruits. Pollination by beekeeping is the key technique to avoid deformed fruits. It is better to place a box of Dutch bumblebee every 1,000 m^2 of greenhouse.

(8) Flower and fruit management. When the strawberry flowers and bears fruits, the flower and fruit thinning shall be carried out in time. First, the higher-order fruitless flowers, abortive fruits, and deformed fruits shall be removed, and then the flower and fruit thinning shall be carried out according to the principle of retaining 3–4 fruits on the first inflorescence, 5–6 fruits on the second or third inflorescence, and 2–3 fruits on the fourth or fifth inflorescence, so as to ensure strawberry yield and quality and improve its commercial value.

4. Harvest

(1) Maturity degree. When the fruit develops to a certain extent, the skin changes from green to white and then to red, and the fruit surface from the top and base of the fruit can be uniformly colored. At the same time, the seeds embedded

in the fruit surface also change from green to brown, indicating that the fruit has entered the mature stage. Colored fruits of 80% maturity or below taste bad, and those of 90% –100% maturity taste good. In winter, the temperature is low, and the fruits are harvested at 100% maturity; after spring, the temperature rises, and the fruits are harvested at 80% –90% maturity to prevent fruit softening.

(2) Harvesting time. Harvesting is available at any time of the day in winter; after the temperature rises in spring, it is better to harvest in the morning when the temperature of the fruit is relatively low. Fruits on both sides of the shelf are separately harvested: For the shelves set in the north-south direction, the fruits not exposed to sunlight in the west are harvested in the morning, and the fruits not exposed to sunlight in the east are harvested in the afternoon.

(3) Harvesting methods. Use the index finger and thumb of the right hand to pinch off the pedicel of a single fruit, leave about 1 cm of pedicel and put the harvested fruits into carry-on shallow plastic baskets. Prevent the pedicel from piercing the skin when placing the fruits.

(4) Post-harvest treatment. The fruits harvested in winter can be directly graded and packed; those harvested in spring when the temperature is high shall be pre-cooled in cold storage at −1°C to 0°C, and then packed and sold after the fruits cool down.

II. Root Restriction Soilless Culture Technique of Grapes

It is generally believed that fruit trees are perennial crops, and that only by cultivating strong and tall tree structures can they achieve high yields and longevity. Therefore, the concepts of "deep roots and big trees" and "deep tillage and more fertilization" dominated people for a long time. However, this traditional method has at least the following deficiencies: ① In order to cultivate tall trees, it is required to form a specific tree shape through re-pruning and cultivation, which will cause young trees overgrowth, with fewer flowers, lower yields, and poorer quality. ② Deep tillage and wide application of fertilizer make the root system widely distributed, making it difficult to determine the exact position of the root system. Fertilization has a certain blindness. Some fertilizer components applied in the early stage sometimes move to the root system position or the root system extends to the fertilizer position only when the fruit quality is formed, causing a delay in the fertilizer effect. At this time, excessive absorption of nitrogen and other fertilizers will inhibit fruit coloration and sugar accumulation. ③ In areas with abundant precipitation or places with high groundwater levels, the root system will excessively absorb water, which will not only induce early overgrowth but also be unfavorable to the sugar accumulation in the later stage, and even cause splitting fruit. People have long recognized the above-mentioned deficiencies of traditional cultivation methods in long-term practice. In order to overcome these deficiencies, researchers engaged in fruit tree production techniques at home and abroad have made long-term and unremitting efforts and made many explorations and attempts. The achievements made in these explorations include applications of root pruning, growth inhibitors, and dwarfing rootstock. However, not all fruit trees have good rootstocks due to the limitation of dwarfing rootstock resources. At the same time, the application of these research results is limited due to the poor practical operability of root pruning and concerns about the product safety of growth inhibitor residues. Moreover, even if the fruit trees like apples to which dwarf rootstocks are most successfully applied, the root system is still distributed in the deep and wide soil range, and the deficiency of difficult regulation and control of fertilizer and water is still not fundamentally solved. Since the 1990s, people have been inspired by root pruning and traditional oriental potted plant art, have explored the root restriction cultivation methods, and then applied them to the soilless culture of grapes. Studies on grapes have shown that root restriction can overcome the shortcomings of the above-mentioned traditional cultivation methods, and has obtained positive results in controlling vegetative growth of aerial parts and improving early yield and fruit quality. Especially, remarkable effects are achieved in improving fruit quality.

Root restriction has been widely applied in grape cultivation in Japan, Australia, and other developed countries. In the process of application of root restriction cultivation, some simple and easy root restriction soilless culture techniques have been developed. For example, Japan has developed a Raised bed soilless root restriction culture method for grape and greenhouse citrus, namely, laying plastic film on the ground surface, stacking organic matter on the film to form ridges, and planting fruit trees on the ridge. The root system is limited in the ridge, and the methods are simple and effective. Wang Shiping et al. further put forward the suitable root volume indicator to alleviate the grapevine vigor, promote fruit setting and improve fruit quality; that is, the effective root volume of 0.05–0.06 m^3 per square meter of leaf area can not only ensure the reasonable growth of the grapevine, but also ensure high yields and quality. This indicator standardizes, quantifies and improves the operability of root restriction cultivation techniques, making it a more mature new technique.

(I) Forms of Restriction Soilless Culture

In the south of the Yangtze River basin where the rainfall is more than 1,000 mm, the high soil moisture content is an important reason affecting the quality of grapes and inducing splitting fruit. Root restriction soilless culture is restricted by root area, and the water absorption area of the root system is strictly limited to a very small range. Transpiration of leaves can timely reduce the root soil moisture content, which can effectively improve quality and overcome splitting fruit. This root restriction mode can adopt ridge-type, trough-type, ridge-trough combination, basket-type, etc.

1. Ridge-type

In the southern region with abundant rain and no frozen soil layer, ridge-type can also be adopted for root restriction culture. The ground is paved with plastic film, on which nutrient soil is piled up to form ridges for growing grapes. In the growing season, the surface of the ridges is covered with black or silver-gray plastic film to maintain the stability of soil moisture and temperature in the ridge. Ridges vary in specification according to culture density. When the row spacing is 4–8 m, the ridges shall have an upper width of 50–100 cm, a lower width of 70–140 cm, and a height of 50 cm (Fig. 8-11). This method has the advantage of simple operation, but the change of root soil moisture content is unstable and the growth is easy to be weakened. Therefore, a good drip irrigation system must be provided.

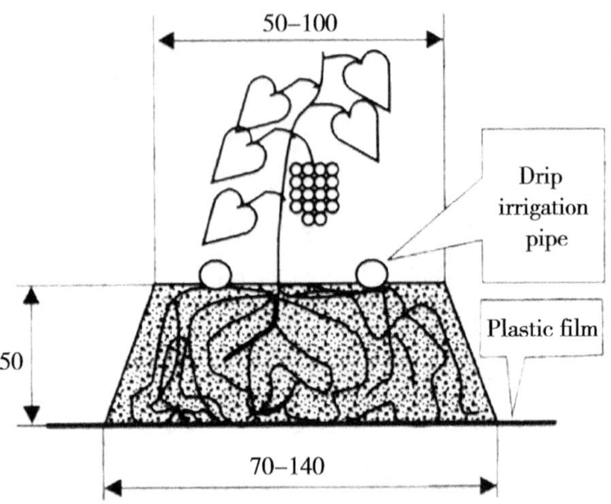

Fig. 8-11 Grape Ridge-type Root Restriction Soilless Culture Mode (Unit: cm)

After paving microporous non-woven fabrics or slightly raised plastic films on the ground (to prevent water accumulation), nutrient soil rich in organic matter is piled up on them to plant fruit trees in the form of ridges or mounds, which is called "Raised bed" in Japanese. The surrounding surface of the ridge is exposed to the air, and the bottom surface has an isolation film, so the root system can only grow in the ridge. This method is easy to operate and suitable for warm regions without soil freezing in winter. However, the root soil moisture content and temperature are not too stable in summer.

2. Trough-type

Good drainage is required in this mode. Excavate a planting furrow with a depth of 50 cm and a width of 100–140 cm, excavate another drainage underdrain with a width of 15 cm and a depth of 20 cm at the bottom of the furrow, pave the bottom and wall of the planting furrow and the drainage underdrain with a thick plastic film (for greenhouses and polytunnels), fill the drainage underdrain with river sand and gravel (if possible, replace river sand and gravel with permeable pipes), and connect the drainage underdrain with the main drainage ditches on both sides (upper width 80–100 cm, lower width 40–50 cm, depth 80–100 cm) to ensure timely and smooth drainage of accumulated water (Fig. 8-12). When the non-woven fabric is used to replace the plastic film and paved on the bottom side wall of the planting furrow, since the non-woven fabric is permeable, it will not accumulate water, and no drainage ditch is necessary. However, the non-woven fabric has a short service life. After 2–3 years, it will lose its restriction effect, and the root system will break through the non-woven fabric and extend to the soil outside the preset root zone. Studies have shown that the change of root soil moisture content is relatively small and excessive stress rarely occurs in the trough-type root restriction soilless culture. The growth of new shoots and leaves of the grapes is moderate and robust, and the fruit quality is good.

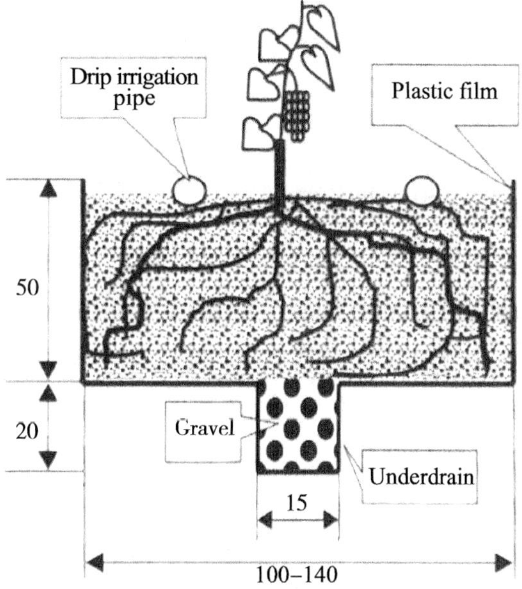

Fig. 8-12 Grape Trough-type Root Restriction Soilless Culture Mode (Unit: cm)

3. Ridge-trough combination

A portion of the root is placed in a trough and a part is placed on the ground in the form of the ridge. Generally, the trough depth is 20–30 cm and the ridge height is 30–20 cm. The specification of the trough and ridge varies according to the row spacing. When the row spacing is 4–8 m, the trough width is 70–140 cm, the lower width of the ridge is 70–140 cm, and the upper width is 50–100 cm (Fig. 8-13). The combination mode of ridges and troughs has the advantages of trough-type of stable root water content, medium growth and good fruit quality, and also has the advantages of ridge-type of simple operation and good drainage.

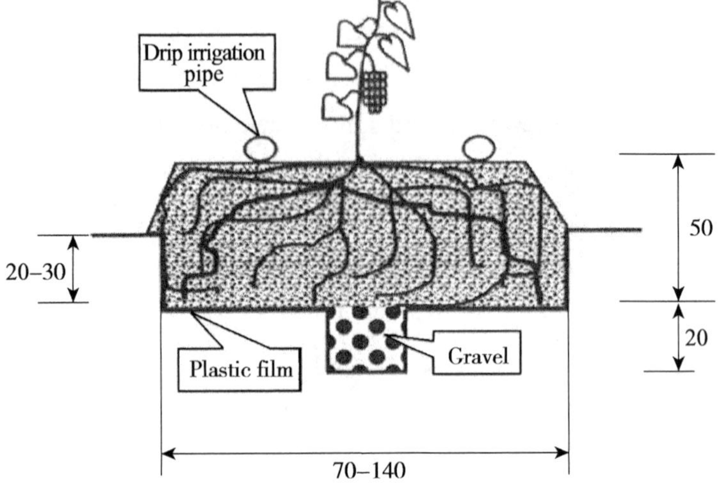

Fig. 8-13 Grape Trough-ridge Combined Root Restriction Soilless Culture Mode (Unit: cm)

4. Basket-type

Fill nutrient substrate in baskets or buckets with a certain volume and plant fruit trees in them. The baskets are easy to move, so it is suitable for application under protected cultivation conditions. The disadvantages still lie in the unstable root moisture and temperature, and the poor resistance to low temperature (Fig. 8-14).

Fig. 8-14 Grape Basket-type Root Restriction Soilless Culture Mode

(II) Production Techniques

1. Variety selection

(1) Summer Black: It is a Euro-American hybrid, and is native to Japan. It

originated from hybridization at the Fruit Tree Experiment Station of Yamanashi-ken, Japan in 1968. After 29 years of adaptability tests in 20 experiment stations in Japan, the product was registered in August 1997 as No. 9732. Zhangjiagang Shenyuan Grape Technology Co., Ltd. introduced the variety from Japan in February 2000.

The fruit cluster is mostly conical, and some have double-split clusters and cluster shoulders, without accessory clusters. The fruit clusters are large, with a length of 16–23 cm, a width of 13.5–16 cm, and an average weight of 415 g. The natural berry weight was 3–3.5 g. After treatment with gibberellin, the fruit berries are large, with a vertical diameter of 2.15–2.67 cm and a transverse diameter of 2.05–2.70 cm. The average berry weight is 7.5 g, the maximum berry weight is 12 g, the average cluster weight is 608 g, and the maximum cluster weight is 940 g. The fruit berries are compact or extremely compact, and the fruit clusters are uniform in size. Fruit berries are nearly round, in dense purplish-black to bluish-black, and very easy to color in southern regions with high night temperatures, consistent in coloration and maturation. The skin is thick, crisp, and non-astringent. The fruit powder is thick. The pulp is hard and crisp, without pulp sac, and the juice is purplish-red. It's sweet and has a strong fragrancc of strawberry. It has no seeds. There are no small green berries. The soluble solid content is 20% –22%. Its table quality is superior.

(2) Giant Rose: Giant Rose is a medium-maturing grape variety successfully selected and bred by the Dalian Academy of Agricultural Sciences, which passed the national official identification in August 2002. This variety is a tetraploid large-berry Euro-American hybrid variety with strong tree vigor and early fruit setting. The plant yield can reach 4–5 kg two years after planting and enters the high-yield period three years after planting. The fruit cluster is large, with an average cluster weight of 675 g and a maximum cluster weight of 1,250 g. The fruit berry is large, with an average berry weight of 9.5–12 g and a maximum berry weight of 17 g. The skin is purplish-red, with good coloration both inside and outside even at high yield, unlike Kyoho, which is not easy to color at high yield. There is no flower or fruit drop, no splitting fruit, neat cluster form, and even berry size. The pulp is

crisp, juicy, and tastes good without pulp sac. It has a pleasant, pure, and strong rose fragrance, with a soluble solid content of 19%–25% and a total acidity of 0.43%. Its quality is excellent and comparable to Muscat Hamburg. It is resistant to high temperature and humidity, has strong disease resistance, and is easy to cultivate and manage, suitable for cultivation in high temperature and humidity regions in northern and southern provinces. It is resistant to storage and transportation, and the quality is even better after storage. The fruits ripen in late August and early September.

(3) Yatomi Rosa: It is a Eurasian variety bred by the famous Japanese horticulturist Yoshimune Yatomi in Tokyo. The first fruit setting was in 1984, and the variety was registered in November 1990 to be very early-maturing. The fruit cluster is large, generally 450–650 g, maximum 1,500 g; the fruit berries are long and elliptical, with a berry weight of 8–10 g, which looks like that of Rizamat, but with a slightly pointed fruit tip. The pulp is pink to purplish-red and tastes crisp and juicy, the skin is thin, and the soluble solid content is 14% –16%, with fragrance, low acid content, sweet and refreshing flavor, and high quality. It has strong growth, high yields, and strong disease resistance, and the fruit has no splitting and is extremely resistant to storage and transportation. The fruits ripen in mid-July in Zhangjiagang. Suitable cultivation area: polytunnel cultivation in high-temperature and humidity regions of southern China.

(4) Shinano Smile: It is a Euro-American hybrid that was introduced from Japan. The fruits ripen in early and mid-September. The average cluster weight is 500–600 g. The ripe fruit berries are golden yellow or red, nearly round, with an average berry weight of 14 g. The pulp is milky yellow and juicy, without splitting fruits and dropping berries, resistant to storage and transportation, has a soluble solid content of 15%, light fragrance, and sweet taste.

(5) Wink: Wink is a Eurasian variety, and is native to Japan. It is a grape variety bred by crossing Kubel Muscat and Kaiji variety by Tomio Shimura from Yamanashi-ken, Japan in 1987. This variety has strong tree vigor, high quality, light coloration in the South, and slightly splitting fruit. It is more resistant to disease than Kaiji, easy to cultivate, and has a longer storage period hanging on the tree

after maturity, which is suitable for delayed cultivation.

The fruit clusters are conical. The fruit clusters are large, with a cluster length of 18–25 cm, a cluster width of 12–14 cm, and an average cluster weight of 450 g. The maximum cluster weight is 575 g, the berries are loose, and the cluster size is uniform. The fruit berries are oval, purplish-red to purplish-black, and ripen at the same time. The fruit berries are large, with a vertical diameter of 2.23–3.63 cm and a transverse diameter of 1.69–2.5 cm. The average berry weight is 10.5 g, and the maximum berry weight is 13.4 g. The skin is medium thick, tough, and non-astringent. The fruit powder is thick, the pulp is crisp with no pulp sac and is rich in greenish-yellow juice, which is extremely sweet. Each fruit berry contains 1–3 seeds, mostly two seeds. Seeds and pulp are easy to separate. There are small green berries. The soluble solid content is over 20%.

(6) Rosario Bianco: Rosario Bianco is a Eurasian variety that is native to Japan. Specifically, it is a pure European variety bred by Uehara Grapes & Vines Research Institute using Rosaki × Muscat of Alexandria in 1976. In August 1987, the variety was registered as No. 1405.

The fruit clusters are large without accessory clusters, with a length of 17–20.5 cm, a width of 12–15 cm, an average weight of 450 g, and a maximum weight of 685 g. The fruit berries are medium compact, and the fruit clusters are uniform in size. The fruit berries are short, oval, yellowish-green, uniformly colored, and ripen at the same time. The fruit berries are large, with a vertical diameter of 2.3–3.7 cm and a transverse diameter of 2.25–2.82 cm. The average berry weight is 8.5 g, and the maximum berry weight is 14 g. The skin is thin and tough with no astringency. The fruit powder is thick. The pulp is thick and crisp with no pulp sac and rich in greenish-yellow juice. It tastes sweet and smells fragrant. Each fruit berry contains 1–4 seeds, mostly two seeds. The pear-shaped seed is medium large and brown, with a medium long and pointed beak. The seeds and the pulp are easy to separate, and there are no small green berries. The soluble solid content is 19%–22%.

2. Key technical points of soilless culture of grapes

(1) Large seedling culture.

Large seedling culture in advance can better reflect the effect of root restriction.

The simplest method is to fill a plastic bag (with a carrying capacity of 10 kg) with a mixture of organic substrate, drill two small holes at the bottom of the bag for water permeation, and plant the seedlings in it. From April to August, apply 30 g compound fertilizer containing 15%–20% nitrogen every month, and fully irrigate the fertilizer in time. After germination, reserve one new shoot for growth, remove the top of other sublateral shoots and reserve one leaf, and remove the top of the main shoot in late August.

(2) Planting and management.

① Culture density: Plant spacing is a multiple of 4–6 m, and row spacing is 6 m.

② Root substrate: Compound mixed organic substrate.

③ Root volume: The projected area of the crown per square meter is 0.05–0.06 m^3, and the root thickness is 40 cm. Assuming that the Kyoho grapes are planted at a plant spacing of 1.8 m and a row spacing of 5.5 m, the projection area of the crown is about 10 m^2, and the root volume should be 0.5–0.6 m^3. To make a 40 cm bed surface, the distribution area of the root is 1.25–1.5 m^2; that is, holes or ridges with a depth of 40 cm, a width of 100 cm, and a length of 150 cm can meet the requirements of tree growth and fruiting. In the same way, if the Kyoho grapes are planted at a plant spacing of 3.6 m and a row spacing of 5.5 m, the projected area of the crown is about 20 m^2, and the volume of the root area should be 1.0–1.2 m^3. When a 40cm bed surface is made, the distribution area of the root is 2.5–3.0 m^2, that is, holes or ridges with a depth of 40 cm, a width of 100 cm, and a length of 300 cm can meet the requirements.

④ Nutrient management: Before the stone hardening stage, 60×10^{-6} comprehensive N liquid fertilizer is sprayed twice a week, with a nutrient solution of 60 L per cubic meter of root volume irrigated each time. The concentration of nutrient solution decreased to 20×10^{-6} after the stone hardening stage, and the application dosage and frequency remained unchanged. When the application of nutrient solution is not convenient, long-acting high-nitrogen organic fertilizers such as fermented bean cakes can be used. About 1.0–1.5 kg of organic fertilizers can be applied in the root zone per cubic meter before germination (for rain-proof culture, around March 20th) and after harvesting (for rain-proof culture, in the mid- and late August).

⑤ Water management: Root restriction soilless culture confines the root system of the grapevine to limited substrate space, and water cannot be freely absorbed from the soil. The water lost by leaf transpiration should be supplemented by irrigation. To what extent should the root substrate start to supplement water is important for the vegetative growth of the tree as well as the development and maturity of the fruit. The drying degree of the substrate is expressed by the water potential, and the substrate water potential that starts to supplement water is called the starting point of irrigation. According to the author's study, the following starting points of irrigation are adopted for different development stages, and the vegetative growth and reproductive growth are well balanced. The pF value is 2.2 (15.1 kPa) before germination and 1.5 (3.1 kPa) after germination to the softening stage of fruit berries; then it becomes 2.2 (15.1 kPa), and the irrigation volume is 60 L/m^3.

(3) Tree shape.

Root restriction soilless culture of grapes has slow growth and is suitable for various tree shapes. There is no low-temperature freezing injury in winter in Shanghai, so it is not necessary to bury soil to overcome the winter. In addition, there are more rainy days and less sunlight. Therefore, it is advisable to use a planar or arched pergola trellis with a high sunlight-receiving rate. The planar pergola trellis is suitable for multi-span polytunnels, and the arched pergola trellis is suitable for training tunnels. Planar pergola trellises mostly adopt H-shaped training or horizontal dragon trunk; Arched pergola trellises mostly adopt the arched dragon trunk or W-shaped training.

Task 5 Soilless Culture of Flowers

The soilless culture of flowers is a new flower cultivation technique that emerged at the end of the 20th century. It refers to using various nutrients required for the growth and development of flower plants to formulate nutrient solutions for their direct absorption and utilization. The growth environment for flowers is suitable, so they grow rapidly, the flower production period is short, and the flower yield per unit area is high. Soilless culture techniques have been widely used in the world for the production of fresh-cut flowers and potted flowers, such as roses, carnations, *Gladiolus*, chrysanthemums,and *Gerbera jamesonii*. China mainly produces high-end fresh-cut flowers, potted flowers, and seedlings.

I. Soilless Culture of Cut Flowers

(I) Chinese Rose

Cut roses, also known as modern roses, refer to a strain formed by the species of roses such as Chinese roses, tea roses, Damask roses, and French roses. The cultivated varieties of the Chinese rose are divided into Hybrid Teas (HT.), Floribundas (FL.), Grandifloras roses (Gr.), Miniature roses (Min.), Climbers (CL.), Ramblers(R.), and Shrubs (S.) of which HT. and FL. are mostly used for cut flowers. Chinese rose flowers throughout all seasons, and has bright colors, wide varieties, and a strong aroma, so it is popular among the people of all countries and is listed as one of the four major cut flowers.

1. Biological characteristics

The Chinese rose is an evergreen or semi-evergreen shrub of the genus *Rosa* in the Rosaceae family, with green small branches, with or without scattered prickles. The plant height is up to 2 m, and the shortest variety is only about 0.3 m. The leaves are alternate, broad-ovate or long-ovate, odd-pinnate leaves, with 3–5 leaflets, taper tips, and sharp serrations. The flowers are at the top of branches or several flowers aggregate to form corymbose cyme. There are sparse unisexual Flowers, and most have multiple corollas, long and caudate sepals, and pinnate lobes at the edge; the styles are separated, extending outside the mouth of the calyx tube, and are as long as the stamens. There is one ovule per ovary, and the fruit is oval or pear-shaped. The sepals will fall off. The anthesis is from April to October. If the conditions are suitable, the flowers can bloom in all seasons. Chinese rose is rich in color. Generally, cut roses can be divided into six color systems, namely, red, scarlet, pink, yellow, white, and other color systems.

Chinese roses have strong adaptability to the climate and soil, but loose, fertile, and slightly acidic soils rich in organic matter are more suitable. Chinese roses like warmth. The optimal temperature of most varieties is 20–27°C during the day and 12–18°C at night. In winter, when the temperature is lower than 5°C, they will enter the dormant state. Generally, they can withstand a low temperature of −15°C and a high temperature of 35°C. In summer, when the temperature is above 30°C, they will enter the semi-dormant state. Although they can still bear buds, the flowers are small with few petals and the color is light without luster, losing their ornamental value. Chinese roses like the leeward, sunny, and well-ventilated environment, and require direct sunlight for at least five hours a day to grow well. They have no strict requirements for soil, and like loose, fertile, and slightly acidic soils rich in organic matter, with a pH value of 6–7. Chinese roses like fertile and water, and are resistant to drought but sensitive to accumulated water. The soil should always be kept moist. Especially from germination to leafing and flowering stages, water should be sufficiently supplied to make the flowers large and bright. After entering the dormant stage, the moisture content should be properly controlled. As the continuous germination, shooting, bearing bud, and flowering during the growth period, it is necessary to timely fertilize to prevent the decline of tree vigor and keep the flowers blooming continuously.

2. Propagation methods

Cut roses are generally produced by cutting and grafting.

(1) Cutting. Softwood cutting or hardwood cutting can be adopted for Chinese roses in both spring and autumn. Spring cutting generally starts in late April and ends at the end of June, and the cuttings can take root about 25 days after cutting, with a high survival rate. Autumn cutting starts in late August and ends at the end of October. However, due to the large temperature difference between day and night, the rooting period is 10–15 days longer than that of spring cutting, and the survival rate is also high. At the time of cutting, quickly dip the lower end of the cutting in 500–1,000 mg/L indole butyric acid or 500 mg/L indole acetic acid to promote rooting. Cutting substrate can adopt rice chaff ash, river sand, vermiculite, rock wool blocks, etc. The depth of cuttings in the cutting substrate is 1/3–1/2 of the cutting length. For softwood cutting, to keep the substrate moist, the cuttings can also be covered with small tunnels for heat insulation and moisture preservation, so

as to prevent the cuttings from losing water and wilting, thus affecting the survival rate.

(2) Grafting. Grafting is the main means of propagating Chinese rose. This method has the advantages of easy material availability, simple operation, rapid growth, high yields in the early stage and long life. Grafting mostly uses cuttings or seedlings of wild species of the genus *Rosa* as rootstock, and the common rootstocks include *Rosa multiflora* and *Rosa multiflora* var. *cathayensis.* Generally, the bud grafting method or stem grafting method is adopted. The bud grafting is carried out from July to August, and the T-grafting method is adopted. Stem grafting is carried out from February to March, and the main method is the cutting grafting method. In general, when the rootstock diameter is 9–13 mm, the grafting can be started.

In addition, tissue culture is also allowed to breed a large number of tissue culture plantlets that maintain the characteristics of the original variety.

3. Nutrient solution management

Nutrient solution management is the most important and critical part of soilless cultivation of cut roses. The concentration and supply of nutrient solution should be adjusted according to the size of the plants and different growing seasons. Generally, in the initial stage of planting, the solution supply can be less, and the nutrient solution concentration should also be slightly lower. Each plant is supplied with about 100 mL of solution every day, and the EC value is controlled at about 1.5 mS/cm. After entering the vigorous vegetative growth stage, the solution supply should be gradually increased. The average daily solution supply is 800–1,200 mL per plant, and the EC value can be increased to 2.2 mS/cm. After entering the anthesis, the solution supply can be increased to 1,200–1,800 mL per day, and the EC value is controlled at 2.2–2.6 mS/cm. The solution is supplied 8–11 times a day in summer and 3–4 times in winter. The solution supply time is mainly from 8：00 to 17：00. The pH value of nutrient solution shall be controlled within 5.5–6.5 during the whole growth period. Rock wool culture or substrate culture is usually adopted for cut roses. See Tables 8-2 and 8-3 for the formula of nutrient solution for soilless culture.

Table 8-2 Basic Formula of Soilless Culture Nutrient Solution for Cut Roses (Unit: mg/L)

Nutritional elements	Substrate trough culture	Rock wool culture	Nutritional elements	Substrate trough culture	Rock wool culture
NO_3^--N	182	144	Fe	1.40	1.40
P	54	46	Cu	0.05	0.03
SO_4-S	48	32	Zn	0.23	0.16
NH_4^+-N	10	7	Mn	0.28	0.28
K	235	225	B	0.22	0.22
Ca	180	120	Mo	0.05	0.05
Mg	24	18	EC(mS/cm)	2.0	1.5

Table 8-3 Formula of Soilless Culture Nutrient Solution for Chinese Roses

Name of compounds	Application rate (mg/L)
Calcium nitrate($Ca(NO_3)_2 \cdot 4H_2O$)	490
Potassium nitrate (KNO_3)	190
Potassium chloride (KCl)	150
Ammonium nitrate (NH_4NO_3)	170
Magnesium sulfate ($MgSO_4$)	120
Phosphoric acid (H_3PO_4, 85%)	130
Chelated iron (Na_2Fe-EDTA)	12
Manganese sulfate ($MnSO_4 \cdot 4H_2O$)	1.5
Copper sulfate ($CuSO_4 \cdot 5H_2O$)	0.125
Zinc sulfate ($ZnSO_4 \cdot 7H_2O$)	0.85
Boric acid (H_3BO_3)	1.24

4. Soilless culture methods

At present, the main soilless culture methods of cut roses in China include substrate culture and rock wool culture.

(1) Substrate culture. Substrate culture is the main soilless culture method in China, including trough-type substrate culture and bag-type substrate culture.

The trough-type substrate culture system consists of the culture trough, drip irrigation pipe system, water pump, reservoir, and timer. Culture troughs can be made of brick, asbestos tile or prefabricated cement board and other materials. The culture trough is 80 cm wide and 25–30 cm high, and its length should not exceed 40 m. Plastic film is paved inside the trough to isolate it from the outside, and then the substrate is filled with a thickness of 20–25 cm. Embedded drip irrigation tape or hose drip irrigation tape is adopted, and 1–2 pieces are laid for each culture trough. An acid-resistant and corrosion-resistant self-priming water pump is used. The plants are planted in two rows, with a plant spacing of 25–30 cm and a row spacing of 30–35 cm. The finished seedlings are removed from the seedling-raising pot and then planted into the substrate, with drip irrigation tapes laid between rows or near the roots of each crop row.

For bag-type substrate culture, the silver or black plastic film bag is used to contain the culture substrate. The bag is 100 cm long, 20 cm wide, and 8 cm high, and a planting hole with a diameter of 10 cm is opened every 20 cm to plant cut roses. Two drainage holes (with a diameter of 0.5 cm) are drilled at both ends of the lower nutrient bag, which is filled with mixed substrate. Each bag can be planted with four plants. The water and nutrient solution are mainly supplied by drip irrigation, and the water and nutrients can also be supplemented by sprinkler irrigation and foliage top dressing.

(2) Rock wool culture. The rock wool culture system consists of a rock wool plantation pad, drip irrigation pipe system, reservoir, timer, etc. For rock wool culture, rock wool plantation pads with a length of 100 cm, a width of 7.5 cm, and a thickness of 30 cm are arranged into a plantation bed, and the rock wool plantation pads are wrapped with black and white double-color film to reduce the loss of nutrient solution. Two rock wool plantation pads are arranged 30 cm apart

in a row, and a shallow trench is excavated between the two rows of plantation pads and covered with plastic film to collect excess nutrient solution discharged from the culture bed.

Before planting, soak the rock wool overnight with the nutrient solution with a pH value of 5.5–6.5 and EC value of 1.0–1.2 mS/cm to reduce its pH value to near neutral. During planting, cut 2–3 incisions on the side of the rock wool plantation pad close to the drainage ditch to discharge excess nutrient solution. Cut a planting hole at the top that matches the seedling block, place the seedling at the planting hole, erect the dropper of the dropper pipe above the seedling block, and drip the nutrient solution directly on the plantation pad. Each 100-cm-long rock wool plantation pad can be planted with 6–7 seedlings.

5. Pruning technique

The pruning of cut roses is an important management measure throughout the production process of cut flowers, which directly affects the yield and quality of cut flowers. The methods, including topping, bud removal, bud picking, branch breaking, and heading cutting, are mainly adopted to enhance the tree vigor, cultivate the flowering basal branch, and promote the formation and development of effective flowering branches. The pruning technique varies greatly in the production of cut roses because of different cultivation methods and growth stages.

(1) Pruning in seedling stage. The main purpose of seedling pruning after planting is to form a robust plant framework and cultivate flowering basal branches. The main method of seedling pruning is to control the flowering of new shoots and promote the germination of lateral buds by topping. Most branches germinating in the early seedling stage are thin and weak, so repeated topping is necessary. Only when the nutrient area reaches a certain degree, the branches with a certain thickness can be germinated. Branches with a diameter of more than 0.6 cm can be used as flowering basal branches after topping (generally, all tender leaves above the first or second compound leaves with five leaflets should be removed). When a plant has more than three flowering basal branches, it can be managed as a flowering plant.

(2) Summer pruning. After a growth cycle, the height of the cut rose plant increases continuously, the growth vigor of the branches decreases, and the yield and quality of the cut flowers also decrease. Especially for plants cultivated in greenhouses for flower production in winter, the plant type must be adjusted to facilitate flower production from autumn to winter and spring. Traditional summer pruning mainly adopts the method of heading cutting and heading back. However, since the plants are still in the growing period in summer, this method is harmful to the tree body and the nutrient area is greatly reduced, which is not conducive to the recovery of growth in autumn. Nowadays, the method of twisting and breaking branches is mostly adopted. The method of twisting branches refers to twisting and bending down branches without injuring the xylem. Breaking branches refers to breaking and bending down branches without separating from the tree body. Twisting and breaking branches can reduce damage to the tree body, ensure sufficient nutrient area, and facilitate the rejuvenation of the tree body. The methods of twisting branch, breaking branch, heading cutting, and heading back may also be combined in production as required.

(3) Winter pruning. Winter pruning is the trimming and pruning of a tree body for rejuvenation after leaves fall and dormancy in the cultivation of Chinese roses after stopping flowering in winter. It is generally carried out after dormancy to one month before germination. Generally, after cutting off the weak branches, branches with pests and diseases and aged branches, the main branches (flowering basal branches) are headed back using the heading cutting method. In general, 3–5 main branches are reserved, and the reserved height of each branch is about 40 cm, which often varies according to different varieties.

(4) Daily pruning. In addition to pruning at the seedling stage and rejuvenation pruning, regular pruning is also important during the growth period and anthesis. Daily pruning includes pruning, bud stripping, bud picking, removal of rootstock sprout tillers of cut flower branches, and pruning of nutrient branches. The pruning of cut flower branches is especially important, because the pruning of cut flower branches not only affects the quality of cut flowers, but also affects the yield and quality of later flowers. Generally, the reasonable cutting position of cut flowers

is the part above 2–3 compound leaves with five leaflets reserved at the base of flowering branches. In addition, it is necessary to conduct bud removal, heading cutting, and topping on weak branches in time to properly reserve leaves and increase the nutrient area.

6. Prevention and control of diseases and pests

Chinese roses are a flower that is prone to pests and diseases, especially in polytunnels and greenhouses. Therefore, the principle of "prevention first" should be implemented in production to strengthen management and enhance the resistance of plants. At the same time, the varieties with strong resistance should be selected according to the characteristics of the cultivation environment. The environment should be clean, the temperature and humidity should be controlled, and pesticides should be applied in time according to the occurrence law of diseases and pests to control their occurrence and spread.

Generally, the diseases that are easy to occur in the production of Chinese roses include black spot, powdery mildew, downy mildew, gray mildew, etc. Tradiecfon, chlorothalonil, thiophanate, carbendazim, tuzet, and metalaxyl can be used for prevention and control. Common pests include mites, aphids, scale insects, *Arge pagana*, *Neosyrista similis*, etc. Dicofol, propargite, amitraz, omethoate, phoxim, fenvalerate, etc. can be sprayed to kill them.

(II) Chrysanthemum

Chrysanthemum is a traditional flower originating from China. It has been documented in writing in China for more than 3,000 years and cultivated artificially for more than 1,600 years. Chrysanthemum was introduced to Japan in the 8th century (Tang Dynasty), to Europe in 1688 through Japan, and to America in the late 18th century through Europe. Chrysanthemum is popular among the people for its elegant color, beautiful gesture, pleasant fragrance, and long anthesis. It is one of the top four cut flowers in the international flower market, accounting for about 30% of the total cut flowers. The traditional cultivation of chrysanthemums in China is mainly potted planting, and the selection and breeding of varieties are mostly potted varieties. Compared with Japan, the Netherlands, the United States, and other countries, breeding and cultivation of cut chrysanthemums started relatively late in China with fewer varieties and lower technology content in cultivation management. However, cut chrysanthemums are in great demand in the international market, especially in the neighboring market of Japan. As long as China makes efforts to improve varieties, cultivation techniques, and product quality, cut chrysanthemums are expected to become an important flower product of China's foreign exchange earnings.

1. Biological characteristics

Chrysanthemum is a perennial root herb of the genus *Chrysanthemum* in the Asteraceae family, sometimes growing into sub-shrubs, with thick and strong stems, many branches, slightly lignified base, and a plant height of 30–200 cm. As a cut flower variety, the plant height is generally 80–150 cm. The leaves are alternate, large, ovate to oblong-lanceolate leaves, with large serrations or incisions, and the depth varies depending on variety, with or without stipules. The flower is solitary or several flowers aggregate at the branch tips to form a capitulum. The capitulum diameter is 2–30 cm, composed of edge tongue-like flowers and central tube-like flowers. The tube-like flowers are mostly yellowish-green, and the tongue-like flowers are extremely rich in color, including yellow, white, pink, red, purple, light green, brownish-yellow, tertiary color, secondary color, etc. There are various types of chrysanthemums, but the flowers of the cut chrysanthemum are mostly in the shape of flat plates, peonies, lotus flowers, or hemispherical shapes, which are neat and round. The seeds (achene) are brown and fine, with a lifespan of 3–5 years.

Chrysanthemum is cryophilic and has a certain cold resistance, especially for small chrysanthemums. The aerial part germinates at above 5°C, the new buds elongate at above 10°C, and 16–21°C is most suitable for growth. Different varieties of chrysanthemums have different requirements on day length and temperature for flower bud differentiation. Chrysanthemum is heliophilous, but also slightly shade-tolerant. In summer, it is appropriately shaded from the sun. It prefers moist and is drought-resistant, but sensitive to waterlogging. It prefers neutral or slightly acidic sandy soil rich in humus, with good ventilation and drainage. It can also grow on weakly alkaline soil, and continuous cropping shall be avoided. The requirement for sunlight duration of the chrysanthemum's flower bud differentiation varies from variety to variety. The autumn chrysanthemum variety requiring short sunlight duration is mainstream, and flower bud differentiation of some varieties is not affected by sunlight duration. The anthesis is from April to December.

2. Propagation methods

Common propagation methods include cuttage propagation, sucker propagation, grafting, tissue culture, and sowing propagation, and sowing is mostly used for breeding. Cut flowers are mostly produced by cuttage propagation, which is mostly carried out from April to August. Cut 7–10 cm on top of robust tender branches, remove the lower leaves for use, and collect the cuttings as needed. If cuttings cannot be made in time after collecting, they can be placed in a moisturizing and permeable plastic bag and stored in the bag at a low temperature of 0–4°C. Cutting substrate mainly include vermiculite, peat, perlite, rice chaff ash, river sand, etc. The temperature of vermiculite, peat, perlite, rice chaff ash, and other substrates rise quickly, so they are suitable for spring cutting, and river sand is suitable for summer cutting. The cutting bed shall be equipped with an automatic intermittent spraying device in full light as far as possible, especially in high-temperature season, which can ensure the survival rate and take root in advance. Rooting is completed 2–3 weeks after cutting. After survival, they shall be planted as soon as possible. If they are in bed for too long, the seedlings will be thin, yellowed, or even rotten and die.

3. Soilless Culture Techniques

There are rich varieties of chrysanthemum, with 20,000–25,000 varieties in the world. According to cultivation and application methods, they can be divided into potted chrysanthemums and cut chrysanthemums; according to the natural anthesis, they can be divided into spring chrysanthemums (late April to mid-June), summer chrysanthemums (late June to early September), early autumn chrysanthemums (early September to early October), autumn chrysanthemums (late October to late November), and winter chrysanthemums (early December to next January); according to the inflorescence diameter, they can be divided into small chrysanthemum strain (less than 6 cm), middle chrysanthemum strain (6–10 cm), large chrysanthemum strain (10–20 cm), and extra-large chrysanthemum strain (more than 20 cm).

Chrysanthemum varieties are often classified according to petal type and flower type. At the Symposium on Variety Classification held by Chinese Society of Horticultural Science and Penjing Branch of China Flower Association in Shanghai in 1982, chrysanthemum varieties were classified into five petal types, namely flat petal, spoon petal, tube petal, fragrans petal, and abnormal petal. The flower types were divided into 30 types and 13 subtypes. Most varieties of cut chrysanthemum have flat petals and spoon petals. A small number of varieties have tube petals and fragrans petals, and most have neat and round flower types.

The cut chrysanthemum varieties for soilless culture should have special requirements, mainly including the following seven aspects:

① The plant grows strong, tall, and upright, and the height should be above 80 cm.

② Flower branches are thick, straight, and hard, with uniform internodes; pedicels (stems) are short, thick, and hard.

The concentration and supply of nutrient solution should be differentiated according to the size of cut chrysanthemum plants and growing seasons. Generally, at the initial stage of planting, the supply of nutrient solution can be less, and the concentration of nutrient solution should also be slightly lower; after entering the vigorous vegetative growth stage, the solution supply shall be increased gradually to supply solution 3–4 times a day and an average of 300–500 mL per plant. The solution supply shall be appropriately reduced on rainy days and increased on sunny days. In addition, the pH Value, EC Value, and NO_3-N of the substrate are determined regularly. According to the results of the determination, the nutrient solution is adjusted. At the beginning of chrysanthemum planting, the concentration of nutrient solution should be at a low level, with an EC value of about 0.8 mS/cm; with the growth of plants, the concentration of nutrient solution can be gradually increased, and the EC value can be increased to 1.6–1.8 mS/cm; at high temperatures in summer, due to a large amount of water evaporation, the concentration of nutrient solution shall be appropriately reduced, and the EC value is 1.2–1.4 mS/cm. In addition, the pH value of the nutrient solution can be adjusted to 5.5–6.5 with a 5% dilute nitric acid solution.

In addition to the supply of fertilizer and water through the nutrient solution, soilless culture of the chrysanthemum can also be carried out by applying basal fertilizer and topdressing during the growth period. For every m^3 of mixed substrate, 5–8 kg of organic fertilizers such as poultry manure after decomposition and disinfection and 0.5–1.0 kg of potassium nitrate can be applied as basal fertilizer. After seedling planting, regular topdressing shall be carried out. The interval and dosage of fertilization shall be determined according to the growth of seedlings. There should be more fertilizers in the vigorous vegetative growth stage. 1.5 kg of organic fertilizers such as poultry manure can be topdressed every 30 days, and urea, potassium nitrate, and diammonium phosphate can also be applied at the same time. In addition, 0.1%–0.5% urea and potassium dihydrogen phosphate can be sprayed for foliar topdressing in combination with pest and disease prevention and control.

4. Plant management

(1) Topping and pruning. For cultivation of multi-head chrysanthemums, they should be topped 1–2 weeks after the seedlings are planted, and only the top buds need to be removed. Pruning shall be carried out about 2 weeks after topping. According to the planting density and variety characteristics, each plant shall reserve 2–4 lateral buds, and the rest shall be removed.

(2) Netting. The cut chrysanthemum requires a straight stem, but it is prone to lodging because of its height. Therefore, when the plant grows to a certain height, a net shall be stretched to support the plant in time so as to prevent the quality from being affected due to the bending of stem caused by plant lodging. The mesh of the supporting net may vary depending on planting density or variety, usually between 10 cm × 10 cm and 15 cm × 15 cm. Generally, 2–3 layers of nets shall be used for support, and the nets shall be tightened flat with supporting rods.

(3) Removal of lateral buds. When the flower bud differentiation of the chrysanthemum begins, its lateral buds begin to germinate and shall be removed in time (except for the variety of multi-head small chrysanthemum). The removal of upper lateral buds will stimulate the germination of the middle and lower lateral buds, so the lateral buds shall be removed several times. With the development of

the flower bud, several lateral buds will be formed around the middle main bud, which should be removed in time to ensure the normal growth of the main bud. The bud should be removed sooner rather than later as long as it is easy to operate. If it is removed too late, the lignification degree of the stem will increase, and it is not easy to operate.

5. Prevention and control of diseases and pests

Chrysanthemum is one of the flowers with many pests and diseases. Although it is less deadly, it greatly affects the quality of cut flowers. Therefore, preventive management should be strengthened in production to enhance the resistance of plants. At the same time, corresponding varieties should be selected according to the cultivation method. The environment should be clean, the temperature and humidity should be controlled, and pesticides should be applied in time according to the occurrence law of diseases and pests to control their occurrence and spread. In addition, crop rotation is also an important means for the prevention and control of diseases and pests in chrysanthemums.

Common diseases of chrysanthemum include spot blight, seedling blight, powdery mildew, etc., and pests include aphids, *Phytoecia rufiventris* Gautier, chrysanthemum leaf miner, Trialeurodes vaporariorum, red spider, looper, grubs, snails, etc. Corresponding fungicides and pesticides should be used in time for prevention and control.

(III) Lily

Lily is a perennial root herb flower of the genus *Lilium*, family Liliaceae. It is one of the world famous flowers. It is a newcomer to the flower market at home and abroad recently and is an important cutting material. Lily can be planted in a pot or made into cut flowers for arrangement in special occasions such as indoor or meeting places; it can be planted in rows, clusters, tufts, or tracts to green courtyards, flower beds, flower nurseries, and gardens; it can be used for edible, and medicinal purposes, integrating ornamental, edible, and medicinal functions, and has high cultivation value. Because lily means "everything is successful" and "a harmonious union lasting a hundred years" in Chinese, symbolizing good connotations of auspiciousness, holiness, reunion, joy, happiness, and blissfulness, it is popular among people.

The soilless culture methods of lily include substrate trough culture, box substrate culture, pot culture, and hydroponics, mainly substrate trough culture.

1. Biological characteristics

Lily is a long-day plant. It prefers cool and humid climates and well-lit environments. It is relatively cold-resistant and does not like high temperatures. If the temperature is higher than 30°C, the growth and development of the lily will be seriously affected: the bud will fall and the flowering rate will decrease; if the temperature is lower than 10°C, the growth will be nearly stagnant. It likes dry conditions and fears of waterlogging. If root humidity is too high, the bulb will rot to death. Continuous cropping should be avoided, and crop rotation should be done every 3–4 years.

2. Variety selection

There are about 100 species of the genus *Lilium*, of which more than 30 species originate from China, and nearly 20 species can be used as ornamental plants. At present, the main varieties cultivated in China include Oriental, Longiflorum, Asia, Longiflorum/Asiatic-hybrid (L/A), and potted varieties, most of which are high-quality first-generation seedballs imported from the Netherlands and New Zealand.

3. Propagation methods

To produce high quality cut lily, the primary requirement is robust and disease-free seedballs. The propagation of lilies can be divided into post-flowering bulb culture, bulblet propagation, scale cutting, bulbil propagation, sowing, and tissue culture. Four main methods of propagation are introduced below.

(1) Bulblet propagation. The stem axis of aged bulbs of lily can sprout many new bulblets. Collect and disinfect bulblets on disease-free plants, and sow them in a complex substrate culture bed or plot made of turf, vermiculite, and fine sand in a ratio of 2 : 2 : 1 at the row spacing of 25 cm × 6 cm. After one year of culture, some of them can reach the seedball criteria (50 g), and the smaller ones can continue to be cultured for another year before being used as seedballs. One year later, plant the grown bulblets in beds or plots. The culture of bulblets requires many fertilizers. The principle of fertilization is less amount and more frequent, and the nutrients

should be complete. Mixing long-acting organic fertilizer in culture substrate is an ideal fertilization method. In the second year of the bulb culture, some flower buds will appear, which should be removed in time to facilitate the culture of the underground bulb. After 2 years of culture, the bulblets can be used as flowering seedballs. After harvest, the seedballs should be graded according to specifications and crated after removing diseased seedballs.

(2) Scale cutting. In autumn, select robust, disease-free, and hypertrophic scales, soak them in 500-fold dilution of benomyl or captan aqueous solution for 30 minutes, take them out and dry in the shade, with the base downward, and insert 1/3–2/3 scales into the substrate bed of peat : fine sand in a ratio of 4 : 1 or pure turf. The density is (3–4) cm × 15 cm. Cover the scales with grass for shading and moisturizing, and avoid water temperature and high temperature to prevent scale rot. The temperature is maintained at 22–25°C, and the air temperature is maintained at about 90%. There is no special requirement for sunlight, but long-day sunlight is more conducive to the formation, growth, and development of bulblets. After 2–3 weeks, 1–2 bulblets will be formed at the incision at the lower end of the scale, 3–5 bulblets at most. After 2–3 years of cultivation, the weight of bulblets can reach 50 g. About 100 kg of scales are needed per mu, and the propagated bulblets can be planted in about 10,005 m^2.

(3) Post-flowering bulb culture. Post-flowering bulb culture is also called big bulb propagation. By the time the lily began to bloom, new bulbs had been formed underground but were not mature. Therefore, when harvesting cut flowers, on the premise of ensuring the length of flower branches, as many leaves as possible shall be reserved to facilitate the culture of new bulbs. The new bulbs are ripe and can be harvested 6–8 weeks after flowering. The success of future forcing culture depends entirely on the maturity of the new bulb.

(4) Tissue culture propagation refers to relevant books on plant tissue culture.

4. Soilless Culture Techniques

(1) Preparation for planting.

① Substrate selection and treatment: At present, the common lily culture substrates in China include sand (diameter less than 3 mm), natural gravel, pumice, volcanic rock (diameter greater than 3 mm), vermiculite, perlite (better when mixed with turf and sand), and turf (can be mixed with slag, etc.). In addition, slag, bricks, charcoal, asbestos, sawdust, fernery, bark, etc. can be used as a substrate for lilies. The substrate shall be disinfected before use.

② Seedball selection and treatment: In production, the daughter bulb with the developed root system, large size, tight scales, white color and regular shape, no wound, no diseases, and pests are mainly selected as seedballs. The seedbulb diameter of the Asiatic lily must be 10–12 cm and that of the Oriental lily must be 12–14 cm. The bigger the seedball, the more flower buds, but different varieties have certain differences in the number of flower buds. After the arrival of the purchased seedballs, the package shall be torn off immediately and the seedballs shall be placed in a cool condition of 10–15°C for slow thawing, and disinfection shall be carried out after complete thawing.

Disinfection method: Soak the seedballs with agricultural streptomycin for 30 minutes or spray 800–1,000-fold dilution of carbendazim for 30 minutes, or soak them with 80-fold dilution of 40% formaldehyde for 30 minutes for disinfection, and dry them in the shade before planting.

(2) Planting.

① Construction of culture trough: Generally, the specification of the culture trough is 96–120 cm wide and 15–25 cm deep, and the length is flexibly determined as appropriate. The trough is lined with film and filled with substrate. Complex substrate should be preferred, e.g. sand : slag = 1 : 2, perlite : vermiculite = 3 : 1, perlite : vermiculite : turf = 2 : 1 : 1, etc.

② Planting: Cut lilies are generally planted in spring and summer. The planting depth requirement is that the top of the bulb is 8–10 cm from the ground surface and 6–8 cm in winter. In spring and summer, they can be densely planted; they should be sparsely planted in winter due to weak sunlight. Shallow holes (8–10 cm) shall be opened for planting, and the plant spacing is generally (25–30) cm × (15–20) cm. See Table 8-4 for planting density of lily seedballs of different populations and specifications.

Table 8-4 Planting Density of Lily Seedballs of Different Populations and Specifications

Population	Specification			
	12–14 cm	14–16 cm	16–18 cm	18–20 cm
Asiatic lily	55–65	50–60	40–50	25–35
Oriental lily	40–50	35–45	25–35	25–30
Longiflorum lily	45–55	40–50	35–45	25–35

Note: Expressed in number of seedballs per square meter.

Potted lilies are usually planted in 12–15-cm-deep pots, with one seed bulb per pot, or 15–18-cm-deep pot, with three scales per pot, which will form a dense flower cluster when flowering. During planting, more pieces of tiles shall be placed at the bottom of the pots, and then the substrate shall be added. The distance between the terminal bud of the bulb and the mouth of the pot shall be 2 cm, and the terminal bud shall be covered with 1-cm-deep soil. Currently, bulb pregermination is adopted in the Netherlands, and the pregermination part must be exposed above the soil surface. If the bulb has germinated before planting, it is not necessary to promote germination. If it has not germinated, the bulb can be arranged in the wood frame containing wood chips to promote germination. The sowing time should be from late September to October.

5. Cultivation management

(1) Nutrient solution.

The formula of the nutrient solution can be selected from the Japanese garden test formula or the Dutch general formula of rock wool culture flowers. In the initial stage of substrate culture and planting, only water can be irrigated. When new leaves grow out after 5–7 days, the nutrient solution shall be applied instead, and 0.5 dose of the standard formula shall be used. After the emerge of aboveground stems, they are irrigated with one dose of the standard formula, and the content of P and K in the nutrient solution is appropriately increased. On the basis of the prescribed dosage of the original formula, the content of P and K is increased by 100 mg/L. During the flowering and fruiting period, 1.5–1.8 doses of the standard formula are used for irrigation. During this period, appropriate foliar fertilization may also be carried out. The adjustment of nutrient solution concentration in hydroponics is similar to that in substrate culture.

During substrate culture, the nutrient solution shall be irrigated once every 2–3 days in winter, and the nutrient solution and clean water shall be irrigated once every 1–2 days in summer. During hydroponics, the DFT intermittent solution supply method is adopted, and circulation may not be necessary. Without circulation, the nutrient solution shall be replaced once every 15–20 days.

(2) Temperature.

Within 3–4 weeks after planting, the substrate temperature must be maintained at 12–13°C to facilitate the development of stem roots, while a temperature higher than 15°C will lead to the poor development of stem roots. After the rooting stage, the optimal air temperature of the Oriental lily is 15–17°C. Air temperature below 15°C will lead to falling buds and yellow leaves. The air temperature of the Asiatic lily is controlled within 14–25°C. The air temperature of the Longiflorum lily is controlled within 14–23°C. In order to prevent petal discoloration, bud deformation, and bud splitting, the day and night temperature shall not be lower than 14°C.

Lilies can tolerate high temperatures to a certain extent, but sustained high temperatures above 30°C are detrimental to their growth and development. In high-temperature summer, ventilation and proper shading shall be strengthened; the temperature difference between day and night shall be controlled at 10°C. If the night temperature is too low, it is easy to cause falling buds, yellow leaves, and bud splitting. If the night temperature is too high, the lily flower stalks are short, the flower buds are few, and the quality is reduced.

(3) Humidity.

The criteria for substrate humidity before planting is that it can be clenched into a lump by hands and dispersed after dropping on the ground. In high-temperature season, if conditions permit, cold water shall be poured once before planting to reduce the temperature of the substrate. After planting, water shall be poured again to make the substrate in full contact with the seedball, thus creating good conditions for the development of stem roots. Subsequent watering shall be based on the criteria of keeping the substrate wet, and it is appropriate to be clenched into a lump by hands but without water. Watering is usually done on sunny mornings. The environmental humidity should be 80%–85%, and too large fluctuation should be avoided; otherwise it will inhibit the growth of lilies and cause leaf burning of some sensitive varieties such as Acapulco. If the night humidity is high in the facility, the facility shall be ventilated in stages in the morning to slowly reduce the temperature.

(4) Plant management.

The root system of lily is relatively shallow and prone to lodging. Therefore, a supporting net or lifting rope shall be used for fixing in due course. When the seedlings are about 50 cm high, the first layer of supporting net shall be erected; or the plant shall be lifted for the first time, and at least one layer of the supporting net shall be erected, or the plant shall be lifted again in the future.

In addition, lilies shall be prevented from bud falling. The prevention method is to spray 0.463 mmol/L silver thiosulfate solution (STS solution), or spray some boric acid when the bud just emerges, which has a certain effect on preventing and controlling falling bud.

6. Prevention and control of diseases and pests

The main diseases include black spot disease, gray mold, and rust, which can be prevented and controlled by spraying with 500-fold dilution of 25% carbendazim wettable powder. Pests include grubs and aphids, which can be killed by spraying with 11,000-fold dilution of 50% dichlorvos emulsifiable concentrate.

7. Harvesting, package, and storage of cut flowers

When three of more than 10 flower buds or two of 5–7 flower buds, or one of less than 5 flower buds are colored, they can be harvested. Early harvesting will affect the color of the flowers; the flowers will be pale and ugly, and some flower buds cannot blossom; late harvesting will bring difficulties to the post-harvest treatment and packaging, the petals will be stained by pollen, and the preservation period of cut flowers will be shortened, which will affect the sales. The best harvest time is in the morning to reduce dehydration. Harvested lilies should be kept in the greenhouse for no more than 30 minutes. After harvesting, lilies are generally graded according to the number and size of flower buds, the length and hardness of the stems, and whether the leaves and flower buds are deformed; then they are bundled, and the yellow leaves, wounded leaves, and leaves 10 cm from the stem base are removed.

Bundles of cut lilies shall be stored directly in clean water or a cold store after the lilies have absorbed sufficient moisture; the cut lilies shall be packaged in a dry perforated box to prevent overheating and fungus reproduction. The seedballs shall be graded and disinfected before storage, and the filling substrate shall be wet sawdust or turf during storage.

(IV) Carnation

Carnation is one of the four major cut flowers widely cultivated around the world because of its beautiful and elegant flowers, long anthesis, high yields, resistance to storage, freshness preservation, water culture, and convenient packaging and transportation.

1. Biological characteristics

The carnation is an evergreen subshrub of the genus *Gypsophila*, family Caryophyllaceae, which is cultivated as perennial root flowers. Plant height is 30–80 cm; the stem is thin and soft; the base is lignified; the whole plant is covered with white powder, with enlarged internodes. The leaves are opposite, linear-lanceolate, entire, thick, and base-clasping. The flower is solitary or several flowers aggregate at the branch tips, with 2–3 layers of bracts closely attached to the calyx tube. The calyx end has five lobes, and most petals are unguiculate. The flower colors are extremely rich, including bright red, pink, light yellow, white, and deep red, as well as compound color and edge color such as agate. The fruit is capsular fruit, and the seed is brown.

Originating from southern Europe, the carnation is widely cultivated around the world. Its main production areas include Italy, the Netherlands, Poland, Israel, Colombia, the United States, etc. Carnation likes warm and cool environments and is not cold-resistant. The most suitable growth temperature is 16–22°C during the day and 10–15°C at night. It likes the circulated air, dry environment, and sunlight, and is a heliophilous and day-neutral flower. However, long sunlight is conducive to flower bud differentiation and development. The soil with good drainage and rich humus is required to be resistant to weak alkali, and continuous cropping shall be avoided. The natural anthesis is from May to October, and the protected cultivation can flower annually.

2. Propagation methods

Cuttage and tissue culture can be used for propagation. Cuttage propagation is mainly in spring or autumn and winter. The lateral branches of 10–14 cm with robust middle parts and short internodes are selected, with 4–5 pairs of cuttings with unfolded leaves. If the cuttings cannot be cut in time, they can be refrigerated at a low temperature of 0–2°C, and generally can be stored for 2–3 months. The cutting substrate is mainly peat, perlite, vermiculite, or rice chaff, which can be used separately or mixed in a certain ratio. Before cutting, treat the cuttings with 500–2,000 mg/L naphthalene acetic acid, indole butyric acid, or their combination, which can promote rooting. The treatment time varies with the concentration. They usually take roots about three weeks after cutting. Tissue culture is mainly used for the detoxification culture of carnation to propagate stock plants. Due to its long seedling stage and thin growth in the early stage, it is rarely used in the production of cut flowers.

3. Soilless Culture Techniques

(1) Variety selection. There are many varieties of the carnation, which can be divided into open-field cultivars and greenhouse varieties according to their cold resistance and ecological conditions. According to the size and number of flowers on the flower stalk, it can be divided into large-flower carnation (also known as single-flower carnation or standard carnation) and scattered-branch carnation (also known as multi-flower carnation). According to the source of the hybrid parents, there are many strains of large-flower carnation varieties, and the common strains in production include Sim and Mediterranean strains. Sim strain is also called American strain, which is a variety group bred after carnation was introduced to the United States in the 19th century, which is characterized by strong adaptability, vigorous growth, long internode, wide leaf, large flowers, round petals with few serrations, but the flowers are easy to split and the cold and disease resistance is weak, and the yield is low, so it is suitable for greenhouse cultivation.

Mediterranean strain is a hybrid variety group bred in European countries, which is characterized by short internodes, long and narrow leaves, rich flower colors and types, strong resistance to cold and disease, and high yield, but the flowers are slightly small.

(2) Culture bed and planting. The soilless culture of carnation mostly adopts the culture bed method using light soilless substrates. The culture substrate is usually peat, vermiculite, rice chaff, perlite, river sand, sawdust, slag, etc. The culture bed is 120–140 cm wide and 20–25 cm high.

The planting time of carnation is mainly determined according to the scheduled anthesis and culture method, and it usually takes about 110–150 days from planting to the first flowering date. Therefore, in general, for the culture method that produces flowers first in autumn and winter, the planting is mostly from May to June in spring, while for the culture method that produces flowers first in spring and summer, the planting is mostly from September to October in autumn. The planting density of the carnation is different according to the habit of the variety: the varieties with strong branchiness can be slightly thinly planted, and the varieties with weak branchiness can be properly densely planted. Generally, the planting density is 30–50 plants/m^2, and the plant and row spacing is mostly 15 cm × 15 cm to 15 cm × 20 cm. Varieties that bloom in spring and summer can be planted densely, and those that bloom in autumn and winter should be planted thinly.

(3) Nutrient solution and its management.See Table 8-5 for the nutrient solution formula for carnation culture. Carnation likes the dry environment, so the nutrient solution and water supply mostly adopt the drip irrigation method.

Table 8-5 Formula of Soilless Culture Nutrient Solution for Carnations

Name of compounds	Application rate (mg/L)
Calcium nitrate[$Ca(NO_3)_2{\cdot}4H_2O$]	950
Potassium nitrate (KNO_3)	500
Potassium dihydrogen phosphate (KH_2PO_4)	170
Ammonium nitrate (NH_4NO_3)	20

continued

Name of compounds	Application rate (mg/L)
Magnesium sulfate ($MgSO_4$)	250
Ammonium molybdate[$(NH_4)_6Mo_7O_{24}\cdot 4H_2O$]	0.15
Chelated iron (Na_2Fe-EDTA)	10
Manganese sulfate ($MnSO_4\cdot 4H_2O$)	2.2
Copper sulfate ($CuSO_4\cdot 5H_2O$)	0.2
Zinc sulfate ($ZnSO_4\cdot 7H_2O$)	1.2
Boric acid (H_3BO_3)	1.9

The concentration and supply of nutrient solution should be determined according to the specific situation. At the initial stage of planting, the concentration is low and the amount is small, and the concentration is high and the amount is large during the vigorous growth Stage. Solution is supplied 4–5 times a day, with an average of 200–400 mL per plant. The pH value and EC value of the substrate shall be measured regularly, and the nutrient solution shall be adjusted according to the measurement results. At the initial stage of planting, the EC value of the nutrient solution is about 1.0 mS/cm, and it gradually increases to 1.8–2.0 mS/cm from the vigorous growth to the anthesis. When the temperature is high in summer, due to the large amount of water evaporation, the concentration of the nutrient solution should be appropriately reduced. Besides, the pH value of nutrient solution shall be adjusted to 6.0–7.0.

In addition, 0.1%–0.5% urea, potassium dihydrogen phosphate, or low-concentration boric acid can be sprayed for foliar topdressing in combination with pest and disease prevention and control.

4. Plant management

(1) Topping. The first topping shall be carried out about 20 days after planting. Topping is a basic technical measure in carnation culture, and different topping methods have different effects on yield, quality, and flowering time. There are three

common topping methods in cut flower production.

① Single topping: Topping is carried out only once on the main stem, which can form 4–5 lateral branches, and the time from planting to flowering is short.

② Half-single topping: When the germinated lateral branches grow to 5–6 nodes after the first topping, the second topping is carried out on half of the lateral branches. Although the yield of the first batch of flowers is reduced, the flower yield is stable.

③ Double topping: After topping of the main stem, when the lateral branches grow to 5–6 nodes, the second topping is carried out for all the lateral branches. This method can ensure a high and concentrated yield of the first batch of flowers, but will weaken the flower stalk of the second batch of flowers.

(2) Netting. After the lateral branches begin to grow, the whole plant will open outside. Columns should be erected as soon as possible to support the net; otherwise, the plants are easy to lodge, affecting the quality of cut flowers. The mesh of the supporting net for carnation may vary depending on planting density or variety, usually between 10 cm × 10 cm and 15 cm × 15 cm. In general, the first layer of the net is generally 15 cm above the bed surface. Usually, 3–4 layers of net are required for supporting, and they shall be tightened flat with supporting rods.

(3) Removal of lateral buds. After the carnation starts the flower bud differentiation, its lateral buds begin to germinate, which shall be removed in time (except for multi-head carnation varieties). Since the removal of upper lateral buds will stimulate the germination of the middle and lower lateral buds, the lateral buds can only be completely removed several times. With the development of flower bud, several lateral buds will be formed around the middle main bud, which should be removed in time to ensure the normal growth of the main bud. If the removal is too late, the degree of stem lignification will be increased, which is inconvenient to operate and will also cause great damage to the plant. The bud thinning should be carried out in time and repeatedly.

5. Prevention & control of diseases and pests

Carnation disease is relatively serious, especially in hot and humid days from

May to September. The main diseases include mosaic disease, stripe disease, mottle disease, ring spot, fusarium wilt, wilt disease, stem rot, rust, etc. The pathogens are viruses, fungi, and bacteria. In addition, the carnation also suffers from aphids, red spiders, and cotton bollworms. The principle of "prevention first" shall be strictly implemented in production to strengthen management and enhance the resistance of plants. Attention shall be paid to keeping the environment clean and controlling the temperature and humidity. The pesticide shall be sprayed regularly according to the occurrence rule of diseases and pests, generally once a week. If diseases and pests occur, the pesticide application frequency shall be once every 3 days; the diseased plants shall be removed and destroyed in time to control the spread of diseases and pests.

II. Soilless Culture of Potted Flowers

(I) Azalea

1. Biological characteristics

Azalea is a flower of the genus *Rhododendron*, family Ericaceae. It is known as one of the top ten famous flowers in China, and has high ornamental value. It has formed different morphological characteristics in different natural environments, including evergreen trees, small trees, shrubs, and deciduous shrubs. The basic forms are evergreen or deciduous shrubs. Azalea has many branches, alternate leaves, and a dark green surface. It has racemes, and the flowers are terminal, axillary, or solitary. The flowers have rich colors, and some varieties have various cultivars.

Azaleas are widely distributed in the cold and temperature zones of the northern hemisphere. There are more than 900 varieties of azaleas in the world and more than 650 varieties in China. They can be vertically distributed from the ground level to the steep mountain with an altitude of 5,000 m, but are most flourishing at an altitude of 3,000 m. They like acidic soil, making them indicator plants of acidic soil, with a suitable pH value of 4.8–5.2. Azaleas are mostly resistant to shade and like warmth. They shall not be exposed to the hot sun and are suitable for growth in scatted light that is not strong. The suitable growth temperature is 12–25°C, 8–15°C for winter-blooming azaleas, 10°C for summer-blooming azaleas, and not lower than 5°C for spring-blooming azaleas. Azaleas like dry conditions and fear of waterlogging, and accumulated water shall be avoided.

2. Propagation methods

(1) Cuttage method. The cuttage shall be carried out during the plum rain season because the survival rate is high under moderate temperature. The new branches of the current year that have been lignified and hardened are selected as the cuttings. The length of each cutting is 7–8 cm, with the lower leaves removed and reserving the top 3–4 leaves. The cuttings are inserted into a moistened substrate, and the cutting bed is placed in a well-ventilated and sheltered place or shaded with a curtain which is removed at night. Spray water only 1–2 times during the day to prevent accumulated water in case of rain. The cuttings can take roots about one month later and can be gradually potted after seedling hardening.

(2) Layering method. The advantage of the layering method is that the seedlings obtained are large. The method is to bend down the branches at the base of the parent and press them into the substrate in the pot, and after 5–6 months, cut them off after rooting. If the branches are at the upper end and cannot be bent down, the air layering method shall be adopted, i.e. filling the soil in the bamboo tube or film for moisture preservation (the same as propagation of the Chinese rose and osmanthus). Pay attention to regular watering, and new roots will come out after July and August.

(3) Grafting method. Some varieties of azaleas have poor effects on cuttage propagation, such as Wangguan, Guixiao, and He Zhizhu, and can be propagated by grafting. The suitable rootstock is selected from robust perennial plants of *Rhododendron pulchrum* with strong vitality and good cold resistance, while the scion is mostly selected from plants of *Rhododendron hybridum* with bright colors and good flower types.

There are three grafting methods: approach grafting, bud grafting, and side grafting.

① Approach grafting: Select one pot of rootstock and scion respectively, arrange them side by side, and cut at the smooth and non-nodular parts with full growth and basically equal branch thickness (rootstock and scion). The cutting surface is 3–4 cm long and deep to the xylem. The cutting surfaces of rootstock and scion should be of the same size. Then align their cambiums against each other, and then bundle them with hemp skin or plastic film tape with moderate tightness. After 5–6 months, the wound is healed and integrated together. Then cut off the scion from the parent plant, and release the bundle for potting in the next spring.

② Bud grafting: Use biennial *Rhododendron pulchrum* as rootstock, remove the buds at the top and cut them flat. Then, cut in the middle, with a depth of about 4 mm, and then cut a tender bud of about 1 cm as the scion. Cut both sides into the same wedge shape, insert the scion into the rootstock, align their cambiums and fit closely, bundle the interface with wire, and place it on a shady shelf. After about 20 days, it can survive and be potted after seedling hardening for one month.

③ Side grafting: Take a scion of 4–5 cm, leave 3–4 leaves at the top, remove all the lower leaves, and cut the two sides of the stem into a wedge shape with a sharp knife, with a length of 0.5–1 cm. The cut surface shall be flat, smooth and clean to prevent contamination. Then cut slantwise at 6–7 cm from the base of the rootstock; the cut depth is slightly longer than the cut surface of the scion. When inserting the scion, their cambiums shall be aligned. Then, bundle their joint with wire, cover the scion and the interface into a small plastic bag, and tie the bag tightly, which is both windproof and moisturizing. After being transferred to the shade, it can survive and be potted after about one month.

3. Variety selection

There are five varieties suitable for soilless culture.

(1) *Rhododendron hybridum*. The flowers and leaves blossom together. The leaves are thick and lustrous. The flowers are large and gorgeous, with multiple petals, and the anthesis is from May to June.

(2) Summer-blooming azaleas. The leaves unfold first and then the flowers blossom. The leaves are small but dense, narrow pointed, with dense fuzz. The flowers are divided into single-petal and double-petal. The flowers are small, and the anthesis is in June.

(3) *Rhododendron simsii*. The flowers blossom before the growth of branches and leaves, cold-resistant, usually clustered in three flowers at the top of branches,

with five bright red petals, and the anthesis is from February to April.

(4) Wangguan. Semi-double petals, red edge on white background, with green spots on the base of three petals. It is very beautiful, and known as the king of azaleas.

(5) *Rhododendron ovatum*. Evergreen, red, or purplish-white flowers with spots. The anthesis is from May to June.

4. Preparation of soilless culture substrate

Mixed substrate is better for the culture of azaleas, and there are various substrate formulas for selection.

(1) Four parts humus soil, three parts humic acid fertilizer, two parts Heishan mud, and one part calcium superphosphate.

(2) Three parts peat, two parts sawdust, three parts leaf mold, one part bagasse, and one part calcium superphosphate.

(3) Five parts dead leaf deposits, two parts vermiculite, one part sawdust, and one part calcium superphosphate.

(4) Four parts lichens, two parts gravels, two parts plastic foam particles, and two parts mountain loess.

The formula substrate must be mixed uniformly and put into a pot for later use after disinfection.

5. Key points of culture

(1) Potting. Potting should be carried out before and after entering the greenhouse in autumn or leaving the greenhouse in spring. The potting method is as follows: ① Cover the drainage hole with several pieces of pot fragments or tiles crosswise, fill a thin layer of particle gravel at the bottom layer, then fill the slag and the coarse soil particles, pave a layer of fine soil at the top layer, and place the seedlings in the center. The root system should be fully extended and the planting depth should be appropriate. ② Hold the seedling with one hand, and add uniformly mixed substrate into the pot with the other hand until the root collar is buried, tamp the substrate in the pot, and then add an appropriate amount of substrate until it is 2–3 cm away from the pot mouth. ③ Water it with a watering can. The first watering should be sufficient until water flows out of the pot bottom. After potting

of the azalea, it needs to go through the 7–10 days recovery stage and be placed in a semi-shaded area in the greenhouse. When leaving the greenhouse, it should be placed under an outdoor sun shelter to avoid direct sunlight, which may cause plant wilting.

(2) Repotting. After potting, the plants become large seedlings through vigorous growth, with dense branches and leaves and a developed root system. The plant should be transferred to a larger pot. Otherwise, the root system cannot stretch in a small pot and intertwine together, which cannot fully absorb fertilizer and water and may affect ventilation and drainage. Plant growth will be in recession. At the same time, after a period of time, the substrate deteriorates and needs to be replaced with a new one. Whether pot replacement is required mainly depends on the growth of the plant; as long as the tree vigor does not decline significantly, the pot may not be replaced. Generally, it is better to replace the pot every three years or so. Large plants tend to require large pots correspondingly, which can also be replaced every five years or so. For super-large plants, as long as the growth does not decline, re-potting may also be unnecessary for many years.

When repotting, cut along the inner edge of the pot with a skewer or blade to separate the roots attached to the inner edge of the pot, then lift the plant to take it out of the pot, remove the broken pieces, or tiles adhered to the bottom of the root, loosen the substrate around the root, and remove some edge soil to make the roots scattered, but the substrate in the center of the top surface cannot be scattered. Cut off excessively long roots, blackened roots, and aged roots to promote new roots. The repotting operation is the same as that of the potting, and the season is similar to that of the potting. However, for plants in the full-blossom stage, repotting should be carried out after flowering.

(3) Watering. Azaleas have weak root systems, and are neither drought-resistant nor flood-resistant. If not watered in time during the growth period, their root system will shrink, leaves will sag or curl, and the tip will become charred yellow. In severe cases, they will not recover for a long time and will gradually die. If watered too frequently, ventilation is blocked, which may cause the root rot. In minor cases, the leaves will become yellow and fall, and stop growth; in severe

cases, they will die. Therefore, watering should not be neglected for azaleas. When the climate is dry, they should be fully watered. During normal growth, they should be properly watered only when the surface soil of the pot is dry. If they are in poor growth and the leaves are grayish-green or yellowish-green, they can be irrigated with 111,000 ferrous sulfate solution alone or in combination with water and fertilizer 2–3 times.

When watering azaleas, pay attention to the water quality. A clean water source is required, and the water temperature should be close to the air temperature. There is bleaching powder in urban tap water, which is harmful to plants and must be stored for several days before use. Alkali-containing water should not be used. The water in the north is slightly alkaline, so sulfuric acid can be added to adjust the pH value before use.

6. Nutrient solution management

The nutrient solution for azaleas is required to be strongly acidic, with an appropriate pH of 4.5–5.5. The components of nutrient solution should be comprehensive and proportional to meet the needs for the growth and flowering of azaleas. Special nutrient solutions for azaleas or general nutrient solutions can be selected. After planting, thoroughly water the plant with nutrient solution (diluted 3–5 times) for the first time. After placing the seedlings in the semi-shaded area for recovery for about half a month, normal management starts. The solution shall be supplemented once every 10 days regularly, with 100–150 mL for medium-sized pots and 200–250 mL for large pots. Supplement water to keep them moist during the period. Azaleas are not alkali resistant. In order to adjust the pH value of nutrient solution, the pH value can be adjusted with acetin or edible vinegar, and measured with pH indicator paper.

In the soilless culture of azaleas, a semi-shaded environment is always required, and shading is required in spring, summer, and autumn. High temperatures and stuffiness in summer often cause leaves to turn yellow and fall off or even cause the death of azaleas. Therefore, attention should be paid to ventilation or water spraying to cool down. The room temperature in winter should be about 10°C.

(II) Cyclamen

Cyclamen, also known as sowbread, Persian violet, or primrose, is a perennial bulbous herb of the genus *Cyclamen*, family Primulaceae. Originating from southern Europe and the Mediterranean, cyclamens now have been widely cultivated around the world.

1. Biological characteristics

Cyclamens have oblate fleshy tubers in dark brown. The leaves are grown at the center of the tip of the tubers. The leaves are heart-shaped and fleshy, and the leaf surface is dark green, mostly with white or light green stripes. The back of the leaves is purplish-red, and the leaf edge is serrated. The flowers are solitary, with slender pedicels and five petals, which roll upward and look like a rabbit ear. The flower colors include red, purplish-red, light red, pink, white, lilac, and compound color, and some varieties have fragrances. The anthesis is in winter and spring. At present, most cultivated cyclamens are horticultural varieties, which are bred and improved from the original seed for many years. They are usually divided into large flower type, flat petal type, wrinkle petal type, silver leaf type, double petal type, rough edge type, aromatic type, etc.

Cyclamens prefer cool, humid, and sunny environments. Autumn, winter, and spring are the growing seasons. In summer, when the temperature is high, they enter the dormant stage. The suitable temperature for growth and development is 15–25°C. The cultivation substrate is required to be loose, fertile, and well-drained. The suitable pH value is 6.0–6.8, and the required air humidity is 60%–70%. The cyclamen is a day-neutral plant that likes sunlight but avoids strong light. The suitable lighting intensity is $(2.8–3.6) \times 10^5$ lx. The full-blossom stage is from December to next April.

2. Propagation methods

Cyclamens can be propagated by sowing, dividing tubers, and tissue culture, and seed propagation is mainly used in production. Sowing is usually carried out from September to November. Perlite, vermiculite, coal cinder, sawdust, and other soilless culture substrates may be used for sowing. Before sowing, the seeds and substrates shall be disinfected. The substrates can be disinfected by high temperatures or chemicals. The seeds shall be soaked in warm water at 30–40°C for one day and night. If the seeds contain viruses, detoxification shall be carried out. For sowing, the seed coat shall be rubbed and removed. Sow the seeds in a shallow pot or sowing bed at a spacing of 1.5–2 cm, cover with a substrate of 0.5–0.7 cm thick, water the substrate thoroughly and keep it wet. Under the temperature of 20–25°C, it can take root after about 20 days, and germinate and grow cotyledons after about one month. At that time, the seedlings can be exposed to light to facilitate the photosynthesis of seedlings. When the emergence rate of seedlings reaches more than 75%, the nutrient solution shall be applied once every 10 days, and the ratio of nitrogen, phosphorus, and potassium is 1 : 1 : 1. When the seedlings grow 2–4 true leaves, the first seedling separation (usually from March to April) shall be carried out, and the seedlings shall be transferred to a flower pot with a diameter of 10 cm, and then start the normal culture and management after seedling recovery. Dividing tuber is carried out when the dormant corms germinate (from September to October). The tubers are cut into several parts according to the number of bud clusters. Each part has buds, and the cut is coated with plant ash or sulfur powder. After that, they are placed in a cool place for drying the cut and then used for the culture of new plants.

3. Soilless Culture Techniques

(1) Substrate Pot Culture. The soilless culture of the cyclamen is mainly by pot culture of substrate. The culture substrate can be made by mixing vermiculite, peat, slag, sawdust, sand, and carbonized rice husk in different proportions, for example, the ratio of vermiculite, sawdust, and sand is 4 : 4 : 2 or the ratio of slag, peat, and carbonized rice husk is 3 : 4 : 3. Clay pot should be used in the seedling stage, with 3–4-cm-thick coarse coal cinder paved at the bottom of the pot, and mixed substrate fills the upper part. Be careful when planting the seedlings, do not damage the root system, spread the fiber roots, add the substrate, gently compact it and expose 1/3 of the corms, and thoroughly water it with nutrient solution (dilute 3–5 times). Cyclamens like fertilizer, but the fertilizer shall be applied evenly. Nutrient solution shall be watered once a week on normal days, and clean water shall be sprayed once every 2–3 days according to the weather conditions. Because the substrate is

loose and permeable, water and fertilizer can be retained, which can meet various needs of seedling growth. For cyclamens, growing ten leaves is an important period, generally from May to June. At that time, they enter the stage of vegetative growth and reproductive growth. In cool regions, the second transplanting can be carried out at this time to a plastic pot, ceramic pot, or porcelain pot with a diameter of 15 cm, and the transplanting method is the same as above. In summer, attention should be paid to humidity reduction and ventilation, so as to preserve existing leaves and control the fertilizer and water to prevent plant overgrowth. In addition, attention should be paid to disease and pest prevention, and bactericides and pesticides such as carbendazim, thiophanate, rogor, and DDVP can be sprayed.

At the end of August, with the cooling of the weather, cyclamens will gradually resume growth and grow many new leaves. At this time, attention should be paid to strengthening lighting and fertilization. Nutrient solution should be applied once a week at the normal concentration, and 0.5% potassium dihydrogen phosphate solution should be sprayed on the leaves every 10 days or so. From October to November, the leaves grow slowly, the development of flower buds is obviously accelerated, and the anthesis starts. At this time, the suitable light is $(2.4–4) \times 10^5$ lx, the suitable temperature is 12–20°C, and the suitable humidity is about 60%. Temperature is the main method to control the anthesis. The anthesis of general varieties can be postponed by 20–40 days at 10°C. Gray mold is easy to occur during the anthesis, so ventilation and chemical control should be strengthened.

The soilless culture of the cyclamen is faster than that in soil, with more large and bright flowers, earlier flowering, and longer anthesis. See Table 8-6 for the recommended formula of soilless culture nutrient solution for cyclamens.

The nutrient solution can be a concentrated solution, and then diluted by different multiples according to different growth stages. Generally, it is a 10-fold concentrate, which can be diluted 3–5 fold for use, and the pH value is adjusted to about 6.5.

(2) Hydroponics.The corm of the cyclamen is put in the upper neck of a special gourd-shaped container, and the roots naturally hang into the large container under

Table 8-6 **Nutrient Solution Formula for Cyclamens**

Name of compounds	Application rate (mg/L)
Calcium nitrate[$Ca(NO_3)_2 \cdot 4H_2O$]	250
Potassium nitrate (KNO_3)	400
Potassium dihydrogen phosphate (KH_2PO_4)	100
Urea [$(NH_2)_2CO$]	200
Magnesium sulfate ($MgSO_4$)	150
Ferrous sulfate ($FeSO_4 \cdot H_2O$)	100
Calcium sulfate ($CaSO_4 \cdot 2H_2O$)	50
Ammonium molybdate[$(NH_4)_6Mo_7O_{24} \cdot 4H_2O$]	10
Zinc sulfate ($ZnSO_4 \cdot 7H_2O$)	10
Boric acid (H_3BO_3)	10

the neck. The whole plant is ornamental, and the green leaves and white roots are complementary.

① Preparation for planting.

a. Preparation of seedlings: In late August, before the cyclamen restore the growth after dormancy, dig out plants with corms of more than 3 cm and more than ten leaves, free from diseases and pests and in good health, and wash the roots for use.

b. Preparation of container: Containers with a diameter of more than 15 cm are generally used for corms above 3 cm. A polystyrene hardboard of 2 cm thickness is used as the cover plate and planting plate.

c. Nutrient solution: Formulate 1/2 dosage level of garden trial formula nutrient solution, with a pH value of 6.0–7.0.

② Planting and management.

Wrap the corms with rock wool or foam plastic, fix them in the planting plate, and immerse the extended root system into the nutrient solution. The nutrient solution shall be replaced once every 30 days, or according to the clarity of the nutrient solution. The rapid growth stage occurs during the high-temperature and high-humidity periods. Pay attention to spraying pesticides such as carbendazim, thiophanate, and rogor once a month.

(III) Arrowroot

Arrowroot is a perennial monocotyledonous herb of the genus *Maranta*, family Marantaceae, which has a beautiful posture. Many varieties have eye-catching stripes on their leaves, which are beautiful, colorful, and strange, and they are the main indoor high-end foliage-ornamental flowers popular in the world. It can be combined with orchid, anthurium, and other flowers to form a variety of green leaf and red flower patterns, which have a better ornamental effect. The soilless culture of arrowroot is mainly substrate pot culture.

1. Biological characteristics

Most varieties of arrowroots have rhizomes or tubers underground, with strong tillering characteristics. There are several growing points directly at the root collar, where leaves can grow out. The leaves vary in shape, with various patterns and fuzz. The leaves have obvious characteristics, that is, there are open leaf sheaths at the base of leaves, and there is a significantly expanded joint at the junction of the leaves and the petiole, which is called a leaf pillow, contains water storage cells and has the function of regulating the direction of the leaves. The leaves stand upright when there is sufficient water at night and unfold when there is insufficient water during the day. This is a characteristic of arrowroot. The flowers are hermaphroditic, bilaterally symmetrical, and often in bracts. They are arranged in spike, head, and sparse panicles, or the inflorescences sprout from rhizomes alone. The fruits are capsule and berry fruits. The flowers are not large but have a graceful posture. The flowers blossom under short sunlight. The growth rate of arrowroot depends on its variety of characteristics and cultivation techniques. It generally takes 6–12 months from seedling to finished product. The height of the finished seedling is 40–80 cm, with plump plant type, lustrous leaves, and free of diseased leaves.

Arrowroot is a thermophilic plant with a suitable growth temperature of 16–28°C. The optimal growth temperature is 22–28°C during the daytime and

18–22°C at night. The growth is slow under a temperature below 16°C, and even stops under a temperature below 13°C. The plants are vulnerable to cold injury under a temperature below 10°C. The suitable lighting intensity is 5,000–20,000 lx. The plant is thin and weak under weak light and the leaves will curl and be burnt under strong light. It likes shade and moist, and the suitable humidity is 65%–70%. The growth is slow under low humidity, and high humidity tends to form spots. The substrate is required to be loose, with strong water and fertilizer retention capacity, a humidity of 60%–70%, a pH value of 4.8–5.5, and an EC value of 0.8–1.2 mS/cm. The water is required to have an EC value of less than 0.1, a pH value of 5.5–6.5, and free of Na^+ and Cl^-. If Cl^- exceeds the standard, it tends to burn the leaves. Arrowroots of different varieties or in different growth stages have different environmental requirements.

2. Variety selection

There are four types of common cultivars of arrowroots:

(1) *Calathea*: *Calathea insignis*, *Calathea zebrina*, *Calathea roseapicta*, etc.

(2) *Ctenanthe*: *Ctenanthe oppenheimiana*, *Calathea orbifolia*, etc.

(3) *Stromanthe*: *Stromanthe sanguinea*, etc.

(4) *Maranta*: *Maranta bicolor*, etc.

Calathea, *Ctenanthe*, and *Stromanthe* plants are tall and easy to plant; *Maranta* plants are short, and some varieties are difficult to plant.

3. Propagation methods

The propagation methods of arrowroot include cuttage propagation, sucker propagation, and tissue culture, and the main methods are sucker propagation and tissue culture. For adult plants that have grown for several years, if the stem is too long and destroys the plant shape, the branches shall be cut in time, and the cut branches and leaves shall be used for cutting. The young stem with 2–3 leaves shall be cut and inserted into the sand bed, and it can take roots in half a month. Sucker division is generally carried out when the temperature reaches above 15°C, and low temperature is easy to damage the root, affecting its survival and growth. When dividing the sucker, remove the original soil to remove the rhizome first, and select strong and neat young plants for potting, respectively. The sucker should not be divided too small, and each block should have many leaves and strong roots; otherwise, it will affect the growth of new plants. The seedlings produced by sucker propagation are easy to be infected by the root-knot nematode, the growth is weak and the seedlings are not uniform, so it can only be used in small-scale culture. Commercially, tissue culture is mostly adopted. The seedlings have the characteristics of good growth, no virus, good plant type, and easy control.

4. Soilless Culture Techniques

(1) Preparation for planting.

① Substrate selection and treatment: Arrowroot prefers a culture substrate with good drainage and air permeability, and a complex substrate with turf : perlite of 2 : 1 or perlite : peat : slag of 1 : 1 : 1 can be used. It is required that the pH value of the substrate is 4.7–5.5, and the EC value of 0.6–0.8 is optimal. The substrate can be used only after full disinfection when the pests and diseases are killed.

② Preparation of culture container: The culture container is determined according to the culture variety and culture method and is generally a hard opaque plastic pot. Generally, the plastic pot has a diameter of 12–14 cm or 17–19 cm.

(2) Planting.

According to the characteristics of arrowroot varieties, it can be divided into two types: single planting and double planting requiring repotting. For large plant varieties with fast growth rates, such as *Calathea zebrina*, *Calathea orbifolia*, *Stromanthe sanguinea*, and *Calathea insignis*, they can be directly planted in pots with a diameter of 17–19 cm; small and dwarf plant varieties can be directly planted in pots with a diameter of 12–14 cm; for large plant varieties with slow growth rates, such as *Calathea ornata*, *Calathea makoyana*, and *Calathea leopardina*, they generally adopt double planting requiring repotting. They are first planted in pots with a diameter of 10–12 cm for about six months and then transferred to pots with a diameter of 17–19 cm.

The root system of *Maranta* plants is relatively shallow, so they are mostly planted in a shallow pot. Pave a layer of ceramsite at the bottom of the pot as the drainage layer when potting, then place the seedlings upright, add the prepared substrate until the pot is 80% full, then compact the substrate by hand, and finally add a layer of ceramsite on the top to prevent from algae growth and washing away the substrate or washing down the seedlings. The new plant should not be planted too deeply as long as the whole root is buried in the soil; otherwise, the growth of new buds will be affected. After planting, the water content of the substrate should be controlled not too high, but water can be sprayed on the leaf surface frequently to increase the air humidity. Only after new roots grow can the plant be fully watered. Spray 1,500-fold dilution of thiophanate methyl and 3,000–4,000-fold dilution of agricultural streptomycin after planting to prevent diseases in the seedling stage.

(3) Cultivation management.

① Temperature. At the initial stage of planting, the daytime temperature is required to be kept at 25–27°C and the night temperature is required to be 17–20°C. The adult plant stage is generally in winter, and the minimum temperature is ensured to be above 16°C; temperature above 35°C or below 10°C is unfavorable to its growth. Therefore, the arrowroot seedlings shall be placed in a cool place in high-temperature seasons like summer; in winter, attention shall be paid to cold prevention, and the plants shall be moved to a warm place without wind to overwinter.

② Light. Arrowroot shall avoid direct sunlight and grow better under indirect radiation light or scattered light. In production, a sunshade net with a shading degree of 75%–80% is used for culture under the sunshade environment. Appropriate sunshading is required at the initial stage of planting, with a lighting intensity of 5,000–8,000 lx. The lighting intensity at the seedling stage is 9,000–15,000 lx, up to 20,000 lx. In summer, direct sunlight and excessive lighting are easy to cause leaf rolling and edge burning. New leaves stop growing and the leaves turn yellow. Attention should be paid to shading, but it should not be too shaded; otherwise, the growth of plants will be weak, and the patterns on the leaves of some variegated-leaf varieties will fade, even disappear, so it's better to place it in a bright place without direct sunlight. The lighting intensity in the adult plant stage can be increased appropriately to reach 10,000–20,000 lx.

③ Humidity. Arrowroot is sensitive to moisture. It should be fully watered during the growth period to keep the substrate in the pot wet, but it is not suitable to accumulate water; otherwise, it will lead to root rot and cause diseases and even death of plant. The suitable relative humidity is 65%–80%, and high humidity is conducive to the unfolding of leaves. In the seedling stage, if the humidity is too high during the daytime (relative humidity > 80%), too much water will accumulate in the leave cells and cause them to break, and brown spots will form on the leaves, which look like disease spots and reduces the ornamental value. The substrate in the adult plant stage shall not be too wet nor watered too often. The humidity shall be kept at 60%–70%. If the environment is too dry during the sprouting period of new leaves, the edges and tips of the new leaves are easy to wither and roll, and even become deformed in the future. The withered leaves cannot recover. Therefore, it is necessary to water frequently in the growing season from March to October every year, and spray water on the leaf surface frequently. In summer, watering shall be carried out three or four times a day in time. The substrate shall be kept slightly dry after autumn. In winter, the plants are in a semi-dormancy state, and the overwintering temperature shall be controlled. The plants shall be placed with sufficient scattered light and the pot soil shall be maintained slightly dry.

④ Fertilizer and water management. In the early growth stage of the arrowroot, N fertilizer can be supplemented properly by applying calcium nitrate or potassium nitrate (alternate application) once a week, with a concentration of 0.1%, and the substrate is kept wet. After the seedlings of 2–4 weeks have grown new leaves, start regular water and fertilizer management, that is, continuous supply under relatively stable conditions of 50%–70% substrate humidity. The general principle of fertilization is “frequent application with little amount” to avoid excessive concentration. The fertilization frequency is generally once or twice a week, depending on the size of the plant and the demand for fertilizer. In order to prevent the burning of new unopened and curled leaves, the leaves shall be washed with clean water after fertilization, and the method of spray fertilization can be adopted. The nutrient solution formula can be selected from formula for foliage-ornamental plants or a nutrient solution with N ∶ P ∶ K of 1 ∶ 0.4 ∶ 1.8 plus trace elements (excessive boron is prone to leaf burning). When the foliage plant nutrient solution is used, the nutrient solution shall be diluted appropriately to water the plant thoroughly for the first time until there is exudate in the tray below the pot. At ordinary times, the solution is supplemented once or twice a week, with 100 mL/plant each time, and water is supplemented to keep the substrate wet. No water shall be supplemented during solution supplement, and no water shall be accumulated in the tray below the pot for a long time to facilitate ventilation and prevent root rot. In the adult plant stage, P and K fertilizers, such as 0.1%–0.2% potassium dihydrogen phosphate solution, are mainly applied to increase plant resistance. The EC value increases with the growth of plants. Generally, the EC value is controlled within 0.5–1.5 mS/cm: 0.5–0.8 mS/cm during the vigorous growth stage and 0.8–1.5 mS/cm during the adult plant stage. Different varieties have slightly different requirements for pH value. For example, *Calathea roseapicta* likes acid and requires a pH value of 4.8–4.9, *Calathea ornata* requires a pH value of 5.1, *Calathea insignis* and *Calathea loeseneri* require a pH value of 5.8–6.0, and most varieties require a pH value of 5.3–5.5. Therefore, the regulation targets for the pH value of the nutrient solution or fertilizer solution in production vary depending on varieties.

⑤ Flower forcing. Most arrowroots are foliage ornamental, but some varieties also have beautiful flowers, such as *Calathea crocata*, *Calathea loeseneri*, and *Calathea zebrina*. There are some varieties such as *Calathea zebrina* and *Calathea roseapicta*, which can also flower under certain conditions, but the flowers are not beautiful. Generally, the flowers are removed or avoided. Arrowroots are short-day plants, so the flower bud induction and formation must be carried out under short sunlight conditions. Key points of flower forcing are as follows.

a. The growth period must last for 3–4 months; otherwise, it will not flower.

b. The plant shall be subject to short sunlight, with lighting hours less than 12 h/day and lasting for 5–6 weeks.

c. The temperature shall be 17–21°C. Too-low or too-high temperatures are not conducive to flowering. To avoid flowering of arrowroot, supplement lighting can be provided in its natural flowering season to make its sunshine hours more than 12 h.

(4) Prevention and control of diseases and pests.

Common diseases of arrowroots include leaf spot and leaf blight, which mainly occur in leaves and may also harm leaf sheath, affecting the ornamental effect. The prevention and control methods include the removal of diseased leaves in time and prevention in advance. For regular prevention, the plants can be sprayed with 800-fold dilution of 75% chlorothalonil wettable powder, 500-fold dilution of 50% captan wettable powder, 800-fold dilution of 70% thiophanate methyl wettable powder, and 3,000–4,000-fold dilution of agricultural streptomycin solution once every 2–3 weeks for two or three times, continuously. Common pests include thrips and red spiders. Thrips mainly harm the leaves of arrowroot, resulting in many small white spots or grayish-white spots on the leaves. In particular, *Calathea zebrina* is very sensitive to thrips. The plants can be sprayed with Chongmanke or 1.8% avermectin once a week for 2–4 times, continuously. The symptom of the red spider is reddish-brown or orange leaves, which may cause an imbalance in water metabolism and affect the normal growth of plants. The plants can be treated with chemicals such as Qimansu once a week for 2–4 times.

[Module summary]

I. Key and Difficult Points

(1) Understand different culture forms of soilless culture of fruit vegetables, and master the methods of the eco-organic soilless culture and rock wool culture of tomatoes, eco-organic soilless culture of cucumbers, trough substrate culture of towel gourds, bag culture of muskmelons, and deep flow culture of eggplants.

(2) Understand different culture forms of soilless culture of leaf vegetables and master the methods of deep flow culture of lettuce, three-layer hydroponics of crown daisy, column substrate culture of *Gynura bicolor*, and deep flow culture of celery.

(3) Understand different culture forms, production sites, and facility preparation of the soilless culture of sprout vegetables, and master the soilless culture techniques of pea seedlings and broad bean sprouts.

(4) Master the propagation techniques of Chinese roses, lilies, carnations, chrysanthemums, etc., as well as the soilless culture techniques of common potted flowers.

(5) Understand the selection and characteristics of excellent strawberry

varieties, master the selection and formulation of nutrient solution and the preparation of culture substrate for strawberry soilless culture, and master the production management techniques of strawberry soilless culture.

II. Summary of Experience and Skills

(1) Eco-organic soilless culture is easy to operate and convenient for management. No inorganic fertilizer is applied during the growth process so that the vegetable produced meets the green vegetable standard.

(2) Soilless culture of leaf vegetables generally has a universal culture facility structure, and the biological characteristics of vegetables and the key points of culture management should be mastered.

(3) As for nutrient solution management, the concentration of the nutrient solution in summer is higher than that in winter, and the concentration in the late growth stage is higher than that in the early growth stage.

(4) The key technical point of sprout vegetable culture is to prevent undesired bacteria and control the temperature, light, and humidity.

[Skill Training]

Skill Training 8-1 Nutrient Film Culture Technique of Mizuna

I. Purpose and Requirements

1. Purposes

Understand the characteristics of NFT hydroponics and master the NFT hydroponics.

2. Requirements

The nutrient solution should be accurately formulated and scientifically managed. The operation should follow the production process. The plant grows normally and the culture effect is good.

II. Preparation for Training

1. Materials and chemical agents

Seedlings of mizuna, 500-fold dilution of carbendazim solution, compounds required for the formulation of nutrient solutions (formula of Japanese Yamazaki lettuce) with the following formula: Calcium nitrate tetrahydrate 236 mg/L, potassium nitrate 404 mg/L, ammonium dihydrogen phosphate 57 mg/L, and magnesium sulfate heptahydrate 123 mg/L.

2. Instruments and tools

Acidimeter, conductivity meter, electronic balance, nutrient film hydroponic facility, tools for formulating nutrient solution, etc.

III. Training Steps

1. Formulation of nutrient solution

See Module 3 for the formulation method.

2. Seedling washing and disinfection

For seedlings raised on the substrate, rinse the substrate in the root system, immerse the seedlings in 500-fold dilution of carbendazim solution for 10 minutes, and then rinse them with clean water before planting.

3. Planting

Select a suitable nutrient solution film culture bed and equip it with an automatic solution supply system for the nutrient solution. Drill holes on the planting plate for planting at the density of 60–70 plants/m^2 in the culture bed.

4. Nutrient solution management

After planting, gradually increase the concentration of nutrient solution from 1/2 dose to 2/3 dose with the growth of plants, and finally to 1 dose. In the daytime, the solution shall be supplied for 15 min every hour, with an interval of 45 min; in the nighttime, the solution shall be supplied for 15 min every 2 h, with an interval of 105 min, which is controlled by a timer. The EC value of the nutrient solution is controlled within 1.4–2.2 mS/cm, and the pH value is controlled within 5.6–6.2.

Supplement the consumed nutrient solution in time at ordinary times, and replace the nutrient solution completely every 30 days.

5. Environmental regulation of aerial part

Mizuna is also cryophilic, so its environmental regulation can refer to lettuce soilless culture.

6. Observation records

Management Records of NFT Hydroponics

Crop name: ____________________ Recorded by: ____________________

Date	Facility environment		Nutrient solution		Growth condition	Treatment measures	Remarks
	Temperature	Humidity	EC value	pH value			

Skill Training 8-2 Soilless Culture of Strawberries

I. Purposes and Requirements

Master the soilless seedling culture technique, the design method of soilless culture trough and the production management technique of soilless culture of strawberries.

II. Plan

1. Materials and tools

Excellent strawberry mother seedlings, seedling substrate, nutrition bowl, culture trough material, etc.

2. Implementation Plans

(1) Preparation of seedling plot, seedling land consolidation, and fertilization.

(2) Planting of mother seedlings.

(3) Formulation and loading of nutrient substrate.

(4) Pruning of stolon seedlings, propagation in nutrition bowl.

(5) Design and assembly of strawberry culture trough.

(6) Formulation of the nutrient solution.

(7) Planting in soilless culture of strawberries.

(8) Daily management in soilless culture of strawberries.

III. Implementation

(1) Refer to relevant materials on soilless culture of strawberries, and formulate the implementation plans for trough soilless culture of strawberries in groups.

(2) Prepare PPTs based on the implementation plans and make reports.

(3) Evaluate each other between groups and amend the implementation plans.

(4) Implement the soilless culture of strawberries in groups

Skill Training 8-3 Root Restriction Soilless Culture of Grapes

I. Purposes and Requirements

Master the design method of soilless culture trough and the production management technique of the soilless culture of grapes.

II. Plan

1. Materials and tools

Grapes seedlings, culture substrates, fertilizers, culture trough materials, etc.

2. Methods and steps

(1) Preparation and fabrication of culture trough.

(2) Formulation and loading of nutrient substrate.

(3) Planting.

(4) Daily management in soilless culture of grapes.

III. Implementation Plans

(1) Formulate the implementation plans for soilless culture of grapes in groups

based on relevant knowledge of soilless culture of grapes.

(2) Prepare PPTs based on the implementation plans and make reports.

(3) Evaluate each other between groups and amend the implementation plans.

(4) Implement the soilless culture of grapes in groups.

[Expansion Task]

I. Review Questions

(1) Analyze the advantages and disadvantages of NFT and DFT hydroponics.

(2) How to prevent the breeding of algae in nutrient solutions?

(3) What is the eco-organic soilless culture technique? What is the significance of applying this technique in agricultural production?

(4) Briefly describe the seed treatment methods in pea seedling production.

(5) Briefly describe the key points of management during the culture of broad bean sprouts.

(6) What are the main management tasks after the grafting of cut roses?

(7) What are the similarities and differences between flower substrate culture and hydroponics?

(8) Think about whether the solution made of compound fertilizer is a nutrient solution, and why?

(9) How to select cuttings of cut roses?

(10) How to select and store lily seedballs?

(11) Briefly describe environmental regulation techniques in the process of soilless culture of arrowroot.

(12) How to formulate the nutrient solution for potted azaleas?

(13) What is the process for H-shaped pruning of grapes? What are the benefits?

(14) How to thin flowers and fruits for grapes?

(15) Try to describe the applications of plant hormones in fruit expansion of grapes.

II. Case Study

1. Scenario

A vegetable grower plans to produce 200 plates of radish sprouts in greenhouses and market them before the Spring Festival. Please design a production plan for radish sprouts.

(1) Scheduling for the production period: ①Sowing period; ②Harvesting period.

(2) Production preparation (seeds, seedling trays, culture racks, etc.).

S/N	Name of material	Specification and model	Quantity	Funds (CNY)

(3) Sowing.

Seed treatment methods	Sowing methods

(4) Management.

Tray stacking	Tray reversing	Temperature	Humidity	Greening	Disease and pest prevention

(5) Harvesting (yield prediction).

2. Production records

Project name: Team: Name:

Name of variety	Sowing period	Germination stage	Harvesting period	Production area	Single-tray yield	Total yield	Total income

3. Summary of production

(1) Exchange successful experiences and summarize measures to be improved.

(2) Write the sprout vegetable production technique in the completed project.

Module 9 Plant Factory

[Learning Objectives]

I. Knowledge Objectives

(1) Master the basic concepts and meanings of the plant factory;

(2) Master the processes and system composition of the plant factory, as well as environmental regulation measures for the plant factory.

II. Skill Objectives

Learn about the history and development of the plant factory from literature, and grasp the environmental regulation measures for the plant factory and the precautions in production.

[Preparation for Learning]

I. Required Resources

(1) Materials about the development of the plant factory;

(2) Understand the development of relevant plant factories both in China and abroad through the Internet;

(3) A multimedia classroom;

(4) Standardized plant factory or picture information of the plant factory.

II. Background Knowledge

(1) Understand the basic knowledge related to the plant factory;

(2) Master the basic knowledge of literature review;

(3) Understand some basic principles of environmental regulation.

[Learning Tasks]

Task 1 Basic Concepts and Meanings of the Plant Factory

I. Concepts of the Plant Factory

The plant factory refers to a highly efficient agricultural system with a commercial facility utilizing artificial light to operate in an industrial production process to realize a standardized, normalized, large-scale, and full-year steady production, and by virtue of high-precision environmental control within the facility, realize full-year uninterrupted production of crops. It represents a labor-saving mode of production in which such environmental factors as temperature, humidity, light, CO_2 concentration, and nutrient solution during the crop growth process are controlled automatically by computers so that the production is not or is rarely restricted by natural conditions. The plant factory is internationally recognized as the highest stage of development of facility agriculture and one of the important indicators to measure the development level of agricultural science and technology in a country, as it makes full use of advanced processes, biotechnology, nutrient solution culture technique, information technology, etc., and is highly technology-intensive.

There is still much debate about the definition and classification of the plant factory. In Europe and America, the factory-like production mode of vegetables and flowers with a greenhouse provided with artificial light sources and using hydroponics or rock wool culture techniques inside is barely referred to as the plant factory utilizing sunlight, but in Asia, especially in Japan, it is classified into the plant factory utilizing sunlight. Professor Toyoki Kozai from Chiba University divides the plant factory into plant factory utilizing artificial light, plant factory utilizing sunlight, and plant factory utilizing both artificial light and sunlight. It is generally accepted that the plant factory can be mainly divided into the plant factory utilizing artificial light and the plant factory utilizing sunlight (further into the plant factory utilizing sunlight with supplementary artificial light and the plant factory utilizing sunlight without supplementary artificial light).

As for the plant factory utilizing artificial light, artificial light serves as the sole light source, and plants are subject to a multi-layer stereoscopic culture. This kind of plant factory is similar to an industrial plant that operates an order-based scale production for which a standardized procedure has been put into practice. The main characteristics of the plant factory utilizing artificial light are as follows: ① The factory building is fully enclosed with a good tightness, where the roof and enclosures are made of light-proof materials for good thermal insulation. ② Artificial light with good characteristics serves as the sole light source, including high-pressure sodium lamp, high-frequency fluorescent lamp, and LED. ③ Online plant detection and network management technologies are employed to detect the plant growth processes and process relevant information. ④ The hydroponics technology with nutrient solution is employed so no soil or even substrate is needed. ⑤ Invasion of pests and pathogenic microorganisms can be effectively controlled to realize pollution-free production. ⑥ Various factors involving the growth of plants can be controlled precisely and regulated at any time to improve stability of plant growth and realize a full-year balanced production. ⑦ High costs for provision of technical equipment and construction of facilities, high energy consumption, and high operating costs.

The plant factory utilizing sunlight refers to a large-sized greenhouse facility with or without supplementary artificial light. It operates in a semi-enclosed greenhouse environment and mainly makes use of sunlight or supplementary artificial light (for a short term) and the nutrient solution culture technique to realize full-year crop production. Characteristics of the plant factory utilizing sunlight are as follows: ① Semi-enclosed greenhouse construction, mostly made of and covered by glass or plastics (fluoroplast, thin film, PC sheet, etc.). ② Mostly sunlight is used, and artificial light can also be supplemented. ③ Various kinds of environmental monitoring and regulation equipment are provided in the greenhouse, including systems for environmental factor data acquisition and automatic environmental regulation. ④ Hydroponics and substrate culture are dominant. ⑤ the production environment can easily be influenced by seasonal and climate changes, leading to a certain restriction to applicable varieties (mostly leaf vegetables and solanaceous vegetables are applicable) and uncertainty in production. ⑥ Costs for the construction of facilities are far lower than, and operating costs slightly are lower than those for plant factories utilizing artificial light.

II. Significance of Developing the Plant Factory

In recent years, the plant factory has attracted the attention of people from all walks of life for a number of reasons. First, under the food crisis caused by the rapid growth of the world population and the reduction of arable land in the world, the plant factory that can efficiently utilize resources has become one of the means available to grapple with this crisis. Second, the problem of out-of-limit residues of pesticides is becoming increasingly prominent and food safety has attracted people's attention. As the plant factory operates in an enclosed production environment in which few or no pesticides are used, it can ensure safety and no pollution in the crops produced by it. Third, the plant factory, boasting a comfortable and pleasant working environment and factory-like production mode, can attract young people with knowledge to engage in agricultural production, thereby improving the quality of the agricultural laborers.

In addition, the plant factory, which represents a mode of agricultural production that is highly technology-intensive and can efficiently utilize resources, also has incomparable advantages over other agricultural modes, mainly including the following aspects.

① Under the highly organized crop production, the full-year production of crops can be realized without being affected by the external environment, where 15–18 crops of leafy vegetables can be harvested a year.

② High yields per unit area and high resource utilization rate: The annual yield of lettuce can be as high as 150 t/1,000 m^2, which is 30–40 times as much as that by open cultivation.

③ With a high degree of mechanization and automation, low labor intensity and a comfortable working environment, the plant factory can attract many young people with knowledge to engage in agricultural production.

④ The products are safe and pollution-free as no pesticides are applied. By means of artificial environmental control, the invasion of diseases and pests can be effectively prevented, and few or no pesticides are applied during production.

⑤ Multi-layer stereoscopic culture: In a plant factory utilizing artificial light, the number of layers for production can reach eight to ten or even higher, thereby significantly improving the land utilization rate.

⑥ Little or no restriction by land as the production is available on non-arable land: With the plant factory, the production of plants can become feasible in suburban regions, barren land, building roofs, deserts, Gobis or even space stations, and other planets.

⑦ The plant factory can be built downtown or in the vicinity of an urban area to shorten the distance between production and sales of vegetables and save the intermediate links, to not only keep the vegetables fresh, but also greatly shorten the distance of logistics, thereby reducing costs and carbon emission related to logistics.

Therefore, the plant factory is considered to be an important way to solve such problems as population growth, shortage in resources, insufficiency in a new generation of labor forces and rising food demands in the future. In particular, vertical farming or skyscraper farming based on the plant factory is a possible way to grapple with the food crisis in the future.

Task 2 Processes and System Composition of the Plant Factory

I. System Overview

Currently, the plant factory is mainly divided into the following two types: ① The plant factory utilizing artificial light, which operates in a fully enclosed environment and employs artificial light sources and nutrient solution culture technique to realize full-year uninterrupted production of plants; ② The plant factory utilizing sunlight, which operates in a semi-enclosed greenhouse environment and employs sunlight or supplementary artificial light (for a short term) and nutrient solution culture technique to realize full-year production of plants. In a broad sense, the plant factory refers to the general term of these two types of plant factories. However, in a narrow sense, the plant factory generally refers to the plant factory utilizing artificial light, i.e. the fully-controlled plant factory. The plant factory utilizing sunlight mainly involves the relevant technologies, including greenhouse engineering, environmental control, nutrient solution culture management, and automatic control. We will focus our study on the plant factory in a narrow sense, i.e. the plant factory utilizing artificial light.

As for the plant factory utilizing artificial light, the enclosed structure is mostly made of light-proof thermal insulation materials, and artificial light serves as the sole light source for the photosynthesis of the plants. Moreover, such factors as temperature, humidity, light, CO_2 concentration, and nutrient solution during the growth and development of plants are controlled automatically by computers to realize the full-year uninterrupted production of the plants. To realize various functional characteristics and meet the production demands, the plant factory utilizing artificial light, in terms of spatial structure, mainly consists of functional rooms, including a cultivation workshop, a seedling culture room, a harvesting and storage room, a mechanical room (nutrient solution tanks, carbon dioxide cylinders, control devices, etc.), and an administration room (office and computer control systems), which, by use of this kind of spatial structure and functional layout, realizes the management throughout the industrial chain (from seeding to harvesting, and to sales). Moreover, in light of the system structure, this plant factory is based on the enclosures and functional units, and with such sub-systems as nutrient solution circulating and control system, multi-layer stereoscopic hydroponics system, air conditioning and purification system, CO_2 fertilizer release system, artificial light source system, and automatic computer control system, it can guarantee the safe operation and intelligent management of the plant factory around the clock.

II. Production Process of Plant Factory

According to the whole-process production requirements (from seeding to harvesting, and to sales) on crops, the basic production process of the plant factory utilizing artificial light includes seeding, germination, seedling culture, cultivation, harvesting, packaging, storage, and sales, which are all under artificial control, and must comply with the requirements on clean and pollution-free processes for vegetables. The system structure is also designed to achieve these objectives.

1. Seeding and germination

As for a large-sized plant factory, both the seeding and germination are carried out in an independent workshop or an independent area delineated in the cultivation workshop of the plant factory. The seedbed plate is generally made of a sponge pad and white plastic foam, and the sponge pad is soaked by special machinery or manually in a nutrient solution that is contained by a boxboard. A seed tray or plate-type seedling planter is used to seed into recesses formed on the sponge pad, at a rate of 300 seeds per sponge pad (25 recesses in 12 rows). The bed plate has a size of 300 mm × 600 mm. Upon the seeding operation, the plantlet tray of the sponge pad that has absorbed enough nutrient solution is placed on the multi-layer seedbed, and then sent to the germination room. The seeds will germinate after two to three days of germination at the temperature and humidity under artificial control. Environmental conditions in the germination room: no light, constant temperature (23℃), and constant humidity (with a relative humidity 95%–100%).

2. Seedling culture

Upon germination, the seedlings of plants are then moved to the multi-layer (three to four layers generally) seedling equipment utilizing artificial light to complete the greening process after about one week of exposure solely to artificial light. The so-called greening refers to a process to promote, by light, the formation of chloroplast in the plants that are germinated in the dark period, so as to prepare for photosynthesis. The seedlings of plants, after the greening process and initiating the photosynthesis, are transplanted to the culture bed containing nutrient solution for growth. The sponge pad used for the seeding operation is evenly cut into pieces, and the plants are separated from each of these pieces and then transplanted to the seedbed for hydroponics to turn into plantlets after about two weeks. The density of transplanting shall depend on the sizes of individual plants upon completion of the seedling operation. On a bed plate of 585 mm × 880 mm, 120 plants can be cultivated (15 plants × 8 rows), i.e. 233.1 plants/m^2.

During the seedling period, environmental control includes the control of temperature, light environment, nutrient solution concentration (EC value), and nutrient solution temperature. Ventilation, cooling, and heating are the main means of temperature control, while fluorescent lamp or LED is used for light environment control (the effective radiation intensity of photosynthesis is 55 μmol/(m^2·s) when light is supplemented).

3. Planting and cultivating

Upon the end of the seedling operation, plantlets together with sponge pieces are transplanted to the cultivation workshop for planting. The plantlets are manually planted on the perforated floating plate for cultivation. After about three weeks of cultivation under artificial light, the lettuce can be harvested. The currently used artificial light sources are dominated by fluorescent lamps or LEDs, i.e. cold light sources, which can not only shorten the spacing between plants being cultivated (to only about 40 cm) and increase the utilization rate of space for cultivation but also eliminate damages of burn to crops. With these cold light sources, generally three to four layers (up to eight to ten layers in some cases) are designed for cultivation in the plant factory. The density of transplanting of the lettuce from planting to harvesting is generally designed to be 12 plants cultivated on each cultivation plate of 585 mm × 885 mm (4 plants × 3 rows), i.e. 23.2 plants/m^2.

4. Harvesting

After about three weeks of cultivation, the plantation trough of vegetables in the maturity stage is moved to the harvesting room for harvesting, packaging, storage, etc. Vegetables are harvested manually or with the assistance of a manipulator. The plantation trough is washed while cutting off the roots of the crops, and then vegetables are placed in plastic boxes, loaded on trolleys, and moved to the packaging workshop for packaging and cold storage.

5. Packaging and storage

The harvested vegetables are moved to the harvesting and storage room and packaged in plastic bags. The packaged vegetables are moved to a fresh-keeping room for precooling. Precooling is an effective process to appropriately cool down the vegetables to be transported or stored. Through the precooling process, the actives in vegetables can be cooled down, and the physiological and biochemical activities in the harvested vegetables can be inhibited to control the invasion of microorganisms and reduce the loss of nutrient substances, thereby improving the fresh-keeping effect. The precooling room is controlled at a constant temperature of 4–5°C and a constant relative humidity close to 100%.

6. Sales

Vegetables produced in the plant factory are subject to scheduled sales, which means that generally a sales schedule is prepared prior to the production, and the supply and sale agreements, as well as contracts specifying the details about quantities, specifications, dates of sales, etc. of the products, are entered into with relevant wholesalers. These agreements and contracts will serve as the basis for the supply and sale of all products. Generally, vegetables are delivered by fresh-keeping refrigerator wagons for sale every day. Sometimes, they can also be delivered by couriers. Vegetables are delivered to auction centers, wholesale markets, or directly to supermarkets, hotels, restaurants, etc.

III. System Composition of Plant Factory

1. Enclosures and materials

The plant factory utilizing artificial light operates in a fully enclosed environment to realize full-year efficient production of plants, and it is designed to minimize the consumption of substances, energy, and resources in the system and maximize the yield of end products. Therefore, in order to reduce the influences of indoor and outdoor substances and energy exchanges as well as outdoor light and heat on the indoor environment, the enclosures shall be made of thermally insulated and light-proof construction materials with good resistance to wind. Currently, polythene-colored steel sandwich panels, polyurethane sandwich panels, and other clean panels are commonly used for enclosures. They are made of two layers of molded metal panels (or panels of other materials) and polymer thermally insulated cores (polyethylene or polyurethane) directly foamed, cured, and formed in these panels, feature good cleanliness, resistance to corrosion and moisture, thermal insulation, etc., and can therefore adapt to the high-humidity environment in the plant factory and meet the needs of cleaning and sterilizing the artificial lights.

It is generally required to provide the enclosures with a concrete structure foundation and a framework of light-gauge steel joist and to comply with the requirements of installation processes for clean panels in terms of splicing and installing works. Fullyenclosed watch windows are provided with dedicated aluminum alloy sections combined with glass panels, with 45° corners. Doors are made of color plates and provided with frames of special aluminum sections, including embedded rubber sealing strips at the edges. Semiarc aluminum sections are used to connect walls with ceilings, and walls with the ground. The base course of the ground is formed by cement mortar, while the top course thereof is formed by self-leveling materials to realize resistance to dust, moisture, wear, skidding, and static electricity, and characteristics of convenience in cleaning operation, rapid construction, convenience in maintenance and good bearing capacities for heavy load and impact load. The *Code for Design of Clean Room* (GB 50073–2001) and the *Code for Construction and Acceptance of Clean Room* (GB 50591–2010) can be used as a reference for the construction and acceptance of enclosures for the plant factory.

2. System composition of the plant factory

The enclosures serve as a basis to ensure a steady environment in the plant factory. However, relevant auxiliary systems are also required in order to realize the full-year uninterrupted production of the plant factory, including nutrient solution circulating and control system, environmental control system, stereoscopic hydroponics system, artificial light source system, and intelligent computer control system.

(1) Nutrient solution circulating and control system.

As for the plant factory utilizing artificial light, currently deep flow technique (DFT) and aeroponics are most commonly used in the nutrient solution culture room. In both of these two modes, the whole-process management of plant culture can be realized by an enclosed automatic nutrient solution circulating system. The enclosed automatic nutrient solution circulating system mainly consists of nutrient solution reservoir (tank), detection sensor (for EC Value, pH value, DO, liquid temperature, etc.), circulating water pump, filtering and sterilization device, solenoid valve and connecting pipeline, culture bed, and automatic controller. During the system operation, such parameters as EC value, pH value, DO, and liquid temperature of nutrient solution reservoir (tank) are detected online in real time by sensors and checked by control software in order to determine whether a regulation to them is necessary. If such regulation is necessary, the blending tanks (including tanks containing major elements, trace elements, acid liquors, alkali liquors, etc.) connected with the nutrient solution reservoir (tank) are controlled by the solenoid valve to realize an automatic blending of nutrient solution. The temperature of the nutrient solution is controlled by heating or cooling equipment, and the increase in dissolved oxygen content is mainly regulated by a stirrer or increasing the circulating rate of the culture solution. The prepared nutrient solution is directly delivered through a circulating pipeline to the culture bed or through an atomizing device to the roots of crops to keep providing nutrients to these crops. The used solution is filtered and sterilized and then returned to the nutrient solution reservoir (tank), and at this time, a complete circulation process is done.

(2) Stereoscopic hydroponics system.

In an early-developed plant factory, there was generally only one layer of culture bed (even for the two-layer structure, the spacing was greater than 1 m) because the artificial light sources with a higher heating value such as high-pressure sodium lamps were used. With the application of cold light sources, including fluorescent lamps and LEDs, such spacing has been reduced to 0.3–0.4 m, so three to four layers, or even up to more than ten layers have become feasible to form a multi-layer stereoscopic hydroponics system in the plant factory. Generally, this kind of stereoscopic hydroponics system consists of a fixed stand, a stand for artificial light sources, a culture trough, waterproof plastic films, foam cultivation plates with holes, a water feeding pipe, an overflow pipe, a circulating pipeline, etc. With the circulating pipeline connected to the automatic nutrient solution circulating system, the multi-layer stereoscopic culture can be realized to greatly increase the utilization rate of space and yield per unit area in the plant factory.

(3) Environmental control system.

As one of the important sub-systems for the plant factory, the environmental control system is designed to comprehensively control the environmental factors above roots, including temperature, relative humidity, CO_2 concentration, lighting intensity and photoperiod, and rhizospheric environmental factors (EC value, pH value, DO, liquid temperature, etc.) in the plant factory. The environmental control system consists of sensors, controllers, and actuators.

Sensors are important tools to acquire environmental information. Generally, the sensors are selected by environmental factors to be collected for the plant factory, which mainly include temperature sensors, humidity sensors, illuminance sensors, CO_2 concentration sensors, pH value sensors, EC value sensors, liquid temperature sensors, and dissolved oxygen content (DO) sensors. The analog signals acquired by these sensors are converted by an A/D converter into digital signals necessary for the control units, and the controlled parameters (or statuses) are compared with the set values to control relevant actuators according to the compared results, thereby achieving the purpose of automatically regulating the controlled parameters (or statuses). The actuators mainly include the air conditioning system, liquid temperature controller, and oxygen-increasing device. This is shown in the schematic diagram of the environmental control system commonly used for plant factories.

(4) Artificial light source system.

As for plants, the light source is the energy source for their basic physiological activities such as photosynthesis, and it can also serve as the information source for the formation of plant morphology and control of the plant growth process. Therefore, it is particularly important to regulate and control the light environment (light intensity, light quality, and photoperiod). In an enclosed plant factory, the growth and development of plants mainly depend on artificial light sources. In an early-developed plant factory, high-pressure sodium lamps, traditional fluorescent lamps, etc. were mainly used as artificial light sources, which generally featured a high energy consumption and operating cost, accounting for approximately 50%–60% of the total operating cost. With the development of LED technology in recent years, it has become feasible to apply LEDs to the plant factory. LED has the advantages of small volume, long service life, low energy consumption, low heating value, availability for short-distance illumination, etc. Moreover, LEDs can be precisely mixed for a light quality (R/B ratio or R/FR ratio) necessary for specific plants to significantly promote the growth and development of plants and therefore improve their yields and qualities. With LEDs, not only energy can be saved, but also layer-to-layer spacing can further be shortened, thereby dramatically improving the utilization rate of space. Generally, the artificial light source system installed in a plant factory consists mainly of lights, voltage regulating and rectifying equipment, and controllers. The light sources can be precisely mixed to meet the needs of growth and development for specific plants.

(5) Computer control system.

The computer control system is the heart of the plant factory. The information acquired by sensors from all processes is transmitted to the computer system for storage and display. Control software is used to analyze and judge this kind of information and then to give commands to relevant actuators to realize control of the system. The computer control system mainly consists of data acquisition units, controllers, and actuators. Such parameters as temperature, humidity, CO_2 concentration, light, and nutrient solution in the plant factory are detected in realtime by sensors, converted by an A/D converter, and then transmitted to single-chip microcomputers so that the process of data acquisition is done. PLC is provided as the core controller, while PC and configuration software are provided as monitoring modules. The core controller and the monitoring modules are connected by serial ports for mutual communication to control the actuators of the system, thereby realizing the intelligent and user-friendly control throughout the production process.

Task 3 Environmental Regulation in the Plant Factory

The environmental control system is one of the major sub-systems and serves as a significant safeguard to realize full-year uninterrupted production of plants in the plant factory. The main environmental elements for the plant factory include air temperature, relative humidity, gas (CO_2 etc.) concentration, light, and rhizospheric environmental factors (EC value, pH value, liquid temperature, and dissolved oxygen content). Moreover, the cleanliness of the air is also included in the environmental control system by some plant factories.

I. Temperature Regulation in the Plant Factory

1. Effects of temperature on physiological activities (photosynthesis) of plants

The growth of crops is closely related to the temperature. Specifically, the growth, development, and final yield of crops are affected by temperature, especially extremely low and extremely high temperatures. As for plants, their physiological activities and biochemical reactions are only available within a certain range of temperature. Generally, the physiological activities and biochemical reactions of plants are proportional to temperature. However, at a temperature lower than or higher than the physiological limits of the crops, the crops will be hindered from growing or even die. Generally, the plants are affected by three critical points of temperature, i.e. the minimum temperature, the optimal temperature, and the maximum temperature. Generally, as for the photosynthesis of crops, the minimum temperature is 0–5°C, the optimal temperature is 20–30°C, and the maximum temperature is 35–40°C. At the optimal temperature, normal growths, and physiological activities of plants can be guaranteed, and the rate of accumulation of photosynthetic products is relatively high. Moreover, the variation in temperature is also accompanied by variation in other factors in the comprehensive environment, such as humidity, which in turn affects the growth and development of crops. Thus, it is very important to regulate the temperature in the plant factory to ensure the efficient production of crops.

2. Temperature regulation in the plant factory

Temperature regulation in the plant factory refers to the artificial regulation of indoor temperature by certain engineering technical means in order to dynamically maintain a temperature suitable for the growth and development of crops and realize a uniform spatial distribution and a gentle temporal variation of temperature, thereby guaranteeing an efficient production in the plant factory. Currently, the main regulation and control measures for temperature in the plant factory include the following aspects.

(1) Cooling control.

The indoor environment can barely be affected by outdoor climate conditions as the plant factory utilizing artificial light is fully enclosed and its roof and enclosures are made of materials with good thermal insulation properties. However, the heat produced by artificial light sources and released from culture beds, structures, etc. can raise the indoor temperature. So, cooling measures are required to forcibly dissipate heat from the indoor area to dynamically maintain a temperature suitable for the growth of the plants.

Generally, air-conditioning units are used for cooling. They are turned on/off by switching on/off a control relay. First, temperature sensors acquire data and output analog signals, which are converted by an A/D converter into digital signals, and then the digital signals are input to single-chip microcomputers and compared with the set values to determine whether a regulation is necessary. Second, the temperature is regulated by the computer system and actuators. When the indoor temperature surpasses the preset upper limit, the single-chip microcomputers send control signals to switch on the relay, so that the air-conditioning units are turned on for cooling, and once the indoor temperature is lowered to the set value, the single-chip microcomputers send control signals to switch off the relay, so that the air-conditioning units are turned off. This is the automatic temperature regulation process for the plant factory. Given that a multi-layer stereoscopic culture technology is applied in the plant factory, it is also very important to control the air flow distribution of the air-conditioning system. In this regard, many tests have been conducted by many organizations both in China and abroad to realize a reasonable air flow distribution, thereby ensuring a uniform spatial distribution of indoor air temperature.

(2) Heating control.

In winter, the outdoor temperature is low, and the indoor temperature in the dark period is generally lower than the temperature suitable for the normal growth of the crops. Thus, heating measures are required to increase the heat to maintain a suitable indoor temperature of the plant factory.

As for the plant factory built in a cold region, generally a hot water heating system is installed for heating. This kind of system consists of a water heating boiler, a heating pipeline, and heat dissipators. Water, after being heated in the boiler, flows into the heat dissipators through the heating pipeline. At the heat dissipators, the air is heated by hot water, and the water with heat dissipated is returned to the boiler for reuse. Generally, heating is performed by low-temperature hot water (water supplied at 95°C and returned at 70°C). Given a smaller altitude difference (less than 3 m) between the boiler and heat dissipators in the hot water heating system, in general cases, the gravity circulation is discarded, but mechanical circulation is adopted; that is, a circulating water pump is installed at the main pipe for water return. Single-pipe connection or dual-pipe connection is provided between the system pipeline and heat dissipators. Hot-dip galvanized round-wing-type heat dissipators with a greater area of heat dissipation and better resistance to corrosion are generally used to adapt to high indoor humidity. Generally, the heat dissipators are laid out at the enclosures, and have an appropriate specification and length to meet the thermal load requirements in heating design. They are evenly distributed indoors to realize a uniform temperature distribution.

As for the plant factory built in temperate or temperate-tropical zones, no greater heating loads are required in winter, and generally, the heating needs can be met by the air-conditioning system; that is, hot air is supplied evenly through the same pipeline as the cooling system, to maintain a temperature suitable for the growth of the plants in the dark period in the plant factory.

In order to maintain a temperature suitable for growth at the roots of the crops, in winter, the nutrient solution is heated by the hot water pipeline or by electric heating to keep the nutrient solution and rhizospheric environment stable for the crops.

II. Humidity Regulation in the Plant Factory

The vapor pressure difference between the leaf surface of the crops and ambient air depends on the relative air humidity in the plant factory and can affect the evaporation at the leaf surface of the crops. Humidity can not only affect the transpiration of crops and surface evaporation but also directly affect the photosynthetic intensity and incidence rate of diseases for the crops. At lower humidity, the evaporation capacity at the leaf surface of crops increases, which, in a severe case, may result in a water deficit in roots, reduced moisture content in crops, shrunken cells, reduced porosity, and reduced photosynthetic products; at higher humidity, the evaporation capacity at leaf surface of crops decreases, which, in a severe case, may cause excessive moisture content in crops, to result in larger stem leaves and therefore affect the yield. Crops can grow normally within a relative humidity range of 25%–80%. Another factor of influence of humidity on crops is diseases and pests. At a humidity higher than 90%, crops can suffer from diseases, while at a too-low humidity, crops can suffer from powdery mildew and pests. Different crops have different requirements for relative humidity in the air, and therefore the air humidity shall be regulated by varieties of crops and based on their specific stages of growth.

1. Dehumidification regulation

Heating, ventilation, dehumidification etc. are available to reduce humidity in the plant factory. Heating can not only raise the indoor temperature but also naturally decrease the relative humidity at constant air moisture. As for the proper ventilation operation, dry air is fed from the outdoor area into the indoor area to gradually replace moisture air in the indoor area, thereby reducing the indoor relative humidity. Alternatively, solid or liquid moisture absorbents can be provided to directly absorb moisture from the air to reduce the air humidity, but they are relatively expensive.

The methods commonly used to reduce humidity and control the excessive indoor relative humidity in the plant factory include the following parts.

(1) Ventilation: As for the indoor area of the plant factory, higher humidity is primarily caused by the enclosed structure of the facility. Forced ventilation is available to reduce indoor humidity and prevent higher indoor temperature and humidity. The criteria to control indoor relative humidity depend on seasons and varieties of crops. But generally, the indoor relative humidity is controlled at 50%–85%. The ventilation volume is associated with the intensities of evaporation and transpiration of crops as well as the indoor and outdoor temperature and humidity conditions.

(2) Heating: Under certain outdoor meteorological conditions and indoor evaporation, transpiration, and ventilation conditions, the indoor relative humidity is inversely proportional to indoor temperature. Thus, appropriately raising the indoor temperature can also be taken as one of the effective measures to reduce indoor relative humidity. In general cases, temperature shall be raised to a level suitable for the growth of crops, at which the leaves of crops are free from dewing (with humidity under proper control).

(3) Heat pump: The refrigerating medium is compressed by a compressor, to absorb latent heat of evaporation from a low-temperature heat source when being evaporated by an evaporator. The compressed refrigerating medium then flows through hot heat dissipators, so that the heat absorbed from the low-temperature heat source and the heat arising from compression work by the compressor are together released to the room to be heated. If the evaporator of the heat pump is placed in the culture room, the evaporator coil can be cooled down to about 5°C, which is far lower than the dewing temperature of indoor air. According to research, the indoor humidity at night can generally be reduced to a level below 85% by the heat pump.

2. Humidification regulation

In a dry season, humidification is necessary when the indoor relative humidity drops below 40%. At a certain air velocity, appropriately increasing the humidity can increase the stomatal apertures and therefore improve the photosynthetic intensity of crops. Spraying humidification and ultrasonic humidification are two commonly used methods of humidification. Ultrasonic humidification, which can avoid the wetting of leaves during the humidification, has already been widely applied to the plant factory.

III. Light Regulation in Plant Factory

1. Requirements of plants for artificial light sources

The requirements of plants for artificial light sources mainly include spectral properties, luminous efficiency, and service life. As for spectral properties, the artificial light sources are required for abundant blue-violet light (400–500 nm) and red-orange light (600–700 nm), an appropriate R/B ratio and an appropriate R (600–700 nm)/FR (700–800 nm) ratio, as well as other appropriate spectral components to meet special requirements (such as to replenish violet light). It is required to not only guarantee a light quality suitable for the photosynthesis of plants but also minimize the useless spectral components and energy consumption.

As for luminous efficiency, a higher ratio of effective radiation for photosynthesis to power consumption is required. The luminous efficiency is expressed by: visible luminous efficiency (luminous efficiency) — Lm/W; effective radiation efficiency for photosynthesis (radiation efficiency) — W/W; effective photon efficiency for photosynthesis (photon efficiency) — (mmol/s)/W or mmol/J. Other desirable performance requirements include longer service life, smaller light attenuation, and lower price.

So far, the artificial light sources used for the plant factory are mostly high-pressure sodium lamps, metal halide lamps, fluorescent lamps, LEDs, laser diodes (LD), etc. LEDs, boasting the advantages in energy conservation, environmental conservation, longer service life, monochromatic lights, cold light sources, etc., are deemed as the ideal light sources for enclosed plant factories. With the application of LEDs to the enclosed plant factory, the energy consumption and operating costs can be reduced, and the efficiency of utilizing light energy and precision to control the light environment can be improved, which can be conducive to promoting the application of enclosed plant factories. Moreover, LEDs can play a significant role in grappling with environmental pollution, increasing the utilization rate of space in plant factories and controlling greenhouse effects, and are anticipated to become the dominant artificial light sources in plant factories.

2. LED light source device and its control methods

By now, multiple types of LED light source devices (tubular, plate-type, etc.) have already been developed to meet requirements for various applications in the plant factory.

(1) Tubular LED light source device: In order to improve the interchangeability with T8 and T5 fluorescent lamp tubes, by now tubular LED light sources that can replace T8 and T5 fluorescent lamps have already been developed. This kind of light source consists of a holder, a tube, and caps, where a special rectifier is installed in the holder to directly convert the 220V AC power into the DC power available for LED; one electrode is embedded at each of two ends of the holder, and the same caps as those for ordinary fluorescent lamp are provided, so that the current fluorescent lamps can be directly replaced by tubular LED light sources without any modification to other structures, and therefore the LEDs can be installed and used conveniently. As for this kind of tubular LED light source, the LED beads that can emit red light (660 nm), blue light (450 nm), and far-red light (730 nm) are evenly distributed to certain proportions on the inner surface of the holder, and can be reasonably mixed to meet specific needs of growth that are determined from researched results about optimal parameters for the light environment, such as R/B/FR of 8 : 1 : 1 for lettuce culture, and R/B/FR of 7 : 1 : 1 for cucumber seedling. The tubular LED light source device can be used independently or jointly with fluorescent lamps for production in the plant factory. It is the most commonly used light source system that is applicable to leaf vegetable culture, seedling, etc. in the plant factory.

(2) LED light source plate and accessories: The plate-type LED light source is also one of the commonly used light sources in the plant factory. It consists of ultra-high-brightness red LEDs (660 nm of peak wavelength) and blue LEDs (450 nm of peak wavelength), which are distributed on the light source plate evenly and

crosswise. In order to maintain a relatively steady temperature, a temperature sensor is installed in the middle of the light source plate to monitor the temperature there in real time. The special cooling fins and axial flow fan are provided for heat dissipation, to improve the efficiency and stability of LEDs. PWM is adopted to control the luminous intensity of the light source, where the luminous intensities of red and blue LEDs can be controlled and regulated separately to meet the needs of the growth of various plants for light environment. The LED light source plate is applicable to leaf vegetable culture, seedling, etc., as well as tissue culture for plants in the plant factory.

(3) LED light source control and effects: The LED light source device is connected with a computer by RS-485 interfaces for mutual communication. The luminous intensity, frequency, and ON/OFF times of the LED light source plate are controlled by the computer. The LED light environment regulation device is also connected to an external control box, with which the luminous intensity and frequency of monochromatic light can be controlled manually. PWM is adopted by the LED light environment control device. Given a linear relationship between the duty cycle and the luminous intensity of the LED light source, the corresponding duty cycle at the time of regulating the R/B ratio can be calculated according to the specific control ratio, thus regulating the luminous intensity of monochromatic light.

IV. CO_2 Concentration Regulation in the Plant Factory

1. CO_2 concentration and photosynthesis of plants

CO_2 is an important raw material for the growth of crops. Photosynthesis refers to a process in which chloroplasts of green plants can, under light conditions, synthesize H_2O, and CO_2 in air into organic matter and release O_2. Through photosynthesis, plants can convert luminous energy into chemical energy (stored in the organic matter). Through respiration (oxidation of carbohydrates), energy can be provided to various biological or chemical reactions in plants.

There are three sources of CO_2 for photosynthesis, namely CO_2 from ambient air, CO_2 produced by the respiration of leaf tissues, and CO_2 absorbed by the roots of crops. Naturally, most of the CO_2 is contributed by the first two sources, while the last source accounts for only 1%–2% of the total mass of CO_2 absorbed by crops. CO_2 diffuses through the epidermis or stomas into the chloroplasts in mesophyll cells. During photosynthesis, CO_2 is continuously consumed by chloroplasts and its concentration in leaves is continuously reduced. In this case, a CO_2 concentration gradient is formed between the inside of the leaf and the ambient environment, and therefore, CO_2 can keep diffusing into chloroplasts in leaves.

2. Sources of CO_2 and technologies of their regulations

In air, the average CO_2 concentration is about 330 mL/L (0.65 g/m^3), which is far below the ideal level for growth of crops. Therefore, CO_2 fertilizer has already become an essential method to guarantee efficient production in the plant factory. Generally, the saturation point of CO_2 is 800–1,000 mL/L or above. The higher the luminous intensity is, the higher the saturation point will be. However, more costs are needed to apply more CO_2 fertilizers. Therefore, an economical fertilization concentration (800–1,000 mL/L) is selected for the plant factory. Currently, there are many methods to apply CO_2 fertilizer, mainly including the following three types.

(1) Liquid CO_2 in cylinders.

Gaseous, liquid, and solid CO_2 with a purity above 99% can be obtained as byproducts from the alcohol-brewing industry. Gaseous CO_2 is compressed into liquid to be stored in cylinders, and these cylinders can be controlled by valves to provide CO_2 at any time in a convenient and safe manner. This method features a convenience in control of CO_2 concentration, abundant sources of raw materials, and lower costs, and is the preferential choice to provide CO_2 sources for plant engineering utilizing artificial light.

(2) CO_2 produced by the combustion of hydrocarbons.

Relatively pure CO_2 can be produced from the combustion of such substances as kerosene, liquefied petroleum gas, natural gas, propane, and paraffin, and can then be piped to the plant factory. Specifically, 3 kg of CO_2 can be produced from 1 kg of natural gas, while 2.5 kg of CO_2 can be produced from 1 kg of kerosene. Given that the concentrations of harmful gases such as SO_2 and CO upon the combustion are not allowed to outnumber the limits harmful to the plants, pure combustion substances,and a dedicated CO_2 generator are required. This method features a convenience in automatic control, but a higher operating cost. It is more used in overseas greenhouses and plant factories utilizing sunlight but less used in plant engineering utilizing artificial light.

(3) CO_2 produced from chemical reactions.

Pure CO_2 can be produced from a chemical reaction of $CaCO_3$ (or Na_2CO_3) and HCl (or H_2SO_4). This method is convenient to use and has abundant and inexpensive raw materials. However, as there are some impurities in the raw materials, attention shall be paid to controlling the pollution of residues and residual liquids (hydrogen sulfide, hydrogen chloride, etc.) after the chemical reaction to the environment, and preventing against strong acids to ensure safety during the operation. With this method, it is difficult to precisely control the CO_2 concentration produced from chemical reactions, so this method is used more in greenhouses and plant engineering utilizing sunlight but less in plant factories utilizing artificial light.

Sources of CO_2 fertilizer for plant factories should be selected to meet the specific needs. Generally, factors such as an abundance of resources, convenience in access to raw materials, purity, harmlessness, price and cost, simple use of equipment, and convenience in automatic control and use should be taken into account.

Task 4 Typical Cases of the Plant Factory

I. Changchun Intelligent Digital Plant Factory

As a brand new agricultural production mode that is highly technically intensive and cannot or can barely be restricted by natural conditions, the plant factory is internationally recognized as the highest stage of development of facility agriculture. The plant factory, featuring no reliance on arable land, safety and no pollution in products, labor-saving operation, high degree of mechanization, and yield per unit area dozens of times or even hundreds of times that in the open field, is treated as the important approach to grapple with the population, resource and environment problems in the 21st Century, and also the important means to realize the in-situ supply of food for the future aerospace engineering as well as lunar exploration and exploration for other planets. This technology is now only mastered by a few developed countries, including Japan, the United States, the Netherlands, etc. On September 7, 2009, China's first LED plant factory centering on intelligent control was successfully developed by the Institute of Environment and Sustainable Development in Agriculture, CAAS, and put into operation in the name of Changchun Intelligent Digital Plant Factory in Changchun Modern Agricultural Park, Jilin Province.

Changchun Intelligent Digital Plant Factory (Fig. 9-1) is China's first productive plant factory,which is centered on intelligent control. The completion of construction of this plant factory denoted China's major technological breakthrough in the field of the plant factory, which made China a country that mastered the core technologies of the plant factory after the United States, Japan, and the EU, and would undoubtedly bring about far-reaching influences on the development of modern agriculture in China.

Fig. 9-1 Changchun Intelligent Digital Plant Factory

This plant factory covers a floor area of 200 m^2, and is divided into a plant seedling factory and a vegetable factory. It is provided with energy-efficient lamps suitable for the growth of plants and LEDs as artificial light sources, and installed with 13 interrelated sub-systems for control including cooling-heating dual-purpose temperature and humidity regulation and control system, illumination system, CO_2 photosynthetic coupling regulation system, online detection and control system for nutrient solution (EC value, pH value, DO, liquid temperature, etc.), environmental data acquisition and automatic control system etc. to achieve the real-time automatic monitoring and intelligent management of such environmental factors as temperature, humidity, light, CO_2 concentration and nutrient solution in the plant factory. The plant seedling factory developed is composed of two rows of five-layer seedling culture beds, and features uniform and robust seedlings, good quality, and seedling culture efficiency per unit area more than 40 times that realized by the conventional seedling culture technology. The vegetable factory is subject to stereo planting with five-layer culture beds, where the duration from planting to harvesting of leaf lettuce cultured in this vegetable factory is only 16–18 days, which is 40% shorter than the period of conventional culture technology, the yield per unit area is more than 25 times that realized by open cultivation, and the products delivered from this factory are clean and free from pollution, with a higher commodity value.

II. Low-carbon and Intelligent Domestic Plant Factory Launched in Expo 2010 Shanghai China

On April 20, 2010, ten days prior to the formal opening of Expo 2010 Shanghai China, the global launch ceremony of the world's first low-carbon and intelligent kitchen and domestic plant factory was held grandly in Shanghai, which denoted the successful launching of the world's first low-carbon and intelligent domestic plant factory.

This domestic plant factory was exhibited at the low-carbon and intelligent kitchen in the exhibition hall, which was themed "We Are the World" of Expo 2010 Shanghai China. It was a fully enclosed intelligent environmental control plant production system powered by a wind-solar complementary renewable energy power generation system that utilized inexhaustible and renewable natural energy sources (wind and solar). Energy-efficient composite LEDs, which could save 60%–80% of energy in comparison to traditional light sources, were provided as the sole light sources for culture. Vegetables were planted on multi-layer MFT hydroponics beds, and the nutrient solution was supplied by the intelligent detection system to precisely meet the growth needs. In the system, the temperature, humidity, light, air velocity, etc. were regulated intelligently by the computer system, and CO_2 was naturally sourced from the human living environment. The domestic plant factory was also provided with the function of the Internet of Things (IoT), so that users were capable of knowing the growths of vegetables, regulating the control parameters, and remotely controlling the system at any time and place by use of cell phones, Internet, and other tools through the remote surveillance system.

The low-carbon intelligent domestic plant factory fully embodied the function of self-sufficiency plant production, and successfully turned the Happy Farm (a virtual game) and Natural Oxygen Bar into reality. The system covered 5 m^2 of vegetable area, with an annual yield of leaf vegetables, fragrant herbs, and other crops of 200–250 kg, to not only meet some family needs for safe, healthy, and green vegetables, but also absorb CO_2 from daily life and release a large amount of oxygen, thereby creating the Natural Oxygen Bar for the domestic environment.

The intelligent control system consisted of sensors, programmable logic controller (PLC), human-computer interface, and actuators, where the sensors, human-computer interface, and actuators were respectively connected with the PLC. The sensors included the pH value sensor, EC sensor, liquid level sensor, liquid temperature sensor, humidity sensor, temperature sensor, CO_2 concentration sensor, light sensor, etc. The human-computer interface was actually an HMI (Human Machine Interface) display panel. The actuators included the artificial lighting device, liquid supply, and return, as well as water supply and drainage device, environmental control device, etc. The intelligent control system was also provided with cameras and a monitor which were connected together through a network. Cameras were installed in the culture room of the intelligent domestic plant factory so that users could perform remote surveillance through the monitor.

This intelligent domestic plant factory utilizes vegetable production equipment in the field of agricultural machinery, and comprises a cabinet to contain a hydroponics and automatic nutrient solution circulating system, an air-conditioning system, an artificial lighting system, and an intelligent control system. The cabinet comprises a light room, a dark room, a bottom layer, and an interlayer on the backside. The artificial lighting system is provided with an LED light source plate consisting of red LEDs, blue LEDs, and green LEDs to provide artificial light sources suitable for the growth of plants. The domestic plant factory is applicable to a variety of plants, including all kinds of leaf vegetables, fruits, sprout vegetables, edible fungi, herbs, spices, ornamental fruits and vegetables, etc., and can dramatically shorten the period of growth and development for leaf plants, and bring forward the initial harvesting time, prolong the harvesting period, and significantly raise the yield for fruit plants. In the room for the growth of leaf vegetables, about 50 plants of leaf vegetables can be cultivated, with a period of growth being 20–25 days. In the room of seedlings, 70 plants of seedlings can be cultivated simultaneously, with a period of seedling being about 20 days. In the room for growth of mushrooms, the period of growth is about 30 days, with a monthly yield of fresh mushrooms being 1.5–2.5 kg.

III. Shandong Shouguang LED Plant Factory

At the 10th China (Shouguang) International Vegetable Sci-tech Fair opened on April 20, 2009, the newly launched, internationally cutting-edge vegetable planting technology named the "plant factory" (Fig. 9-2) attracted the attention of visitors, media, and journalists. It was then rapidly reported and publicized and at that time became the central focus of this fair.

Fig. 9-2 Shandong Shouguang LED Plant Factory

The plant factory operated in the soilless culture mode. Beneath the plates used to keep steady the root systems of vegetables, the nutrient solution was provided precisely under the control of a water-fertilizer system to develop the root systems of each vegetable. Moreover, by the use of a remote control system, technicians, with only cell phones or computers, could realize such operations as watering and fertilization. The plant factory was powered by a PV solar power generation system, which consisted of solar cell modules and an inverter, to serve as the regular power supply to provide reliable power to artificial LEDs for the growth of plants in greenhouses and polytunnels, and also provide surplus power for the daily operation of agricultural equipment or to the power grid. The plant factory utilized PV solar energy. As for the plant factory, the power generated by the PV solar power generation system could not only be provided to the artificial LED light system for the growth of plants, but also to the nutrient solution circulating system, biological insecticidal lamps, automatic control system, etc., thereby effectively lowering the costs in the plant factory.

Basically, the technology of the plant factory was subject to indoor operations, and the plant factory operated in a relatively enclosed environment, subject to physical sterilization and disinfection, with few diseases and pests. Therefore, the plant factory had great advantages in food safety. Moreover, in the plant factory, temperature, humidity, light, and pH value of the nutrient solution associated with the growth and development of plants were monitored in real time and controlled automatically by the IoT, so that the growth and development of plants in the plant factory could not or could barely be restricted by the natural conditions. Full-year uninterrupted production of multiple batches of crops could be realized under the illumination of LEDs.

At this fair, an operation system and demonstration platform for the application of technologies of the IoT were exhibited. With an intelligent operation system for the environment in the greenhouse, the indoor temperature and humidity, soil temperature and humidity, and CO_2 concentration could be monitored, and for any anomalous case found, an alarm could be sent automatically by a function called "greenhouse monitor". Moreover, any things that happened in the exhibition area were recorded in the surveillance system for safety in the production, including the growth of plants. The images about the growth of plants for a certain time period could be zoomed in by the PTZ (Pan/Tilt/Zoom) system and called out at any time, and also transmitted through the Internet. It was convenient to switch to the screen of surveillance by logging into a specific website.

The plant factory was a forward-looking technology that broke through the bottleneck of "vegetable planting without sunlight", and once applied to aerospace engineering, would conquer the difficulty in providing fresh vegetables to astronauts.

[Module Summary]

I. Key and Difficult Points

(1) Master the composition and types of main systems of the plant factory.

(2) Master the main types of plant factories in the world at present.

(3) Understand the typical cases of plant factories in China.

II. Review Questions

(1) What is the plant factory? What are the characteristics of the plant factory? What are the main types of plant factories?

(2) How is the current level of development and application of plant factories in the world? What is the development prospect of plant factories?

(3) What are the typical cases of plant factories in China?

(4) How is the environmental control system of plant factories realized?

(5) How are the problems of light sources for artificial light source plant factories solved?

Module 10 Application of Soilless Culture Techniques for Home Gardening

[Learning Objectives]

I. Knowledge Objectives

(1) Understand the development history of home gardening;

(2) Master the soilless culture techniques for home gardening;

(3) Master the main forms of soilless culture for home gardening.

II. Skill Objectives

(1) Be able to comprehensively apply the theoretical knowledge and skills learned, and independently engage in soilless culture for home gardening;

(2) Be able to master the management of soilless culture for home gardening.

[Preparation for Learning]

I. Required Resources

(1) Knowledge of soilless culture of vegetables, flowers, and fruit trees;

(2) A digital reading room and an electronic resource library;

(3) A multimedia classroom;

(4) A formulation laboratory of nutrient solution.

II. Background Knowledge

(1) Master the formulation of nutrient solution for soilless culture and the relevant properties of culture substrate;

(2) Master the construction and management of common production equipment for soilless culture;

(3) Understand the knowledge and key management points of the soilless culture of common vegetables, flowers, and fruit trees.

[Learning Tasks]

Task 1 Introduction to Soilless Culture for Home Gardening

Home gardening refers to the activities of horticultural plant culture and decoration in indoor spaces, balconies, or courtyards. Home gardening conveys the concept of healthy life, and increases enjoyment of life. According to the research, the shoots and leaves of indoor green plants have the functions of purifying indoor air pollution, as well as absorbing living waste gas, regulating air humidity, and reducing noise.

Although the rise of home gardening in China is not long, the trend of home gardening has long been developed abroad. In 1964, the UK conducted the activities of creating a beautiful street landscape with flowers and green plants, called "Blooming in the Beautiful UK". In 2008, several institutions in the UK co-sponsored a "Grow Your Own" campaign, which is themed on how to turn even a tiny balcony into a productive plot and is called the "London Food Up Front Scheme". Currently, in France, "Orchard on the Balcony" has become a buzzword in recent years. In Sweden, about 65% of adults spend part of their leisure hours in gardening, and home gardening has become one of the most common recreational activities for Swedish residents. In the USA, gardening is one of the most popular outdoor leisure activities. According to the survey, among the households engaged in gardening in the USA, the gardening area of each household covers an area of 55 m^2 on average; according to the *Organic Gardening* journal in the USA, 78 million people in the USA are keen on gardening, accounting for over 40% of the adult population in the USA.

The gardening marketing channels abroad are also quite developed. In Europe, the gardening centers have developed into national chains in recent years, and the home gardening industry chain has been developing and improving, with a high degree of technology, a high level of scientific research, and a very stable industrial order. Due to the national territorial area being far larger than each European country, the USA has not formed a nationwide chain of gardening centers similar to those in Europe; instead, it has formed hundreds of independently operated local multi-purpose gardening centers. These gardening centers have similar functions to those in Europe: they also provide various gardening supplies such as hand tools, irrigation tools, gardening information, and garden furniture, and offer simple meals. Home gardening products, such as seeds, seedlings, trees, soil, fertilizers, garden furniture, sculptures, or even small greenhouses, can be readily purchased in gardening centers in each European country and the USA.

China is in an economic transitional period, and rapid industrialization and urbanization have caused many environmental problems. In addition to outdoor pollution, indoor pollution also becomes one of the biggest threats to human health. International environmental experts have listed indoor air pollution as the third generation of air pollution following air pollution due to coal combustion and photochemical smog air pollution. Assessed according to the *Ambient Air Quality Standards* (GB 3095–2012), the number of days of heavy pollution and above in the Beijing-Tianjin-Hebei region in 2014 accounts for up to 17%, and hazy weather is becoming increasingly serious, making it impossible for the people to open their windows for indoor and outdoor air circulation as usual. According to the research, the shoots and leaves of indoor green plants have the functions of purifying indoor pollution, increasing oxygen content, reducing CO_2 content, and generally improving indoor environmental conditions.

With the continuous and rapid economic growth, people's living standards are improving constantly. People's order of consumption regarding clothing, food, housing, and mobility has changed obviously, and the consumption structure has transformed from the previous food and clothing consumption to development-centric consumption and enjoyment-centric consumption. In addition, housing conditions are constantly improving, so people are not just satisfied with the simple life of having a house to live in, and more and more people are longing for close contact with nature. As a result, the demands for indoor potted flowers and trees grow sustainably. In addition to beautifying the indoor environment and purifying the air, indoor potted flowers and trees can also cultivate sentiment and make people spiritually pleasant. Therefore, the demand for home gardening products has been increasing in recent years, and more and more enterprises have engaged in the development, application, and promotion of home gardening products.

Regarding the development requirements of urban agriculture, urban agriculture is an inevitable trend of economic and social development. It was only in some economically developed metropolises, such as Beijing and Shanghai, that urban agriculture was first incorporated into the urban development planning. However, in the past 10 years, urban agriculture has been developing rapidly in China and has now become a theme of development of the metropolises in China, showing a good development trend. As a part of urban agriculture, home gardening can not only increase urban employment, but also play a role in greening the urban environment. In the current society with high stress, high pollution, and fast pace, people's physical and mental health problems are increasingly prominent. Utilizing the vacant space in modern homes and offices for home gardening can deliver the functions of protecting the eco-environment, reducing dust and noise, absorbing hazardous substances in the air, and adding the content of negative oxygen ions, meeting the development requirements of urban agriculture.

In Beijing, Tianjin, Shanghai, and other cities, home vegetable planters, balcony agriculture, and other home gardening products have emerged. In January 2008, UGarden Life Guangzhou Flagship Shop, the first high-end customer-oriented theme mall in South China providing home gardening products and services, operated on a trial basis, and the event "First Year of Home Gardening

Life 2008 in China" started synchronously, marking the home gardening in China towards the industrial era. The efforts in making landscaped and interesting vegetable gardens also provide more directions for the development of home gardening. Currently, studies on planting vegetables on balconies have been explored in terms of variety, sowing date, planting pattern, planting vessel, substrate screening, fertilizers, occurrence, and prevention of plant diseases and pests, and irrigation facilities for the vegetables to be grown. The developed devices for vegetable farms on balconies include the ladder type device, wall-mounted device, and upright column type device, as well as small-sized products such as the domestic vegetable cultivator, which can not only save space but also achieve the purpose of beautifying.

In recent years, many enterprises in China have developed and produced home gardening products. For example, the home vegetable farm of Kingpeng focuses on the concept of the plant factory, and advocates vacating a space at home, so that relevant equipment can be used to provide the required environment for vegetables without considering the domestic illumination, temperature, and soil conditions; AgriGarden launched hydroponic vegetable racks, vegetable desk lamps, and humidifiers that focus more on design fashion and use convenience; Guangzhou Miugarden launched the vegetable planting rack with the grow light converting solar light to LED energy; Beijing Photon developed the smart vegetable planter that uses a microcomputer system to control the water supply and fertilization to vegetables, making the management easier; some other enterprises started to develop aquaponic products, namely, integrated culture trough installed over the fish tank, which follows the principle of "no watering for growing vegetables and no replacement of water in the fish tank". Such products have been publicized and promoted at garden fairs and vegetable expos in China, and there is a constant stream of consumers of such products.

Task 2 Main Forms of Soilless Culture for Home Gardening

Soilless culture has the advantages of high yields, good quality, safety and no pollution, no hogweeds, cleanness and sanitation, and unrestricted culture places. It is an ideal culture method of home gardening and the development direction of future urban home gardening. Home gardening includes hydroponics and substrate culture. Hydroponics is a soilless culture method where the plant roots directly contact nutrient solution. Substrate culture is a soilless culture method where plants are planted in the substrate with good physical structure and stable chemical properties, and nutrient solution is supplied to meet the growth requirements of the plants. Substrate culture is characterized by being simple, economical, and easy to manage, and it is the main method of soilless culture in home gardening. The substrate has the functions of fixing plants, retaining water and fertilizers, ventilating, and buffering ion concentration. There are wide varieties of substrates for soilless culture. Generally, the substrates for home gardening shall be light, beautiful, safe, and clean and shall have sufficient strength and appropriate structure to meet the growth requirements of the root system. The organic substrates with peculiar tastes that are likely to breed mosquitoes should not be used for home gardening. Therefore, the suitable substrates for soilless culture in home gardening mainly include rock wool, pumice stones, perlites, turf, and other inorganic substrates. The success of soilless culture lies in the formula and formulation of nutrient solutions. The formula of nutrient solutions required by each plant is not exactly the same, and it is different even for the same plant in different growth periods. However, restricted by conditions for soilless culture in home gardening, it is impossible to formulate a special nutrient solution for each variety of plants, and generally the general formula (Japanese garden test formula) or the common formula for vegetable nutrient solution (Yamazaki nutrient solution formula) is used.

I. Mode of Substrate Container Cultivation for Home Gardening

The planting areas for soilless culture in home gardening are located in indoor spaces, balconies, courtyards, and other urban buildings, which are generally cement structures on which plants cannot grow. In indoor gardening activities, containers are the most commonly used for planting. Container cultivation is the main planting mode of soilless culture in home gardening. Container cultivation is a production mode for planting with containers. The biggest difference between container cultivation and open cultivation is that container cultivation is not affected by the land, and the root system basically grows in the container. There are great varieties of containers for home gardening, ranging from pots and baskets to bag containers. The containers can be selected according to the type and size of the plant to be cultivated and the features of the balcony space. The eliminated domestic basins, earthen bowls, and other containers can be recycled as good balcony cultivation containers. The ideal cultivation containers for home gardening should be economical, portable, easy to handle, durable, breakage-proof, permeable, well-drained, etc.

1. Types of substrate cultivation containers for home gardening

(1) Pot containers.There are great varieties of pot containers with diverse sizes. They are usually named according to the materials used, such as clay pots, ceramic pots, purple sand pots, plastic pots, and wooden pots. Clay pots, also called plain burning pots or earthenware pots, are formed by burning clay, and there are red and grey clay pots, which are coarse in texture, economical, and durable, making them very suitable for balcony cultivation. Ceramic pots and purple sand pots have good luster, beautiful appearance, and fine texture, but have poor permeability and drainage. Plastic pots are portable and cheap, but have poor permeability and drainage and are likely to age. Wooden pots have good decorative effects and good permeability, but the parts applied with preservatives are likely to become mildewed and rot.

Pot containers are mostly round and have diverse sizes and specifications. Suitable containers for flower and tree planting can be selected according to the sizes of the plants. Generally, pot containers with a diameter of over 20 cm should be selected for planting vegetables and potted fruit trees.

(2) Case and tank containers. Case and tank containers are generally made of waste-packing wooden cases, plastic frames, or styrofoam cases. Alternatively, cultivation cases (tanks) can be specially prepared for planting vegetables on balconies, and they can be made of plastic boards, wooden boards, bamboo sheets, etc. Styrofoam cases are usually used to load fish, shellfish, vegetables, etc. on the market. They are portable and sturdy and have good thermal insulation performance, making them very suitable for container cultivation. Wooden cases (tanks) shall be subject to preservative treatment inside or lined with a layer of plastic film to reduce the corrosion by the moisture in the soil. Case and tank containers are generally rectangular and can save space and area if placed on the balcony or hanged, making them very suitable for vegetable planting. Case and tank containers should be about 20 cm wide and 15–20 cm high, and the length depends on the size of the balcony.

(3) Bag containers. Bag containers are plastic bags that contain various culture substrates for cultivation. The greatest advantage of bag containers lies in their economical efficiency, simplicity, and flexibility. The size, shape, and placement mode of plastic bags can be changed with the space and site. They are particularly suitable for multi-layer and multi-combination of balcony gardening through three-dimensional space utilization. For example, small bag containers can be hung on the walls and supports of the balcony, or placed in gaps between other containers to make full use of light and space. Small bag containers are also suitable for edible fungus cultivation on balconies.

2. Mode of multi-layer substrate culture

The indoor planting area of most urban buildings is limited, generally only several square meters. It is the primary consideration of how to make rational use of the space to increase the planting area for greater harvest in home gardening. The mode of multi-layer cultivation is the most direct and scientific utilization of space for planting. Multi-layer cultivation is, on the basis of indoor container cultivation, a multi-layer cultivation combination by using various supports to erect the tanks, cases, pots, bags, and other containers for plant cultivation in multiple-layer or staggered special containers. Multi-layer cultivation includes, among other things, bookshelf-type multi-layer cultivation, stepped multi-layer cultivation, wall-attachment multi-layer cultivation, column-type multi-layer cultivation, and staggered container multi-layer cultivation.

(1) Bookshelf-type multi-layer cultivation. The rack for bookshelf-type multi-layer cultivation is structurally simple and easy to make. It can be made of aluminum alloy, wood, and plastic, and generally has 3–6 layers, each with an adjustable height (Fig. 10-1). It is suitable for the cultivation of vegetables, fruits, and ornamental plants in balcony gardening, as well as edible fungus cultivation on balconies. Bookshelf-type multi-layer cultivation has the advantage of small footprint and large space to increase. This mode can make maximum use of the balcony space. However, it has the disadvantage of weaker illumination at the lower layers if the cultivated plants are dense, which will affect the growth of the plants at the lower layers. Therefore, light-demanding plants can be planted at the upper layers, and shade-enduring plants can be planted at the lower layers to avoid poor growth of the lower-layer plants.

Fig. 10-1 Multi-layer Cultivation Rack

(2) A-shaped (Half A-shaped) multi-layer cultivation. The rack for A-shaped multi-layer cultivation is a variant of bookshelf-type multi-layer cultivation. It looks like a capitalized letter A from the cross-section, so it is called A-shaped multi-

layer cultivation. The advantage of the rack for this type of cultivation is that each layer is staggered outward so that each layer can receive sufficient illumination; the disadvantage is that it occupies a large land compared with bookshelf-type multi-layer cultivation. Half A-shaped multi-layer cultivation only has racks on one side, and it is suitable for gardening on small balconies (Fig. 10-2). In addition, the rack for half A-shaped multi-layer cultivation made of PVC (polyvinyl chloride) tubes and stainless steel supports has also been widely used, which is very practical (Fig. 10-3).

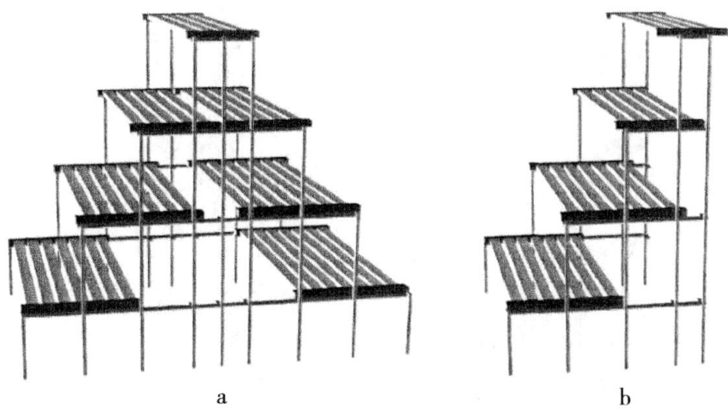

Fig. 10-2 Racks for A-shaped (Half A-shaped) Multi-layer Cultivation

a. Rack for A-shaped multi-layer cultivation b. Rack for half A-shaped multi-layer cultivation

Fig. 10-3 Rack for Half A-shaped Multi-layer Cultivation Made of PVC Tubes and Stainless Steel Supports

(3) Column-type multi-layer cultivation. Column-type multi-layer cultivation, also known as tower-type multi-layer cultivation, is a multi-layer cultivation method that is fixed in the middle with a strut, and the surrounding round, square, polygonal, and other shapes of cultivation plates are arranged in layers (Fig. 10-4). This method of multi-layer cultivation has the advantage of small footprint, large space to increase, and utilization of the space to the maximum extent. Column-type multi-layer cultivation is mostly used for soilless culture, and the rack is generally made of polyfoam and PVC tubes.

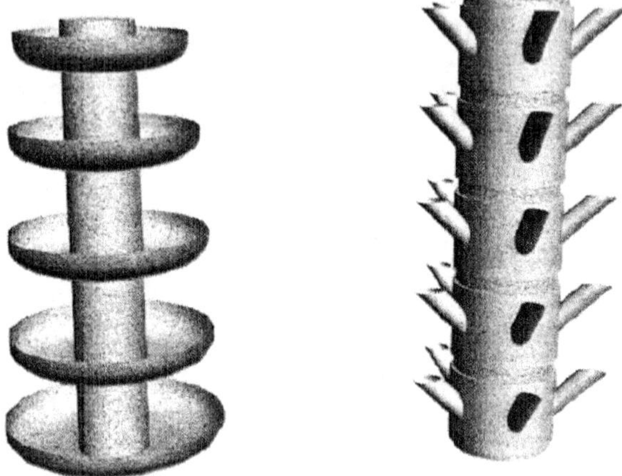

Fig. 10-4 Sketches of Column-type Multi-layer Cultivation

(4) Staggered container multi-layer cultivation. Staggered container multi-layer cultivation is to stagger specially made containers on each layer in a well-arranged manner, and each layer of containers can be exposed to plant vegetables and fruits (Fig. 10-5). The advantage of this form of multi-layer cultivation is that the number of layers can be adjusted at will and it is easy to move; the disadvantage is that the containers at the lower layers are easily shaded by the containers at the upper layers and cannot be fully utilized. Staggered container multi-layer cultivation includes staggered protruding pot multi-layer cultivation and wall-type staggered container multi-layer cultivation. Currently, there are patented products for staggered container multi-layer cultivation on the market.

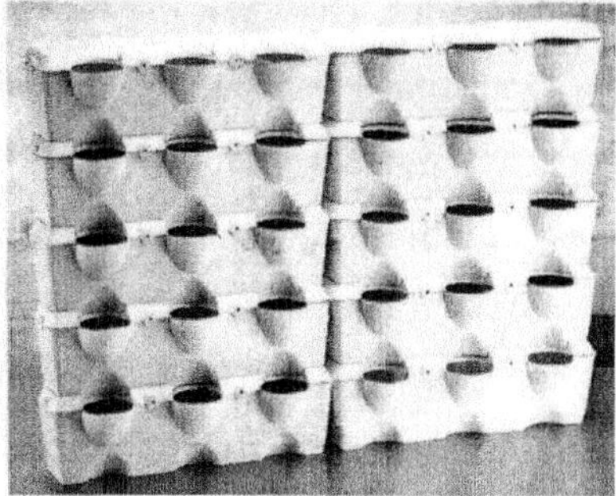

Fig. 10-5 Sketches of Staggered Container Multi-layer Cultivation

II. Mode of Hydroponics in Home Gardening

1. Simple soilless culture in still water

Simple soilless culture in still water is the simplest soilless culture method. Generally, in this soilless culture method, the nutrient solution does not flow, and the roots of the cultivated plants go deep into the nutrient solution to absorb the nutrient solution so that the plants can obtain mineral nutrition and water. There are many forms of simple soilless culture in still water. Suitable devices can be made according to the actual conditions, and the simple soilless culture on a floating bed is a typical representative (Fig. 10-6).

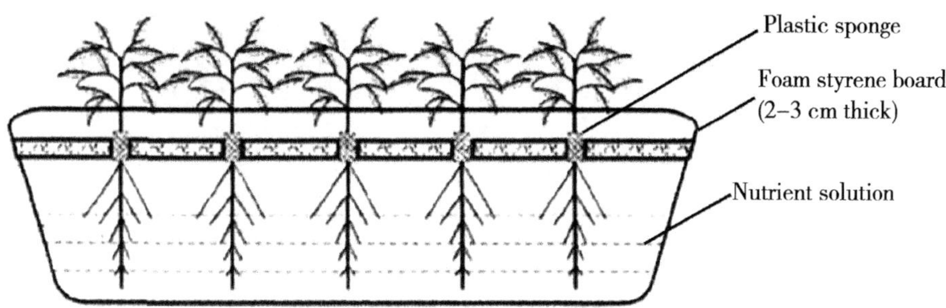

Fig. 10-6 Sketch of Simple Soiless Culture in Still Water

2. Column-type soilless culture

Column-type soilless culture belongs to multi-layer cultivation. In the middle of the device is an upright column, through which the nutrient solution can pass. There are many cultivation holes around the upright column, which can be used to plant leaf vegetables, strawberries, flowers, etc. Column-type soilless culture can provide a large space for massive planting in a small area, making it suitable for balcony gardening. Standard devices for column-type soilless culture are commercially available on the market. The standard column is assembled from several foam polystyrene pots, each having five cultivation holes made of PVC tubes, and containing foam materials and substrates such as pumice stones. The standard column is erected in the liquid collecting pot with a small motor. After being energized, the motor can deliver the nutrient solution to the standard column to form a loop, thus ensuring sufficient nutrients to the plants on the column. (Fig. 10-7)

Fig. 10-7 Device for Column-type Soilless Culture

3. Tubular soilless culture

Tubular soilless culture belongs to multi-layer cultivation. Several layers of tubes at different heights are connected, with nutrient solution passing through the tubes. There are holes on the tubes that can be used to plant leaf vegetables, strawberries, and other plants. The soilless culture device called vegetable planter available on the market is a simple device for tubular soilless culture. It is made by connecting PVC tubes with different thicknesses. The outlet nozzle is equipped with a liquid collecting pot with a small motor. After being energized, the motor can deliver the nutrient solution to the uppermost inlet nozzle to form a loop, thus ensuring nutrient supply to the plants in the holes on the horizontal tubes (Fig. 10-8).

Fig. 10-8 Device for Tubular Soilless Culture

Task 3 Daily Management of Soilless Culture in Home Gardening

I. Soilless Culture Substrates for Home Gardening

1. Substrate requirements

(1) Safe and hygienic soilless substrates may be organic or inorganic, but they generally require pollution-free surroundings. Some chemical substances constantly give off unpleasant odors or release some substances harmful to humans and plants. Such substances must not be used as soilless culture substrates. One of the disadvantages of soil is dust pollution. The selected substrates must overcome this defect of soil. Both flower producers and flower consumers should understand the

basic chemical and physical properties of some soilless culture substrates, and select absolutely safe and hygienic substrates for flower and vegetable planting.

(2) Portable and beautiful soilless culture is an elegant technology and art. Soilless cultivated flowers must meet the need for indoor decoration. Therefore, it is necessary to select light and well-structured materials that are easy to handle and have an appearance in harmony with the flower shape and arrangement and the environment to overcome the defects of heavy and sticky soil and difficulty in handling.

(3) As the substrate shall also support the plant body of appropriate size and maintain a good root system environment, the substrate shall also have sufficient strength and a suitable structure. Only when the substrate has sufficient strength can the plants not reel right and left. Only when the substrate has a suitable structure can it have a suitable ratio of water, air, and nutrient, thereby ensuring the root system is in the best environmental state, which in turn leads to lush foliage and beautiful flowers. Some substrates can provide appropriate nutrients to the plants and are certainly good. However, without this capacity, the substrates are still appropriate as long as they have the capacity of water retention, fertilizer retention and ventilation, and provide a good environment for the root systems. The reason is that the nutrients required by plant growth can be completely supplied by the nutrient solution formulated according to the scientific formula. The different plant root systems require different best environments, and different substrates provide different water/air/nutrient ratios. Therefore, suitable substrates can be selected, or mixed substrates can be formulated according to the physiological needs of the plant root systems.

2. Substrate selection

The following three factors shall be considered in the selection of soilless culture substrates for home gardening: adaptability to the root system, i.e. the substrate can meet the growth needs of the root system; practicability, i.e. the substrate shall be light in weight, good in performance, and safe; economic efficiency, i.e. the substrates can be sourced from the local place, or are cheap on the market.

(1) Adaptability to the root system. One of the advantages of soilless culture

substrates is that they can create the best environmental conditions required by the growth of plant root systems, i.e. they can provide the best water and air ratio. Aerial roots and fleshy roots require excellent ventilation and humidity surrounding the root systems to be over 80%. Thick root systems require good ventilation and the humidity to be over 80%. Thin root systems, such as the root systems of azaleas, require the ambient humidity of the root systems to be over 80% or even 100%, as well as good ventilation. Some substrates with good air permeability, such as pine needles and sawdust, are very suitable for regions with great air humidity. However, the air permeability of such substrates is so great that the root systems are likely to become air dry in the north with dry air.

(2) Practicability. The water in North China is generally alkaline, requiring the substrates to have a certain capability of regulating the hydrogen ion concentration. Therefore, better effects can be achieved by using peat-mixed substrates. The substrate bulk density shall be small to make it easy to handle soilless cultivated flowers. The preferred substrates include ceramsite, vermiculite, perlite, rock wool, sawdust, urea formaldehyde, and peat, and the substrates of their mixtures.

No matter which substrate is used, the substrate must meet the environmental conditions required by the plant root systems. Therefore, the following shall be noted in substrate selection. First, if there is only one type of substrate available in the local place, select flowers that are suitable for growing in this substrate based on this substrate. Second, if a certain flower already exists, select the substrate suitable for planting it based on its biological characteristics. For example, if *Clivia miniata* already exists, select the complex substrate formulated by peat and vermiculite, or by perlite and sand. Soilless culture substrates must also be unharmful to human health. They must first be non-toxic and odorless, preferably natural inorganic substrates. Although some organic substrates are good for plant growth, it is difficult to predict the substances they would release during decomposition, and it is impossible to ensure that such substances are unharmful. In particular, this should be noted when families with children select soilless culture substrates. Although some synthetic substrates have good performance, they should not be used as soilless culture substrates for home gardening if they give out unpleasant odors.

Because of this, more attention should be paid to the safety and hygiene of the substrates for flower planting than those for vegetable planting.

(3) Economic efficiency. One of the most important in selecting soilless substrates is to spend as little money as possible, and it is best for the substrate to be sourced from the local place. This is not only for the purpose of reducing costs but also highlighting the unique features.

II. Application of vegetables in home gardening

Growing vegetables through soilless culture in courtyards or indoors can not only purify the air and beautify environment but also provide fresh and pollution-free vegetables for families.

1. Vegetable varieties

The vegetable varieties that are suitable for soilless culture in home gardening can be divided into fruit vegetables and leaf vegetables. Fruit vegetables include tomatoes, cucumbers, bitter gourds, strawberries, cherries, radishes, etc. Leaf vegetables include lettuce, celery, Chinese chives, garlic sprouts, coriander, pakchoi, small rape, water spinach, *Crowndaisy coronarium*, etc. These vegetables have different lengths of maturity stages, generally 3–5 months for fruit vegetables and 2–3 months for leaf vegetables. Fruit vegetables require a temperature of 10–25°C and light throughout the day; leaf vegetables have relatively fewer requirements and are suitable to be planted by general families. Vegetable seeds can be bought from institutes of agricultural science and seed companies everywhere.

(1) Green onion.

Green onions can be bred by sowing or bulb seedling cultivation. However, for home planting, it is more convenient and simple to use bulbs for seedling cultivation.

① Seedling cultivation with bulbs: Generally, it is conducted from April to May or from September to October. The bulbs reserved in the previous year or the bulbs with roots bought from the market can be used.

② Cut off the leaves before planting, but not to the white leaf sheaths.

③ Divide 2–3 bulbs into a group, and plant them into the pot with soil, with a space of about 10 cm. Do not bury them too deep. Just bury the root in the soil, with the bulbs slightly exposed from the soil surface. Water the soil thoroughly, and keep the soil wet. After the bulbs survive, they should be taken care of normally.

④ Green onions can be planted when six principal leaves grow out. They are generally planted in spring. All the roots should be buried underground and watered thoroughly.

(2) Chinese chive.

Chinese chives are generally suitable to be planted in spring. As long as they are properly cared for, they can be harvested all the time.

① Seeding: First, soak the seeds in water at 40°C to remove wizened seeds and wash the Seeds. Second, broadcast sow the seeds evenly in the soil, cover the seeds with 1cm-thick soil, and water the soil with a sufficient amount of water.

② Top dressing: One week later, the plantlets start to germinate. Keep the soil moist. After the plantlets grow to about 15 cm, apply decomposed organic fertilizer once. When the plantlets grow to about 18 cm, water them once every week. Note that there shall be no accumulated water, and hogweeds (if any) shall be immediately removed. At this time, the plantlets can be planted.

③ Planting: Thoroughly water the plantlets one day before planting, generally when there is no sunlight. When planting, put several plantlets together with a slightly larger plant spacing to ensure there is sufficient growth space.

(3) Lettuce.

The outward leaves of the lettuce can be picked to eat. After the outward leaves are picked, the middle leaves will grow gradually.

① Evenly sow healthy seeds free from diseases and pests in the soil, and cover the seeds with ±1-cm-thick soil. The seeds will germinate at 20°C about 5 days after sowing.

② Do not water the seeds too much, just keep the soil moist.

(4) Water spinach.

Water spinach is easy to plant as there is basically no concern about disease and pest attacks on the water spinach. After the uppermost crop of the water spinach is picked, the lateral buds can grow and extend horizontally. Therefore, the growth of water spinach needs large horizontal space. If water spinach is not picked in time, the stems and leaves will become thick and hard, but it can be picked all the time as long as it is properly cared for.

The water spinach bought from the market can be directly planted in the soil for cutting. If it is too long, it shall be cut into small sections of 10 cm long for cutting. During cutting, bury three sections of the stem in the soil, expose three sections on the soil, and compact the soil. Water the plant every day. It can survive about four days after cutting.

(5) Malabar spinach.

① Seeding: First, put the soil in a flowerpot (or foam box), dig a hole, sow the seeds evenly in the flowerpot (or foam box), and completely cover the seeds with soil. Second, water the seeds with sufficient water.

② Germination: Thoroughly water the seeds to keep the soil moist when the temperature is kept at 20°C. The seeds will start to germinate six days later. The shells of the plant are very hard. If there are hard shells on the top of the plant, remove them by hands; otherwise, they will affect the growth. Generally, plantlets basically grow out about one week later. After the plantlets grow out, start to remove bad plants and hogweeds (if any), and apply decomposed organic fertilizer once.

③ Planting: Plant the Malabar spinach one month later after five principal leaves grow out, and apply decomposed organic fertilizer once.

(6) Spinach.

Spinach is also suitable to be planted on balconies.

① Evenly sow healthy seeds without diseases and pests in the soil.

② Plantlets grow out about one week later. If the interval between the plantlets is too close, thin the seedlings by removing the weak plantlets to keep the plant spacing at about 3 cm.

③ Two weeks later, when 2–3 leaves grow out, apply fertilizers appropriately by spreading the mixture of fertilizers and soil at the roots of the plantlets. About three weeks later when the plants grow to a height of about 10 cm, apply fertilizers again.

(7) Amaranth.

Amaranth can be sown all year round, and it only takes four to eight weeks from sowing to harvest.

① Sowing: Flat the surface of the soil in the pot, and evenly sow the amaranth seeds in the soil. As the seeds are very small, the seeds may be mixed with sand for sowing, and thoroughly water the seeds to keep the soil moist.

② Seedling thinning: About 10 days later, plantlets will break through the soil. Where there is dense seedling emergence, remove some weak plantlets to prevent crowding from affecting the growth.

③ Final singling: About 45 days later, when 5–6 leaves grow out, thin the seedlings for the second time. Just remove the plantlets alternatively, but be sure not to remove the adjacent plantlets. The removed plantlets can be eaten.

④ Top dressing: After final singling, the plants are in the best period of growth. At this time, fertilizers need to be supplemented in time, and a great amount of nitrogenous fertilizers is needed. Top dressing is required once every time of picking in the future.

2. Principles for application of ornamental vegetables

The application of ornamental vegetables can produce a landscape that both meets scientific law and has art appreciation value only by following a certain allocation principle.

(1) Timely application in suitable places. The healthy growth of vegetables needs a suitable environment. Only by understanding the growth characteristics of various ornamental vegetables can they be planted in suitable places in due time. For example, the slightly shade-tolerant vegetable varieties that like heat but cannot bear high temperatures can be planted on northwards balconies, while the drought-resistant vegetable varieties that like light can be planted on southwards balconies. In addition, plants of vegetables must be adjusted, and excessive shoots must be removed to ensure large, regular and bright-colored fruits and achieve the optimal ornamental effect.

(2) Highlighting the key points while considering the diversity of plants. The beauty of fruits shall be highlighted in the creation of the aesthetic environment.

For this purpose, it is required to consider the reasonable mix and systematic combination of an ornamental vegetable with other ornamental vegetables in the selection of types and varieties. In the early period, the green leaves are the main ornamental parts, and some bright yellow or cherry-red ornamental plants can be added to improve the highlights. In the later period where there are colorful fruits, it can be more casual to match colors. However, it shall be particularly noted that the matching in different antheses and of different flower colors shall be reasonable and harmonious so as to achieve flowers in three seasons, landscape in four seasons, continuous blossom and harmonious color tone, and avoid temporary blossom and single flower color.

(3) Principle of harmony. Ornamental vegetables shall be selected so that they are suitable for the surroundings. Harmony produces beauty. The ornamental vegetables can be appropriate and pleasing to the eye, natural, and generous and deliver the best ornamental effect only when they are in harmony with the surroundings.

3. Application of vegetables in home gardening

(1) Application of vegetables in courtyards and terraces.

In modern home gardening, courtyards and terraces are relatively close to the natural environment. In order to highlight the ornamental effect, the space shall be utilized efficiently for plane or stereo planting, or various placement effects can be designed and a unique landscape can be created according to personal preference. The courtyards and terraces of different families may not be exactly the same, and suitable modes can be designed according to the particular conditions. Generally, there are rectangular and square courtyards and terraces. In order to better exhibit the ornamental effects, different ornamental vegetables can be systematically combined with the courtyards and terraces.

① Rectangular courtyards and terraces. A pergola can be erected in the middle of the courtyard with the road as the axis so as to plant the vine plants with a strong ornamental effect. On both sides of the road, ornamental vegetables such as fruit- and foliage-ornamental vegetables can be planted. This design has a sinuous effect. After passing through the shelter, we can see colorful fruit- and foliage-ornamental vegetables on both sides.

② Square courtyards and terraces. Square shelters can be built in the middle of the courtyard and terrace to plant ornamental vines. Fruit- and foliage-ornamental vegetables can be planted around courtyards and terraces.

The above two design methods of the courtyards and terraces not only bring a cool place in the hot summer, but also beautify our courtyards and terraces, and provide us with edible vegetables or fruits while enjoying the beautiful scenery.

(2) Application of vegetables on balconies and indoors.

Balcony and indoor growing of ornamental vegetables are mainly to beautify our living space and sometimes can provide fresh and edible pollution-free products, bringing endless vitality and energy to leisure time. Ornamental vegetables planted on balconies and indoors should be small and suitable for pot cultivation.

Due to the special balcony and indoor environment, it is difficult to plant ornamental vegetables indoors. Therefore, the balcony and indoor growing are often effectively combined to facilitate the growth of plants. Most ornamental vegetables are heliophilous and shade-resistant, and grow poorly under weak indoor light conditions. The most prominent manifestation includes defoliation, flower drop and fruit drop. In order to solve the problem of growing ornamental vegetables indoors, balcony planting and indoor appreciation can be combined. Specific method: Grow various ornamental vegetables as small potted plants, move them indoors when they enter the appreciation period, and place them in the living room, dining room, window sill, etc. Note that when placed indoors, they should be changed every 1–2 days. This can ensure the good growth of various ornamental vegetables.

III. Soilless Culture Techniques for Home Gardening

1. Ceramsite culture of plants

In addition to traditional hydroponics, there has been a new popular soilless culture method of using ceramic nutrient soil as cultivation substrate (also called ceramsite culture) in recent years. This culture substrate contains many mineral elements such as potassium, sulfur, calcium, and magnesium, which provide necessary fertilizers for plant growth, and is a real soilless culture substrate. It is suitable for fiber root plants, fleshy root plants (orchids, *Clivia miniata, Cymbidium hybrid-um*, *Zamioculcas zamiifolia*, etc.) and woody plants (peony, jasmine, *Spathiphyllum floribundum 'clevelandii'*, etc.). It can be widely used for potted flower planting and as the culture substrate in the roof garden, which opens a new path for soilless culture. Ceramic nutrient soil is lighter than soil, permeable and water-drainable, without hardening and dust. It will not disintegrate after soaking and has no pests or diseases. Although many nutrient elements are added to ceramic nutrient soil, plants consume a large amount of elements such as nitrogen, phosphorus, potassium, sulfur, calcium and magnesium, which shall be continuously supplemented. Therefore, the nutrient solution shall be added in the process of soilless culture.

Generally, ceramsite soilless culture is easy for the following plants: *Monstera deliciosa*, *Aglaia odorata, Clivia miniata*, camellia, Chinese rose, jasmine, azalea, *Rohdea japonica, Matthiola incana*, *Phalaenopsis aphrodite*, fuchsia, *Pinus parviflora, Philodendron, Ficus elastica, Dracaena fragrans,* begonia, fern, Palmaceae plants, etc.; various foliage-ornamental plants, such as clustered *Philodendron selloum, Spathiphyllum floribundum clevelandii, Anthurium andraeanum, Aglaonema modestum, Monstera deliciosa, Spathiphyllum kochii, Aglaonema commutatum, Syngonium podophyllum* of the Araceae; *Commelina communis* and *Tradescantia zebrina* of the Commelinaceae; *Aloe vera, Chlorophytum comosum* and *Rohdea japonica* var. *variegata* of the Liliaceae; *Aeonium arboreum* of the Crassulaceae, and over 100 other species including *Clivia miniata, Paphiopedilum, Zygocactus truncatus, Phyllanthus,* silverleaf chrysanthemum, Brazilwood, *Hedera helix*, *Coleus scutellarioides*, etc.

The culture methods are as follows:

(1) Detachment: Use your fingers to push out the roots and soil from the bottom hole of the pot.

(2) Wash roots: Soak the root system with soil in water at a temperature close to the ambient temperature, and wash the soil of the root system. During the

cleaning process, ensure the root integrity.

(3) Soaking: Soak the washed root in the prepared nutrient solution for about 10 minutes to allow it to fully absorb nutrients.

(4) Potting and filling liquid: First wash the glass container, cover the bottom of the glass container with a small amount of soilless culture ceramsite, and then place the flowers in the corresponding position of the glass container. Before covering the soilless culture ceramsite, small ceramsites with a diameter of about 2–4 mm shall be used to reinforce the root system, and then small ceramsites with a diameter of about 4–8 mm shall be used to cover it until it reaches the top of the main root system. During the covering process, sufficient ceramsites shall be filled between the main root system and other sub-root systems. According to the height of the main root system and the glass container, 50% of the capacity of the whole glass container shall be filled with ceramsites with a diameter of about 2–4 mm, and 2/3 of the remaining part shall be filled with ceramsites with a diameter of about 4–8 mm. The top of the glass container is covered with ceramsites with a diameter of about 8–12 mm. After all ceramsites are paved, water shall be added to 1/3 of the container height.

(5) Daily management: The requirements for light, temperature, and other conditions of flower soilless culture are the same as those of soil culture. During the growth stage of the plant, the nutrient solution shall be watered once a week, and the dosage shall be determined according to the size of the plant. The dosage for flowers with slow leaf growth shall be appropriately reduced. In winter or dormant stage, the plant shall be watered once every half month to once a month. Indoor foliage-ornamental plants can survive in weak light conditions, and the dosage of nutrient solution should be reduced. Nutrient solution can also be used for foliar spraying, and attention should be paid to timely watering.

Ceramic nutrient soil is free of mud, dust, odor, mosquitoes, and flies, and it is clean and sanitary, convenient for maintenance, in which plants can grow well. Therefore, it is an ideal culture substrate for flowers, especially indoor flowers, and is also the best choice for modern fashion life.

2. Crystal mud culture of plants

Crystal mud is crystal clear and bright after absorbing water, and looks like crystal. It is mainly colorless and transparent, red, blue, yellow, and green, and can be used alone or in combination to form various color effects, with high ornamental value. At the same time, crystal mud is also a new type of soilless culture substrate instead of soil to plant plants. It is a highly absorbent carrier for storing water, nutrients, and trace elements, which is processed on the basis of highly absorbent resins for agriculture and forestry. Crystal mud can absorb water up to 50–100 times its own weight in several hours and can slowly release the water to support plant growth. When crystal mud is used to grow various shade plants indoors, even if they are not watered for more than a month, the plants will still be alive. It also contains nitrogen, phosphorus, potassium, and trace elements, which can ensure the growth of plants for several months. Crystal mud is non-toxic, odorless, clean and environmentally friendly, simple and convenient to use, and can maintain the moisture and nutrients required for plant growth of several weeks by absorbing water to a saturated state.

The effect of planting flowers with crystal mud is very obvious. It can not only avoid the trouble of frequent watering and fertilization in soil culture, but also is clean and easy to create an indoor garden belonging to oneself. Placed on the dressing table, desk, computer table, tea table, toilet, and other places, it can add beautiful scenery to home life. Crystal mud is a colloidal substance synthesized by high-tech methods. According to the characteristics of plants, the colloid can be made into particles of different shapes and sizes. Crystal mud has no smell, can provide sufficient nutrients, moisture, and oxygen for plant growth, and has no environmental pollution. The crystal mud colloid is transparent and glossy after being soaked in pure water or distilled water for 24 hours.

[Module Summary]

I. Key and Difficult Points

(1) Master the soilless culture methods for home gardening;

(2) Master the daily management in the soilless culture for home gardening.

II. Summary of Experience and Skills

The soilless culture technique can solve the problems that cannot be solved in home gardening, make the courtyards planting cleaner, improve the microclimate of the courtyards, green the urban space, and improve the air quality.

[Skill Training]

Skill Training 10-1 Crystal Mud Culture of Flowers

I. Purposes and Requirements

(1) Purposes: Investigate the characteristics of the crystal mud culture of flowers;

(2) Requirements: Master the steps of crystal mud culture of flowers.

II. Preparation for Training

Materials: flowers, crystal mud, etc.

III. Preparation for Planting

(1) Select suitable plant varieties.

(2) Remove rotten roots and leaves, and wash it.

Wet the soil-cultured plants with water first, then remove the plants and wash away the soil, cut off rotten roots and leaves, and wash the plants and dry them for later use. Do not damage the main root during operation. Most shade plants can be planted with crystal mud after washing and drying the roots. For a few plants with undeveloped root systems, they can be placed in clean water and cultured until the root system has developed before transplanting (the time of hydroponics depends on the air temperature and plant conditions).

Description of hydroponics: Clean the soil at the root of the plant for hydroponics, then place it in water. The root shall be immersed in water, and the water shall be changed once every four days. The plant can be planted after hydroponics until the new root system grows out.

(3) Combination of colors.

It is better to separately soak crystal muds in 1–2 colors for combination in one vase, and the crystal muds shall have similar colors, for example, red, orange, yellow, and purple is a series (two optional); cyan, blue, and green is a series (two optional). Do not mix multiple colors.

(4) Method of soaking.

Dry crystal mud particles shall be soaked in 50–100 times of water for 2–4 hours (Note: Do not stir when soaking).For xerophytic plants such as orchids, cacti, succulents, etc., the soaking proportion can be appropriately reduced to about 1 : (50–60). The crystal mud shall be soaked in clean water (such as distilled water, mineral water, purified water, and cold boiled water). Before soaking, a few dry crystal mud particles shall be reserved and placed at the bottom of the pot to absorb excess water.

IV. Planting Steps

(1) Wash and dry the vase, place the reserved dry crystal mud particles at the bottom of the pot to absorb excess water, and then lay a layer of crystal mud at the bottom according to the designed color. Put the plant into the vase, spread the root system of the plant as far as possible, and evenly put crystal mud around the root (if different colors are used, place them in the order of designed colors).

(2) After all crystal muds are filled, the height is better to cover the root completely to the vase mouth. Lift the plant slightly to stretch the root, and then gently compact the crystal mud with your hands. If the plants are stable without inclination after planting, only a small amount of water is necessary.

(3) During planting, it is better to soak it with a flower disinfection solution before planting. Disinfect the root with 3‰ potassium permanganate solution for 10 minutes, rinse the root with water, and then remove the water on the surface of the root by airing. Select a suitable vase, plant the flower, and spread

the roots.

(4) During the preparation of some plants *Philodendron xanadu, Monstera deliciosa, Zamioculcas zamiifolia, Dieffenbachia* cv. *Camilla, Eucharis grandiflora,* etc., attention should be paid to preventing water from dripping on the leaves. The leaves will rot with water and should not be sprayed with water at ordinary times.

(5) The container for crystal mud culture should not be too large, generally with a volume of 250–2,500 mL. It is suitable for planting dwarf indoor evergreen foliage-ornamental plants.

V. Nursing

1. Phenomenon: Damaged and rotten root system or yellowing of some leaves

After transplanting, some root systems of the plant will be damaged and rotten or some leaves will turn yellow regardless of soil culture, hydroponics, or crystal mud culture. Yellow leaves shall be cut off. Rotten roots will affect growth and shall be cut off in time before planting in crystal mud. If there are too many rotten roots, pour out the crystal mud, cut off the rotten roots and clean and filter the crystal mud before planting.

2. Phenomenon: Dust or dirt on the surface of crystal mud

Take out the surface crystal mud, rinse it with water, soak it for recovery, dry the surface water and put it into the vase.

VI. Others

During use, crystal mud can be added or replaced in time according to the plants and their growth, and an appropriate amount of nutrient solution suitable for plant growth can also be added regularly. Most plants need to be sprayed with water on the leaf surface every 1–2 weeks (the sprayed water on the leaf surface can be properly added with leaf surface fertilizer), and it is inadvisable to spray frequently; otherwise, the leaves are easy to yellow. The crystal mud itself has a strong ability to keep moisture, so it shall not be watered as much as possible at ordinary times. When the height of the crystal mud drops obviously, water less than half of the normal amount of nutrient solution can be added to allow the crystal mud to absorb and expand again. However, do not fill it up, which may reduce the ventilation, and the excess water should be poured and filtered.

[Expansion Task]

I. Review Questions

(1) What are the methods of soilless culture for home gardening?

(2) What are the advantages of home soilless hydroponics?

II. Case Study

Mini Garden

Since the beginning when the modern home gardening products of Mini Garden were developed, the company had taken "uniqueness and proficiency" as the design goal, and proposed the slogan of "Home Gardening" first, which emphasizes the new concept of planting, takes the fashionable family as the object, creatively breaks through the traditional concept of flower planting and makes flower planting simple and convenient. The products are mainly compact and exquisite, and encourage people to cultivate themselves and enjoy the fun of planting. Up to now, the company has developed many series of products. The Mini Garden canned flower series applies the new concept of modern cultivation and develops canned flowers, which highly integrates seeds, fertilizers, substrate, flower cans, and other planting elements, and uses advanced soilless culture techniques. It only needs watering every day, requires no fertilization, has no odor, and is clean and convenient, making planting truly a leisure activity. The small potted flower series, consisting of various flower pots, succulent plants, foliage-ornamental plants, and gardening accessories, is a brand-new horticultural brand series product in one-stop purchasing mode. This series of products has rich varieties, exquisite shapes, unique styles, rich vitality, easy maintenance, and other characteristics, and is decorative or artistic, making it a new fashion product for home decoration and personal gifts.

The newly launched magic mat and soil for plant growth are processed by special production techniques, and the seeds are deftly placed in the magic fiber layer developed by the company. The physical structure is loose, which can fully absorb water and has good air permeability and heat preservation. It can be naturally decomposed into organic fertilizer, and is an ideal breeding ground for seed germination and plant growth.

Appendixes

Appendix 1　Properties and Requirements of Major Element Compounds and Auxiliary Materials Used as Plant Nutrients

S/N	Name	Molecular formula	Relative molecular mass	Color and shape	Solubility ①	Alkalinity and acidity		Element content (%)	Required purity ② (%)
						Chemical	Physiological		
1	Calcium nitrate tetrahydrate	$Ca(NO_3)\cdot 4H_2O$	236.15	White, small crystals	129.3	Neutral	Alkaline	N11.86,Ca16.97	90 for agricultural use
2	Potassium nitrate	KNO_3	101.10	White, small crystals	31.6	Neutral	Weak alkaline	N13.85,K38.67	98 for agricultural use
3	Sodium nitrate	$NaNO_3$	85.01	White, small crystals	88.0	Neutral	Highly alkaline	N16.50,Na27.00	98 for agricultural use
4	Ammonium nitrate	NH_4NO_3	80.05	White, small crystals	192.0	Hydrolytically acidic	Acidic	N35.0	98.5 for agricultural use

continued

S/N	Name	Molecular formula	Relative molecular mass	Color and shape	Solubility	Alkalinity and acidity		Element content (%)	Required purity (%)
						Chemical	Physiological		
5	Ammonium sulfate	$(NH_4)_3SO_4$	132.15	White, small crystals	75.4	Hydrolytically acidic	Highly acidic	N21.20,S24.26	98 for agricultural use
6	Ammonium chloride	NH_4Cl	53.49	White, small crystals	37.2	Hydrolytically acidic	Highly acidic	N26.17,Cl66.27	96 for agricultural use
7	Urea	$CO(NH_2)_2$	60.03	White, small crystals	105.0	Neutral	Acidic	N46.64	98.5 for agricultural use
8	Ammonium dihydrogen phosphate	$NH_4H_2PO_4$	115.05	Grey, powder	36.8	Hydrolytically acidic	Not obvious	N12.18,P26.92	>90 for agricultural use
9	Diammonium hydrogen phosphate	$(NH_4)_2HPO_4$	132.07	Grey, powder	68.6	Hydrolytically acidic	Not obvious	N21.22,P23.45	>90 for agricultural use
10	Potassium dihydrogen phosphate	KH_2PO_4	136.07	White, small crystals	22.6	Hydrolytically acidic	Not obvious	N22.76,K28.73	96 for agricultural use

continued

S/N	Name	Molecular formula	Relative molecular mass	Color and shape	Solubility	Alkalinity and acidity		Element content (%)	Required purity (%)
						Chemical	Physiological		
11	Dipotassium phosphate	K_2HPO_4	174.18	White, small crystals	167.0	Hydrolytically acidic	Not obvious	P17.78,K44.90	98 for industrial use
12	Sodium dihydrogen phosphate	$NaH_2PO_4 \cdot 2H_2O$	119.97	White, small crystals	85.2	Hydrolytically acidic	Not obvious	P25.81,Na19.16	98 for industrial use
13	Disodium hydrogen phosphate	$Na_2HPO_4 \cdot 2H_2O$	141.96	White, small crystals	80.2(50)	Hydrolytically acidic	Not obvious	P21.82,Na32.39	98 for industrial use
14	Triple superphosphate	$Ca(H_2PO_4)_2 \cdot H_2O$	252.02	Grey, powder	15.4(25)	Highly acidic	Not obvious	P24.6,Ca15.9	92 for agricultural use
15	Potassium sulfate	K_2SO_4	174.26	White, small crystals	11.1	Neutral	Highly acidic	K44.88,S18.40	95 for agricultural use
16	Potassium chloride	KCl	74.55	White, small crystals	34.0	Neutral	Highly acidic	K52.45,Cl47.55	95 for agricultural use
17	Calcium chloride	$CaCl_2$	110.98	White, small crystals	74.5	Neutral	Acidic	Ca36.11,Cl47.55	98 for industrial use

continued

S/N	Name	Molecular formula	Relative molecular mass	Color and shape	Solubility	Alkalinity and acidity		Element content (%)	Required purity (%)
						Chemical	Physiological		
18	Calcium sulfate	$CaSO_4 \cdot 2H_2O$	172.17	White, powder	0.204	Neutral	Acidic	Ca36.11,S18.62	98 for industrial use
19	Magnesium sulfate	$MgSO_4 \cdot 7H_2O$	246.48	White, small crystals	35.5	Neutral	Acidic	Mg9.86,S13.01	98 for industrial use
20	Ammonium bicarbonate	NH_4HCO_3	79.04	White, small crystals	31.0	Alkaline	Weak acidic	N17.70	95 for agricultural use
21	Potassium carbonate	K_2CO_3	138.20	White, small crystals	110.5	Highly alkaline	Negligible	K56.58	98 for industrial use
22	Potassium bicarbonate	$KHCO_3$	100.11	White, small crystals	33.3	Highly alkaline	Negligible	K39.06	98 for industrial use
23	Calcium carbonate	$CaCO_3$	100.08	White, powder	6.5×10^{-13}	Alkaline	Negligible	Ca40.05	98 for industrial use

continued

S/N	Name	Molecular formula	Relative molecular mass	Color and shape	Solubility	Alkalinity and acidity		Element content (%)	Required purity (%)
						Chemical	Physiological		
24	Calcium hydroxide	$Ca(OH)_2$	74.10	White, powder	0.165	Highly alkaline	Negligible	Ca54.09	98 for industrial use
25	Potassium hydroxide	KOH	56.11	White, block	112.0	Highly alkaline	Negligible	K69.69	98 for industrial use
26	Sodium hydroxide	NaOH	40.00	White, block	109.0	Highly alkaline	Negligible	Na57.48	98 for industrial use
27	Phosphoric acid③	H_3PO_4	97.99	Light yellow, liquid	Soluble	Acidic	Negligible	P31.60	98 for industrial use
28	Nitric acid	HNO_3	63.01	Light yellow, liquid	Soluble	Highly acidic	Negligible	N22.22	98 for industrial use
29	Sulfuric acid	H_2SO_4	98.08	Light yellow, liquid	Soluble	Highly acidic	Negligible	S57.48	98 for industrial use

Notes: ① Solubility: the maximum grams dissolved in 100 mL water (calculated as an anhydrous compound) at 20°C, and the number in brackets is another temperature.② Required purity: grams of this substance in every 100g of solid substance, that is, wt%. Matters other than this substance are impurities.③ With three acids (H_3PO_4, HNO_3, and H_2SO_4) specified as liquids, grams of this substance in every 100 g of liquids, that is, wt%. Matter other than this substance are mainly water, which may also contain traces of impurities, where the limit of hazardous substances is the same as ②.

Appendix 2 Properties and Requirements of Trace Element Compounds Used as Plant Nutrients

S/N	Name	Molecular formula	Relative molecular mass	Color	Shape	Solubility ①	Alkalinity and acidity	Element content (%)	Required purity ② (%)
1	Ferrous sulfate	$FeSO_4 \cdot 7H_2O$	278.02	Light green	Small crystals	26.5	Hydrolytically acidic	Fe 20.9	98 for industrial use
2	Ferric trichloride	$FeCl_3 \cdot 6H_2O$	270.30	Yelloish-brown	Crystal block	91.9	Hydrolytically acidic	Fe 20.66	98 for industrial use
3	Na_2-EDTA	$Na_2C_{10}H_{14}O_8N_2 \cdot 2H_2O$	372.42	White	Small crystals	11.1(22)	Slightly alkaline		99 chemically pure
4	Na_2Fe-EDTA	$Na_2FeC_{10}H_{12}O_8N_2$	389.93	Yellow	Small crystals	Soluble	Slightly alkaline	Fe 14.32	99 chemically pure
5	NaFe-EDTA	$NaFeC_{10}H_{12}O_8N_2$	366.94	Yellow	Small crystals	Soluble	Slightly alkaline	Fe 15.22	99 chemically pure
6	Boric acid	H_3BO_3	61.83	White	Small crystals	5.0	Slightly alkaline	B 17.48	99 chemically pure
7	Borax	$Na_2B_4O_7 \cdot 10H_2O$	381.37	White	Powder	2.7	Alkaline	B 11.34	99 chemically pure
8	Manganese sulfate	$MnSO_4 \cdot H_2O$	223.06	Pink	Small crystals	62.9	Hydrolytically acidic	Mn 24.63	99 chemically pure

continued

S/N	Name	Molecular formula	Relative molecular mass	Color	Shape	Solubility	Alkalinity and acidity	Element content (%)	Required purity (%)
9	Manganese chloride	$MnSO_4 \cdot H_2O$	197.09	Pink	Small crystals	73.9	Hydrolytically acidic	Mn 27.76	99 chemically pure
10	Zinc sulfate	$ZnSO_4 \cdot 7H_2O$	287.54	White	Small crystals	54.4	Hydrolytically acidic	Zn 22.74	99 chemically pure
11	Zinc chloride	$ZnCl_2$	174.51	White	Small crystals	367.3	Hydrolytically acidic	Zn 37.45	99 chemically pure
12	Copper sulfate	$CuSO_4 \cdot 5H_2O$	249.69	Blue	Small crystals	20.7	Hydrolytically acidic	Cu 25.45	99 chemically pure
13	Copper chloride	$CuCl_2 \cdot 2H_2O$	170.48	Bluish-green	Small crystals	72.7	Hydrolytically acidic	Cu 37.28	99 chemically pure
14	Sodium molybdate	$Na_2MoO_4 \cdot 2H_2O$	241.95	White	Small crystals	65.0	Hydrolytically acidic	Mo 39.65	99 chemically pure
15	Ammonium molybdate	$(NH_4)_6Mo_7O_{24} \cdot 4H_2O$	1235.86	Light yellow	Crystal block	Soluble		Mo 54.34	99 chemically pure

Notes: ①Solubility: the maximum grams dissolved in 100 mL water (calculated as an anhydrous compound) at 20°C, and the number in brackets is another temperature.② Required purity: grams of this substance in every 100g of solid substance, that is, wt%.

Appendix 3 Solubility Product Constant of Some Insoluble Compounds

(Ksp, 18–25°C)

Chemical formula	Ksp	Chemical formula	Ksp
$CaCO_3$	2.8×10^{-9}	$MgNH_4PO_4$	2.5×10^{-13}
CaC_2H_4	2.6×10^{-9}	$Mg(OH)_2$	1.8×10^{-11}
$Ca(OH)_2$	5.5×10^{-8}	$MnCO_3$	1.8×10^{-11}
$CaHPO_4$	1.0×10^{-7}	$Mn\ (OH)_2$	1.9×10^{-13}
$Ca_3(PO4)_2$	2.0×10^{-29}	MnS crystal	2.0×10^{-13}
$CaSO_4$	9.1×10^{-6}	$ZnCO_3$	1.4×10^{-11}
CuCl	1.2×10^{-6}	$Zn(OH)_2$	1.2×10^{-17}
CuOH	1.0×10^{-14}	$Zn(PO_4)_2$	9.1×10^{-33}
Cu_2S	2.0×10^{-48}	ZnS	2.0×10^{-22}
CuS	6.0×10^{-36}	FeCO	3.2×10^{-11}
$CuCO_3$	1.4×10^{-10}	$Fe(OH)_2$	8.0×10^{-16}
$Cu(OH)_2$	2.0×10^{-20}	$Fe(OH)_3$	4.0×10^{-38}
$MgCO_3$	3.5×10^{-8}	$FePO_4$	1.3×10^{-22}
$MgCO_3\cdot3H_2O$	2.1×10^{-5}	FeS	6.3×10^{-18}

References

[1] CAO W R. Course of Soilless Culture [M]. Beijing: China Agricultural University Press, 2015.

[2] CHEN X Y. Cultivation of Rare & Special Vegetables [M]. Beijing: China Agricultural University Press, 2011.

[3] CHEN Z H. Plant and Plant Physiology [M]. Beijing: China Agriculture Press, 2010.

[4] DONG Q H, ZHU D X. Questions and Answers on Strawberry Cultivation Techniques [M]. China Agricultural University Press, 2008.

[5] DUAN Y D, FAN L Q, WU Z G, et al. Current Situation and Development Prospects of Vegetable Soilless Culture [J]. Northern Horticulture, 2008. (8): 63–65.

[6] FENG S Z, ZHAO S T. Fruit Tree Production Techniques (Northern Edition) [M]. Beijing: Chemical Industry Press, 2007.

[7] GAO G R. Operation Procedure for Soilless Culture Technique of Vegetables [M]. Beijing: Jindun Press, 2007.

[8] GUO S R. Soilless Culture [M]. Second Edition. Beijing: China Agriculture Press, 2011.

[9] HAN S D. Vegetable Production Techniques (Northern Edition) [M]. Beijing: China Agriculture Press, 2001.

[10] JIANG G D. On the Basic Steps and Methods of Investigation Report Writing [J]. Journal of Kaili University, 2004, 22 (5): 94–95.

[11] JIANG W J. New Techniques for Soilless Culture of Vegetables [M]. Revision. Beijing: Jindun Press, 2008.

[12] LI C J. Advanced Plant Nutrition [M]. Second Edition. Beijing: China

Agricultural University Press, 2008.

[13] LIAN Z H. Principles and Techniques of Soilless Culture [M]. Beijing: China Agricultural Press, 1994.

[14] LIU B. Production Principles and Techniques of Plug Seedlings [M]. Beijing: Chemical Industry Press, 2007.

[15] LIU S Z. Modern Practical Soilless Culture Techniques [M]. Second Edition. Beijing: China Agricultural University Press, 2004.

[16] LIU Z X. Practical Soilless Culture Techniques of Common Vegetables [M]. Beijing: China Agriculture Press, 1997.

[17] LU J L. Plant Nutrition (I) [M]. Second Edition. Beijing: China Agricultural University Press, 2003.

[18] MA J. Fruit Tree Production Techniques (Northern Edition) [M]. Beijing: China Agriculture Press, 2009.

[19] MA N, Bao S S. Current Situation of Home Gardening Development [J]. Agricultural Engineering Technology, 2016,36 (4).

[20] PEI X B. Collection of Soilless Culture Techniques for Organic Vegetables [M]. Beijing: Chemical Industry Press, 2010.

[21] STYER, R C. Production Principles and Techniques of Plug Seedlings [M]. Beijing: Chemical Industry Press, 2011.

[22] TANG X S. Study on Balcony Horticultural Planting Mode in Chengdu [D]. Chengdu: Sichuan Agricultural University, 2012.

[23] QIN X H, LI H B, LI P, et al. Soilless Culture Techniques [M]. Chongqing: Chongqing University Press, 2015.

[24] WAN J. Current Situation and Development Trend of Soilless Culture Techniques at Home and Abroad [J]. Science and Technology Innovation Herald, 2011 (3).

[25] WEI S L. 2001. Soilless Culture of Flowers [M]. Beijing: China Forestry Press.

[26] WANG H R, RU S J, BEI Y W, et al. Disease Surveys of Fruit Vegetables in Hydroponic Systems in Zhejiang Province [J]. China Vegetables, 2001, 1 (1): 33–34.

[27] WANG H S. Soilless Culture Techniques of Flowers and Vegetables [M]. Changsha: Hunan Science & Technology Press, 1993.

[28] WANG H F. Soilless Culture of Flowers [M]. Beijing: Jindun Press, 1997.

[29] WANG J X. Illustration of Soilless Culture of Vegetables [M]. Beijing: Jindun Press, 2011.

[30] WANG Y P. Soilless Culture Techniques [M]. Beijing: China Agriculture Press, 2014.

[31] WANG Z L.Course of Soilless Culture [M]. 2nd Edition. Beijing: China Agricultural University Press, 2014.

[32] WANG X H. Questions and Answers on Soilless Culture Production Techniques of Vegetables [M]. Beijing: China Agriculture Press, 1998.

[33] XING Y X. New Edition of Principles and Techniques of Soilless Culture [M]. Beijing: China Agriculture Press, 2002.

[34] XU W H, WANG H X. Soilless Culture Techniques for Home Vegetables [M]. Beijing: Chemical Industry Press, 2013.

[35] YANG J S. Practical Techniques of Soilless Culture [M]. Shenyang: Liaoning Science & Technology Press, 1995.

[36] YANG Z C, ZOU Z R. New Soilless Culture Techniques in Greenhouse [M]. Yang Ling: Northwest Agriculture and Forestry University Press, 2005.

[37] ZHANG W Q. 500 Questions on Soilless Culture of Home Flowers [M]. Beijing: China Agriculture Press, 1999.

[38] ZHAO Y F. Illustration of Strawberry Cultivation Techniques [M]. Nanjing: Jiangsu Science & Technology Press, 2005.

[39] ZHOU H C. Standardized Production Techniques of Strawberry [M]. Beijing: Jindun Press, 2008.

[40] LIU J, TAO J P, MENG L L, et al. Design of a Greenhouse Environment Monitoring System Based on Internet of Things Technology [J]. Journal of Chinese Agricultural Mechanization, 2016, 37 (12), 179–182.

[41] YAN Z M, SUN J, GUO S R. Effects of Exogenous Proline on Seedling Growth, Photosynthesis, and Photosynthetic Fluorescence Characteristics in Leaves of Melon under Salt Stress [J]. Jiangsu Journal of Agricultural Sciences, 2013, 29

(5): 1125–1130.

[42] YAN Z M, SUN J, GUO S R. Effects of Exogenous Proline on Growth and Reactive Oxygen Substance Metabolism of Melon Seedlings under Salt Stress [J]. Jiangsu Journal of Agricultural Sciences, 2011, 27(1):141–145.

[43] YAN Z M, SUN J, GUO S R, et al. Effects of Exogenous Proline on the Ascorbate-Glutathione Cycle in Roots of Cucumis Melo Seedlings under Salt Stress [J]. Plant Science Journal, 2014, 32(5):502–508.

[44] YAN Z M, SHI H L, CAI S Y, et al. Development and Application of Two Types of Balcony Hydroponic Contraptions [J]. Southern Horticulture,2016, 27 (1): 35–39.

[45] ZHANG Y, XU J H, LI W L. Present Situation and Development Trend of Soilless Culture [J]. Agricultural Outlook, 2008, 4 (5).

[46] YANG F J, LI T L, ZANG Z J, et al. Effects of Timing of Exogenous Calcium Application on the Alleviation of Salt Stress in the Tomato Seedlings [J]. China Agricultural Science, 2010, 43 (6): 1181–1188.

[47] ASHRAF M, FOOLAD M A. 2007. Improving Plant Abiotic-stress Resistance by Exogenous Application of Osmoprotectants Glycine Betaine and Proline [J]. Environ Exp Bot, 2007 (59): 206–216.

[48] RODRIGO M J, ALQUEZAR B, ZACARIAS L. 2006. Clonging and Characterization of Two 9-cis-epoxycarotenoid Dioxygenase Genes, Differentially Regulated during Fruit Maturation and under Stress Condtions from Orange(Cotrus sinensis L.Osbeck)[J].Journal of Experimental Botany, 2006, 57 (3): 633–643.

[49] CAKIR B, AGASSE A, GAILLARD C, et al. A Grape ASR Protein Involved in Sugar and Abscisic Acid Signaling[J].Plant Cell, 2003, 15 (9): 2165–2180.

[50] EHSANPOUR A A, FATHAHIAN N. Effects of Salt and Proline on Medicago sativa Callus [J].Plant Cell Tissue and Organ Culture, 2003, 73(1):53–56.